Darwin

On Evolution

To Karoline Kohn

Darwin

On Evolution

The Development of the Theory of Natural Selection

Edited by

Thomas F. Glick

and

David Kohn

Hackett Publishing Company, Inc.
Indianapolis/Cambridge

07 06 05 04 03 02 01 00 99 98 97 96
1   2   3   4   5   6   7   8   9

For further information, please address

Hackett Publishing Company, Inc.
P.O. Box 44937
Indianapolis, Indiana 46244-0937

Cover and text design by Dan Kirklin

Library of Congress Cataloging-in-Publication Data

Darwin, Charles, 1809–1882.
    On evolution: the development of the theory of natural selection
    edited by Thomas F. Glick and David Kohn.
        p.   cm.
    Includes bibliographical references.
    ISBN 0-87220-285-2 (pbk.)   ISBN 0-87220-286-0 (cloth)
    1. Evolution (Biology)   2. Natural selection.   I. Glick, Thomas F.
II. Kohn, David, 1941–   .   III. Title.
QH365.Z9D37   1996
575.01'62—dc20   96-9388
                                                          CIP

The paper used in this publication meets the minimum requirements of
American National Standard for Information Sciences—Permanence of
Paper for Printed Library Materials, ANSI Z39.48-1984.

CONTENTS

103494

ACKNOWLEDGMENTS

Selection from *Charles Darwin's "Natural Selection,"* 1975, edited by R. C. Stauffer. Reprinted by permission of Cambridge University Press.

Charles Darwin to Emma Darwin, July 5, 1844 [*Darwin Correspondence* 3:43–44]. This letter is located in the General Library Manuscripts of the Natural History Museum, London. Reprinted by permission of the Natural History Museum, London, and George Pember Darwin.

Charles Darwin to Joseph Dalton Hooker, May 10, 1848 [*Darwin Correspondence* 4:140]. Reprinted by permission of the Syndics of Cambridge University Library and George Pember Darwin.

Charles Darwin to Asa Gray, 1857. Reprinted by permission of Gray Herbarium Archives, Harvard University, Cambridge, Massachusetts, U.S.A., and George Pember Darwin.

Selections from Notebooks B, C, D, and E, in the context of *Darwin on Evolution*, are reprinted by permission of David Kohn. Selections from Notebook RN are reprinted by permission of Sandra Herbert. Selections from Notebooks M and N, edited by Paul Barrett, are reprinted by permission of Wilma Barrett.

The editors wish to thank Frederick Burkhardt for his help in clarifying the proper rights holders to selections from Darwin's correspondence, Gretchen Wade, Gray Herbarium Archives, for her help with the cover art, and the Natural History Museum, London, for supplying the photograph.

INTRODUCTION

In the evolutionary worldview, human origins and human nature are aspects of the history of all living beings. For this reason, the theory of evolution, with its global dissemination since the mid-nineteenth century, marks a watershed in human history. Because evolutionary thought has challenged the ways in which humanity has viewed itself, it has been—and remains—controversial. The natural progression of species, involving the transformation of one into another along the "great chain of being," has engaged the interest of numerous naturalists, from the ancient Greek philosopher Anaximander, who believed that men evolved from fish, to eighteenth-century theorists such as Jean-Baptiste Lamarck and Erasmus Darwin. By the early 1800s, evolutionism, though it contradicted the biblical account of man's creation, had attained some respectability as a corollary of the historical thinking prevalent in Europe. Nevertheless, by the 1830s, serious scientists from the French anatomist Georges Cuvier to Darwin's own mentor Charles Lyell rejected evolution and made the origin of species a focus of biological controversy. Because evolutionists could not point to a believable naturalistic mechanism that supported their ideas, evolutionary speculations remained outside the realm of natural knowledge. That state of affairs was changed by Charles Robert Darwin, whose theory of natural selection provided a compelling naturalistic mechanism for species change that reorganized all of natural history around the concept of evolutionary descent.

Natural selection, the notion that competition for survival ultimately favors the "selection," through differential reproduction, of favorable variations, provided an answer for those naturalists who were looking for a reasonable explanation for the origin of adaptations.

Darwin's breakthrough occurred in 1838, but no account of it was published until 1859, by which time the secularization of society was well underway. During the subsequent century, natural selection became a handy shibboleth, over which secularists and their religious antagonists could attack one another, although many Christians sought to reconcile themselves to evolution, which became a keystone of modern thought. The religious issues have, at the "popular" level at least, tended to obscure the real significance, both philosophical and scientific, of Darwin's theory. The author of a recent letter in *The New York Times* suggested that, inasmuch as both Marx and Freud had lost much of their intellectual prestige in the 1980s and 1990s, Darwin's time for obsolescence had also come. But Darwin's natural selection, like the heliocentrism of Copernicus and Kepler, will not go away. Darwin's account explains vast areas of form and physiology, structure and behavior in the animal and plant worlds, and it has generated, and continues to generate, a plethora of empirical studies by thousands of natural scientists. Like all forms of robust science, the theory of evolution has continued to evolve. We are now in the midst of the fourth strong neo-Darwinian period since 1859. Since Darwin's time, however, there has also been scientific opposition to natural selection among evolutionists, and that situation has not changed. In this volume, we trace the development of Darwin's idea of natural selection from its early formulation through its manifold applications. Our intent is to make his ideas, and the course of their elaboration by Darwin, more accessible to the general reader as well as to give some indication of the broad and solid evidentiary base on which those ideas were constructed.

Charles Darwin was born in Shrewsbury, England, in 1809 and was buried in Westminster Abbey in 1882. His father was a Whig—that is, liberal—physician, and his grandfather, Erasmus Darwin, was a physician-poet who wrote a famous work called *Zoonomia*, which promoted an evolutionary worldview. His mother Susannah was the daughter of Josiah Wedgwood, the wealthy, enlightened, and scientifically inclined founder of the Wedgwood pottery firm. As a young man, Charles pursued the typical life of a country gentleman—hunting was his strongest interest—and one of the few signs he showed of his vocation as naturalist was his avidity as a beetle collector. He studied medicine in Edinburgh from 1825 to 1827, but, unable to bear the bloody anatomical demonstrations, he spent much of his time attending meetings of the Plinian Society, a natural history society founded by political radicals, and learning invertebrate zoology informally with Robert Grant, himself a Lamarckian evolutionist.

Darwin's father then sent the young man to Cambridge University, expecting him to take Anglican orders upon completing his studies. It was commonplace in early Victorian England for a young man like Darwin with an inclination toward natural history to seek a rural vicarage and combine (in Darwin's case) beetle collecting with a secure income and a respectable position. One of Darwin's most significant acts during his time at Cambridge was his forging of a friendship with the Reverend Professor John Stevens Henslow, with whom he botanized on walks in the environs of the University and dreamed of a tropical journey in the footsteps of the voyager-naturalist von Humboldt. Darwin also read William Paley's *Natural Theol-*

ogy (1802), which attracted his keen interest. Paley was best known for his doctrine of "perfect adaptation," according to which God had created each animal and plant species perfectly adapted to its particular environment. The theological significance of perfect adaptation is that it appeared to offer empirical support for the existence of purposeful—teleological—design in nature. That was the key assumption of Paley's "argument from design," which held that belief in God could be founded on rational study of "the book of nature." Natural theology enjoyed wide popularity in the vicar-naturalist class in which Darwin moved, members of which delighted in expounding on particularly interesting or vivid cases of adaptation in their articles and sermons in order to point out God's design. The result was the creation of a body of scientific data to which Darwin would later return in constructing his own concepts of adaptation and its role in evolution.

Before he had to seek a professional position, Darwin received an invitation, secured on his behalf by Henslow, to serve as an unpaid naturalist aboard H.M.S. *Beagle*. The voyage of the *Beagle* (1831–1836) became legendary because Darwin's observations in South America were crucial in the formulation of the theory of evolution. In fact, however, it was a standard "hydrographic" expedition of the British Navy, ordered to survey the coastlines of Argentina, Chile, and Peru as part of England's policy of charting continental coasts and islands in regions that were considered to be of strategic importance. The ship carried no fewer than twenty-four chronometers with which to compute longitudes, and the captain, Robert FitzRoy, was a talented navigator. Near the end of the voyage, in February 1836, Darwin reported ruefully that from Tasmania on, the trip was to be devoted to chronometric measurements, with little time left for natural history.

In his popular account of the voyage in the *Journal of Researches*, Darwin demonstrates his excellence as a field naturalist. In South America, as well as collecting many plant and animal specimens, he made geological observations that led his contemporaries to view him as a naturalist of great promise. His work with the mammalian fossil megafauna of Patagonia, his studies of the endemic birds and flowering plants of the Galapagos, and his convincing explanation of the origins of coral islands are impressive—even spectacular—examples of Darwin's public science shortly after the *Beagle* returned to England. During that period, however, a number of his observations caused him to wonder whether the accepted view of the fixity of species was correct. The first of these observations was the existence in the same locality of fossil species and living species that were closely related but not the same. The example he used compared the extinct megatherium and the living armadillo. The second observation was that on the vast expanse of the pampas, with no intervening topographical barriers, the range of one species would grade off into that of a different but closely related species.

Here his example was that of two ostriches, the greater rhea and a smaller one that is now called Darwin's rhea. He became obsessed with the analogy between the cases, one of which suggests change over time (that is, descent), the other, change over space (that is, geographical distribution). The third observation had to do with species he observed in the Galapagos Islands. According to the doctrine of perfect adaptation, the flora and fauna of the Galapagos should have resembled those of the Cape Verde Islands, off the African coast, which he had visited on the journey from England. According to the theory current at that time, places with similar climates and habitats should have had the same species. Clearly they did not. The Cape Verde species resembled those of continental Africa, and those of the Galapagos resembled those of South America. Darwin began to wonder how plants and animals might migrate from one place to another.

Before he boarded the *Beagle*, Darwin had never questioned the truth of Christianity. He was quick to admit, later on, that he had never been much attracted by theological speculation. Notwithstanding familiarity with his grandfather's freethinking, which remained a family tradition for his father and his uncle John Wedgwood, and his exposure to Robert Grant's materialism, he began the voyage with a conventional view of species. It is unlikely that his view of species changed dramatically during the voyage, although Darwin did make a few frankly speculative notes as early as 1835. In 1836, some months before the end of the voyage, he seemed to question the stability of species in his Ornithological Notes. A few months after his return, he began to record evolutionist ideas in what he called the Red Notebook, a practice he continued in a series of notebooks over the next two years.

Darwin's collections of living and fossil animals were too extensive for the young naturalist to study them all personally, and he was not competent to do so. Therefore, he distributed them by subject to experts residing in London. Fossil mammals fell to the comparative anatomist Richard Owen (1804–1892); living birds, to ornithologist John Gould (1804–1881). When these experts began to report their findings, Darwin's doubts about the fixity of species began to crystallize. By the middle of February 1837, Owen had ascertained that virtually all of Darwin's specimens were extinct prototypes of smaller forms currently inhabiting South America.[1] Then, sometime between March 7 and March 12, Gould informed Darwin in a chance meeting that some of the Galapagos birds replaced one another on different

1. Here we follow the convenient chronology compiled by Frank Sulloway in "Darwin and the Galapagos," *Biological Journal of the Linnean Society* 21 (1984): 29–59, page 46, which is consistent with the dating first offered by Sandra Herbert in "The Logic of Darwin's Discovery," Ph.D. dissertation, Brandeis University, 1968, 78.

islands. Darwin had been so convinced that the Galapagos finches were all members of the same species that he had not even bothered to label his specimens by island; Gould had to track down more precisely labeled finches from other members of the *Beagle* crew, including the evangelical Captain FitzRoy. Gould's news astounded Darwin, who had already admitted that such a conclusion would cast doubt on the fixity of species (see Ornithological Notes, chapter 2, 49).

On March 14, Darwin and Gould each read a paper on the two rhea species that replaced each other geographically.[2] On March 15, approximately, Darwin wrote his first clearly evolutionary note in the Red Notebook. In July of the same year, he began Notebook B, which was devoted solely to the species question. About a year later, he wrote in a pocket diary:

> In July [1837] opened first note Book on "transmutation of Species"—Had been greatly struck from about Month of previous March—on character of S. American fossils—& species on Galapagos Archipelago. These facts origin (especially latter) of all my views.[3]

Darwin filled four such transmutation notebooks (B–E) and two more on human evolution (M–N) from 1837 to 1839. In these notebooks, Darwin's wide-ranging speculations explored every branch of natural history that the theory of evolution would subsequently illumine. Darwin had laid the foundations for the coherent evolutionary research edifice that is represented by the chapters of the present book. On 28 September 1838, after having read Malthus's *Essay on Population*, Darwin first formulated his theory of natural selection to explain the origins of adaptation. Paley's perfect adaptation was surpassed by Darwin's relative adaptations arising from the "warring of species as inference from Malthus" (Notebook D:134). As Darwin would write in his *Autobiography* decades later, at last he "had a theory by which to work."

The "origin of the *Origin*" can be followed in this volume from Darwin's early views of species as noted in his *Journal of Researches* (chapter 1), in the notebooks (chapter 2) and in marginal annotations in his copy of Lamarck's *Zoological Philosophy* (chapter 3), and then in two early sketches blocking out his theory (chapter 4). After completing his 1844 Essay, Darwin launched into a major project of invertebrate classification and morphology, his multivolume study of cirripedes, or barnacles; we include a passage dealing with an exquisitely intricate example of adaptation (chapter 5) of the kind that

2. Darwin, "Notes upon the *Rhea Americana* and upon *Rhea Darwinii*," *Proceedings of the Zoological Society of London* 5 (1837): 35–36; Gould, "On a New Rhea (*Rhea Darwinii*) from Mr. Darwin's Collection," *Proceedings of the Zoological Society of London* 5 (1837): 35.

3. *Correspondence*, 2: 430–431, Appendix 2, Chronology 1837–1843.

particularly pleased Darwin because natural selection provided such a simple explanation for it. In the late 1850s, he began to write up his theory in definitive form (called by Darwin the "Big Species Book" and published in 1975 as *Natural Selection*). A section from that work on the principle of divergence is reproduced here to illustrate the progress of Darwin's theorizing during that period (chapter 6). He never finished that volume, however, because in June 1858 he received an essay from Alfred Russel Wallace containing the essence of his own theory, arrived at independently. Wallace's paper (appendix 2) and an abstract of Darwin's theory sent to the American botanist Asa Gray in 1857 (chapter 7) were presented together at a meeting of the Linnaean Society to establish Darwin's priority. Darwin then set to work to produce an "abstract" of the "Big Species Book," which is the volume we know as *On the Origin of Species*. Its first edition sold out on the very day of its publication, 22 November 1859. Included here (in chapter 8) are the famous chapters "Struggle for Existence" and "Natural Selection." We have also excerpted key passages from the chapters "Difficulties on Theory" and "Instinct." These selections are framed by Darwin's introduction and conclusion.

Darwin named his evolutionary mechanism natural selection to contrast it with the artificial selection practiced by plant and animal breeders. He set out his views on the latter in *Variation of Animals and Plants Under Domestication* in 1868. The book is most famous, however, not for its discussion of selection, but for Darwin's bold proposal of a mechanism to explain heredity, which he named pangenesis (chapter 9).

In *On the Origin of Species*, Darwin had deliberately avoided the question of human evolution, perhaps in order to avoid a bitter confrontation with religionists. In 1871, however, after T. H. Huxley (1863), A. R. Wallace (1864), and the German Ernst Haeckel (1868) had all published works on human evolution, Darwin presented his views on the subject in *The Descent of Man*. Our selections (chapter 10) include the famous chapter "Moral Faculties," which is in fact an evolutionary reinterpretation of English natural-man theory of the eighteenth century, and part of the section on sexual selection, which Darwin thought was particularly important for human evolution.

In the latter part of his career, Darwin was a botanist as well as a zoologist. We present excerpts from three of his great botanical works (chapter 11), those on orchids, on cross-fertilization, and on sexual polymorphism, all of which contributed to Darwin's view of adaptation.

The final chapter (chapter 12) is extracted from the autobiography that Darwin prepared for members of his own family, wherein he recounts some of the steps in the development of his theory, including his famous account of reading Thomas Malthus's *Essay On Population* (excerpts in Appendix 1).

We have also included some passages in which Darwin or his correspondents commented on particular works. These selections are presented at the ends of the appropriate chapters.

Where specific themes are reiterated or embroidered, we have cross-referenced them in the text so that the reader can easily track the development of specific ideas or themes. In a similar fashion, we have organized the index around the central themes of Darwin's thought: adaptation, competition, divergence of character, natural selection, speciation, and so forth. Footnotes in Darwin's texts have been omitted, except for those cross-referencing others of his works.

Chapter 1

JOURNAL OF RESEARCHES

INTRODUCTION: FIRST EDITION (1839)

The *Beagle* was a hydrographic vessel—that is, it was equipped to chart coastlines—and it was first sent to South America on a hydrographic mission in 1826 under the captainship of Pringle Stokes. Sir Francis Beaufort, hydrographer to the British Admiralty, deemed the first mission incomplete and dispatched a second in 1831 under Robert FitzRoy. The second voyage was extremely successful, producing eighty-two coastal charts and eight harbor plans.

As the day of departure approached, Darwin's enthusiasm mounted. "Every body, who can judge," he wrote to his cousin William Darwin Fox, "says it is one of the grandest voyages that has almost ever been sent out. Everything is on a grand scale. 24 Chronometers. The whole ship is fitted up with Mahogany, she is the admiration of the whole place." Darwin viewed himself as part of the grand tradition of English exploration: "When I think of all that I am going to see & undergo, it really requires an effort of reasoning to persuade myself, that all is true. That I shall see the same land, that Captain Cook did."[1]

Cook, who explored the South Seas of the Pacific Ocean in three voyages between 1768 and 1779, cast a long shadow in the field of British exploration. If he did not begin the custom of carrying naturalists on voyages of exploration, he certainly developed it considerably. Joseph Banks and Daniel Solander were the botanists on his first expedition, and Darwin was keenly aware that, in his own exploration of Patagonia, he was following their tracks.

---

1. Darwin to W. D. Fox, 17 November 1831; *Correspondence*, 1: 182–183.

Darwin's observations on the pampas of Uruguay and Argentina set in motion a train of thought that led him to question the received view that species were fixed. Insofar as life history was concerned, Darwin made an early and crucial step in his own thinking when he unearthed fossil mammals in the same places where closely related living species were now found, which suggested that the latter were descended from the former. In chapter 9, he first writes of the Guanaco—the wild camelid, closely related to the domestic llama—and what he thought was its fossil ancestor, as well as of the similar cases of fossil edendates (the megatherium being the most spectacular example) and current armadillos. These relationships fascinated him, and later his fascination turned into obsession, as we will see from entries in his notebooks (see, for example, Red Notebook 130, p. 50, and Notebook B:20, p. 54).

The first edition of the *Journal of Researches* was published in 1839.

PATAGONIA

December 6th, 1833. The *Beagle* sailed from the Rio Plata, never again to enter its muddy stream. Our course was directed to Port Desire, on the coast of Patagonia. Before proceeding any further, I will here put together a few observations made at sea.

Several times when the ship has been some miles off the mouth of the Plata, and at other times when off the shores of Northern Patagonia, we have been surrounded by insects. One evening, when we were about ten miles from the Bay of San Blas, vast numbers of butterflies, in bands or flocks of countless myriads, extended as far as the eye could range. Even by the aid of a glass it was not possible to see a space free from butterflies. The seamen cried out 'it was snowing butterflies,' and such in fact was the appearance. More species than one were present, but the main part belonged to a kind very similar to, but not identical with, the common English *Colias edusa*. Some moths and hymenoptera accompanied the butterflies; and a fine Calosoma flew on board. Other instances are known of this beetle having been caught far out at sea; and this is the more remarkable, as the greater number of the Carabidae seldom or never take wing. The day had been fine and calm, and the one previous to it equally so, with light and variable airs. Hence we cannot suppose that the insects were blown off the land, but we must conclude that they voluntarily took flight. The great bands of the Colias seem at first to afford an instance like those on record of the migrations of *Vanessa cardui*; but the presence of other insects makes the case distinct, and not so easily intelligible. Before sunset a strong breeze sprung up from the north, and this must have been the cause of tens of thousands of the butterflies and other insects to have perished.

On another occasion, when seventeen miles off Cape Corrientes, I had a net overboard to catch pelagic animals. Upon drawing it up, to my surprise, I found a considerable number of beetles in it, and although in the open sea, they did not appear much injured by the salt water. I lost some of the specimens, but those which I preserved, belonged to the genera, colymbetes, hydroporus, hydrobius (two species), notaphus, cynucus, adimonia, and scarabaeus. At first, I thought that these insects had been blown from the shore; but upon reflecting that out of the eight species, four were aquatic, and two others partly so in their habits, it appeared to me most probable that they were floated into the sea, by a small stream which drains a lake near Cape Corrientes. On any supposition, it is an interesting circumstance to find live insects, quite alive, swimming in the open ocean, seventeen miles from the nearest point of land. There are several accounts of insects having been blown off the Patagonian shore. Captain Cook observed it, as did more lately Captain King of the *Adventure*. The cause probably is due to the want of shelter, both of trees and hills, so that an insect on the wing with an off-shore breeze, would be very apt to be blown out to sea. The most remarkable instance I ever knew of an insect being caught far from the land, was that of a large grasshopper (*Acrydium*), which flew on board, when the *Beagle* was to windward of the Cape Verd Islands, and when the nearest point of land, not directly opposed to the trade-wind, was Cape Blanco on the coast of Africa, 370 miles distant.

On several occasions, when the vessel has been within the mouth of the Plata, the rigging has been coated with the web of the Gossamer Spider. One day (November 1st, 1832) I paid particular attention to the phenomenon. The weather had been fine and clear, and in the morning the air was full of patches of the flocculent web, as on an autumnal day in England. The ship was sixty miles distant from the land, in the direction of a steady though light breeze. Vast numbers of a small spider, about one-tenth of an inch in length, and of a dusky red colour were attached to the webs. There must have been, I suppose, some thousands on the ship. The little spider when first coming in contact with the rigging, was always seated on a single thread, and not on the flocculent mass. This latter seems merely to be produced by the entanglement of the single threads. The spiders were all of one species, but of both sexes, together with young ones. These latter were distinguished by their smaller size, and more dusky colour. I will not give the description of this spider, but merely state that it does not appear to me to be included in any of Latreille's genera. The little aeronaut as soon as it arrived on board, was very active, running about; sometimes letting itself fall, and then reascending the same thread; sometimes employing itself in making a small and very irregular mesh in the corners between the ropes. It could run with facility on the surface of the water. When disturbed it lifted up its front legs, in the attitude of attention. On its first arrival it appeared very thirsty, and with exserted maxillae drank eagerly of the fluid; this same circumstance has been observed by Strack: may it not be in consequence of the little insect having passed through a dry and rarefied atmosphere? Its stock of web seemed inexhaustible. While watching some that were suspended by a single thread, I several times observed that the slightest breath of air bore them away out of sight, in a horizontal line. On another occasion (25th) under similar circumstances, I repeatedly observed the same kind of small spider, either when placed, or having crawled, on some little eminence, elevate its abdomen, send forth a thread, and then sail away in a lateral course, but with a rapidity which was quite unaccountable. I thought I could perceive that the spider before performing the above preparatory steps, connected its legs together with the most delicate threads, but I am not sure, whether this observation is correct.

One day, at St. Fe, I had a better opportunity of observing some similar facts. A spider which was about three-tenths of an inch in length, and which in its general appearance resembled a Citigrade (therefore quite different from the gossamer), while standing on the summit of a post, darted forth four or five threads from its spinners. These glittering in the sunshine, might be compared to rays of light; they were not, however, straight, but in undulations like a film of silk blown by the wind. They were more than a yard in length, and diverged in an ascending direction from the orifices. The spider then suddenly let go its hold, and was quickly borne out of sight. The day was hot and apparently calm; yet under such circumstances the atmosphere can never be so tranquil, as not to affect a vane so delicate as the thread of a spider's web. If during a warm day we look either at the shadow of any object cast on a bank, or over a level plain at a distant landmark, the effect of an ascending current of heated air will almost always be evident. And this probably would be sufficient to carry with it so light an object as the little spider on its thread. The circumstance of spiders of the same species but of different sexes and ages, being found on several occasions at the distance of many leagues from the land, attached in vast numbers to the lines, proves that they are the manufacturers of the mesh, and that the habit of sailing through the air, is probably as characteristic of some tribe, as that of diving is of the Argyroneta. We may then reject Latreille's supposition, that the gossamer owes its origin to the webs of the young of several genera, as Epeira or Thomisa: although, as we have seen that the young of other spiders do possess the power of performing aerial voyages.

During our different passages south of the Plata, I often towed astern a net made of bunting, and thus caught many curious animals. The structure of the Beroe (a kind of jelly fish) is most extraordinary, with its rows of vibratory ciliae, and complicated though irregular systems of circulation. Of

Crustacea, there were many strange and undescribed genera. One, which in some respects is allied to the Notopods (or those crabs which have their posterior legs placed almost on their backs, for the purpose of adhering to the under side of ledges), is very remarkable from the structure of its hind pair of legs. The penultimate joint, instead of being terminated by a simple claw, ends in three bristle-like appendages of dissimilar lengths, the longest equalling that of the entire leg. These claws are very thin, and are serrated with teeth of an excessive fineness, which are directed towards the base. The curved extremities are flattened, and on this part five most minute cups are placed, which seem to act in the same manner as the suckers on the arms of the cuttle-fish. As the animal lives in the open sea, and probably wants a place of rest, I suppose this beautiful structure is adapted to take hold of globular bodies of the Medusae, and other floating marine animals.

In deep water, far from the land, the number of living creatures is extremely small: south of the latitude 35°. I never succeeded in catching any thing besides some beroe, and a few species of minute crustacea belonging to the Entomostraca. In shoaler water, at the distance of a few miles from the coast, very many kinds of crustacea and some other animals were numerous, but only during the night. Between latitudes 56° and 57° south of Cape Horn the net was put astern several times; it never, however, brought up any thing besides a few of two extremely minute species of Entomostraca. Yet whales and seals, petrels and albatross, are exceedingly abundant throughout this part of the ocean. It has always been a source of mystery to me, on what the latter, which live far from the shore, can subsist. I presume the albatross, like the condor, is able to fast long; and that one good feast on the carcass of a putrid whale lasts for a long siege of hunger. It does not lessen the difficulty to say, they feed on fish; for on what can the fish feed? It often occurred to me, when observing how the waters of the central and intertropical parts of the Atlantic, swarmed with Pteropoda, Crustacea, and Radiata, and with their devourers the flying-fish, and again with *their* devourers the bonitos and albicores, that the lowest of these pelagic animals perhaps possesses the power of decomposing carbonic acid gas, like the members of the vegetable kingdom.

While sailing in these latitudes on one very dark night, the sea presented a wonderful and most beautiful spectacle. There was a fresh breeze, and every part of the surface, which during the day is seen as foam, now glowed with a pale light. The vessel drove before her bows two billows of liquid phosphorus, and in her wake she was followed by a milky train. As far as the eye reached, the crest of every wave was bright, and the sky above the horizon, from the reflected glare of these livid flames, was not so utterly obscure, as over the rest of the heavens.

As we proceed further southward, the sea is seldom phosphorescent; and off Cape Horn, I do not recollect more than once having seen it so, and then it was far from being brilliant. This circumstance probably has a close connection with the scarcity of organic beings in that part of the ocean. After the elaborate paper by Ehrenberg, on the phosphorescence of the sea, it is almost superfluous on my part to make any observations on the subject. I may however add, that the same torn and irregular particles of gelatinous matter, described by Ehrenberg, seem in the southern as well as in the northern hemisphere, to be the common cause of this phenomenon. The particles were so minute as easily to pass through fine gauze; yet many were distinctly visible by the naked eye. The water when placed in a tumbler and agitated gave out sparks, but a small portion in a watch-glass scarcely ever was luminous. Ehrenberg states, that these particles all retain a certain degree of irritability. My observations, some of which were made directly after taking up the water, gave a different result. I may also mention, that having used the net during one night I allowed it to become partially dry, and having occasion twelve hours afterwards, to employ it again, I found the whole surface sparkled as brightly as when first taken out of the water. It does not appear probable in this case, that the particles could have remained so long alive. I remark also in my notes, that having kept a Medusa of the genus Dianaca, till it was dead, the water in which it was placed became luminous. When the waves scintillate with bright green sparks, I believe it is generally owing to minute crustacea. But there can be no doubt that very many other pelagic animals, when alive, are phosphorescent.

On two occasions I have observed the sea luminous at considerable depths beneath the surface. Near the mouth of the Plata some circular and oval patches, from two to four yards in diameter, and with defined outlines, shone with a steady, but pale light; while the surrounding water only gave out a few sparks. The appearance resembled the reflection of the moon, or some luminous body; for the edges were sinuous from the undulations of the surface. The ship, which drew thirteen feet of water, passed over, without disturbing these patches. Therefore we must suppose that some animals were congregated together at a greater depth than the bottom of the vessel.

Near Fernando Noronha the sea gave out light in flashes. The appearance was very similar to that which might be expected from a large fish moving rapidly through a luminous fluid. To this cause the sailors attributed it; at the time, however, I entertained some doubts, on account of the frequency and rapidity of the flashes. With respect to any general observations, I have already stated that the display is very much more common in warm than in cold countries. I have sometimes imagined that a disturbed electrical condition of the atmosphere was most favourable to its production. Certainly I think the sea is most luminous after a few days of more calm weather than ordinary, during which time it has swarmed with various animals. Observing that the water charged with gelatinous particles is in an impure state, and that the luminous appearance in all common cases is produced by the agitation of the fluid in contact with the atmosphere, I have always been inclined to consider that the phosphorescence was the result of the decomposition of the organic particles, by which process (one is tempted almost to call it a kind of respiration) the ocean becomes purified.

December 23rd. We arrived at Port Desire, situated in lat. 47°, on the coast of Patagonia. The creek runs for about twenty miles inland, with an irregular width. The *Beagle* anchored a few miles within the entrance in front of the ruins of an old Spanish settlement.

The same evening I went on shore. The first landing in any new country is very interesting, and especially when, as in this case, the whole aspect bears the stamp of a marked and individual character. At the height of between two and three hundred feet, above some masses of porphyry, a wide plain extends, which is truly characteristic of Patagonia. The surface is quite level, and is composed of well-rounded shingle mixed with a whitish earth. Here and there scattered tufts of brown wiry grass are supported, and still more rarely, some low thorny bushes. The weather is dry and pleasant, and the fine blue sky is but seldom obscured. When standing in the middle of one of these desert plains, the view on one side is generally bounded by the escarpment of another plain, rather higher, but equally level and desolate; and on the other side it becomes indistinct from the trembling mirage which seems to rise from the heated surface.

The plains are traversed by many broad, flat-bottomed valleys, and in these the bushes grow rather more abundantly. The present drainage of the country is quite insufficient to excavate such large channels. In some of the valleys ancient stunted trees, growing in the very centre of the dry watercourse, seem as if placed to prove how long a time had elapsed, since any flood had passed that way. We have evidence, from shells lying on the surface, that the plains of gravel have been elevated within a recent epoch above the level of the sea; and we must look to that period for the excavation of the valleys by the slowly-retiring waters. From the dryness of the climate, a man may walk for days together over these plains without finding a single drop of water. Even at the base of the prophyry hills, there are only a few small wells containing but little water, and that rather saline and half putrid.

In such a country the fate of the Spanish settlement was soon decided; the dryness of the climate during the greater part of the year, and the occasional hostile attacks of the wandering Indians, compelled the colonists to desert their half-finished buildings. The style, however, in which they were commenced, showed the strong and liberal hand of Spain in the old time.

The end of all the attempts to colonize this side of America south of 41°, have been miserable. At Port Famine, the name expresses the lingering and extreme sufferings of several hundred wretched people, of whom one alone survived to relate their misfortunes. At St. Joseph's bay, on the coast of Patagonia, a small settlement was made; but during one Sunday the Indians made an attack and massacred the whole party, excepting two men, who were led captive many years among the wandering tribes. At the Rio Negro I conversed with one of these men, now in extreme old age.

The zoology of Patagonia is as limited as its Flora. On the arid plains a few black beetles (Heteromera) might be seen slowly crawling about, and occasionally a lizard darted from side to side. Of birds we have three carrion hawks, and in the valleys a few finches and insect feeders. The *Ibis malanops* (a species said to be found in central Africa) is not uncommon on the most desert parts. In the stomachs of these birds I found grasshoppers, cicadæ, small lizards, and even scorpions. At one time of the year they go in flocks, at another in pairs: their cry is very loud and singular, and resembles the neighing of the guanaco.

I will here give an account of this latter animal, which is very common, and is the characteristic quadruped of the plains of Patagonia. The Guanaco, which by some naturalists is considered as the same animal with the Llama, but in its wild state, is the South American representative of the camel of the East. In size it may be compared to an ass, mounted on taller legs, and with a very long neck. The guanaco abounds over the whole of the temperate parts of South America, from the wooded islands of Tierra del Fuego, through Patagonia, the hilly parts of La Plata, Chile, even to the Cordillera of Peru. Although preferring an elevated site, it yields in this respect to its near relative the Vicuna. On the plains of Southern Patagonia, we saw them in greater numbers than in any other part. Generally they go in small herds, from half a dozen to thirty together; but on the banks of the St. Cruz we saw one herd which must have contained at least five hundred. On the northern shores of the Strait of Magellan they are also very numerous.

Generally the guanacoes are wild and extremely wary. Mr. Stokes told me, that he one day saw through a glass a herd of these beasts, which evidently had been frightened, running away at full speed, although their distance was so great that they could not be distinguished by the naked eye. The sportsman frequently receives the first intimation of their presence, by hearing from a long distance, the peculiar shrill neighing note of alarm. If he then looks attentively, he will probably see the herd standing in a line on the side of some distant hill. On approaching them, a few more squeals are given, and then off they set at an apparently slow, but really quick canter, along some narrow beaten track to a neighbouring hill. If, however, by chance he should abruptly meet a single animal, or several together, they will generally stand motionless and intently gaze at him; then perhaps move on a few yards, turn round, and look again. What is the cause of this difference in their shiness? Do they mistake a man in the distance for their chief enemy the puma? Or does curiosity overcome their timidity? That they are curious is certain; for if a person lies on the ground, and plays strange antics, such as throwing up his feet in the air, they will almost always approach by degrees to reconnoitre him. It was an artifice that was repeatedly practised by our sportsmen with success, and it had moreover the advantage of allowing several shots to be fired, which were all taken as parts of the performance. On the mountains of Tierra del Fuego, and in other places, I have more than once seen a guanaco, on being approached, not only neigh and squeal, but prance and leap about in the most ridiculous manner, apparently in defiance as a challenge. These animals are very easily domesticated, and I have seen some thus kept near the houses, although at large on their native plains. They are in this state very bold, and readily attack a man, by striking him from behind with both knees. It is asserted that the motive for these attacks is jealousy on account of their females. The wild guanacoes, however, have no idea of defence; even a single dog will secure one of these large animals, till the huntsman can come up. In many of their habits they are like sheep in a flock. Thus when they see men approaching in several directions on horseback, they soon become bewildered, and know not which way to run. This greatly facilitates the Indian method of hunting, for they are thus easily driven to a central point, and are encompassed.

The guanacoes readily take to the water: several times at Port Valdes they were seen swimming from island to island. Byron, in his voyage, says he saw them drinking salt water. Some of our officers likewise saw a herd apparently drinking the briny fluid from a salina near Cape Blanco. I imagine in several parts of the country, if they do not drink salt water, they drink none at all. In the middle of the day, they frequently roll in the dust, in saucer-shaped hollows. The males fight together; two one day passed quite close to me, squealing and trying to bite each other; and several were shot with their hides deeply scored. Herds sometimes appear to set out on exploring-parties: at Bahia Blanca, where, within thirty miles of the coast, these animals are extremely unfrequent, I one day saw the tracks of thirty or forty, which had come in a direct line to a muddy salt-water creek. They then must have perceived that they were approaching the sea, for they had wheeled with the regularity of cavalry, and had returned back in as straight a line as they had advanced. The guanacoes have one singular habit, which is to me quite inexplicable; namely, that on successive days they drop their dung in the same defined heap. I saw one of these heaps which was eight feet in diameter, and necessarily was composed of a large quantity. Frezier remarks on this habit as common to the guanaco as well as to the llama; he says it is very useful to the Indians, who use the dung for fuel, and are thus saved the trouble of collecting it.

The guanacoes appear to have favourite spots for dying in. On the banks of the St. Cruz, the ground was actually white with bones, in certain circumscribed spaces, which were generally bushy and all near the river. On one such spot I counted between ten and twenty heads. I particularly examined the bones; they did not appear, as some scattered ones which I had seen, gnawed or broken, as if dragged together by beasts of prey. The animals in most cases, must have crawled, before dying, beneath and amongst the bushes. Mr. Bynoe informs me that during the last voyage, he observed the same circumstance on the banks of the Rio Gallegos. I do not at all understand the reason of this, but I may observe, that the wounded guanacoes at the St. Cruz, invariably walked towards the river. At St. Jago in the Cape de Verd Islands I remember having seen in a retired ravine a corner under a cliff, where numerous goats' bones were collected: we at the time exclaimed, that it was the burial-ground of all the goats in the island. I mention these trifling circumstances, because in certain cases they might explain the occurrence of a number of uninjured bones in a cave, or buried under alluvial accumulations; and likewise the cause why certain mammalia are more commonly embedded than others in sedimentary deposits. Any great flood of the St. Cruz, would wash down many bones of the guanaco, but probably not a single one of the puma, ostrich, or fox. I may also observe, that almost every kind of waterfowl when wounded takes to the shore to die; so that the remains of birds, from this cause alone and independently of other reasons, would but rarely be preserved in a fossil state.

One day the yawl was sent under the command of Mr. Chaffers with three days' provisions to survey the upper part of the harbour. In the morning we searched for some watering-places mentioned in an old Spanish chart. We found one creek, at the head of which there was a trickling rill (the first we had seen) of brackish water. Here the tide compelled us to wait several hours; and in the interval I walked some miles into the interior. The plain as usual, consisted of gravel, mingled with soil resembling chalk in appearance, but very different from it in nature. From the softness of these materials it was worn into many gullies. There was not a tree, and, excepting the guanaco, which stood on the hill-top a watchful sentinel over its herd, scarcely an animal or a bird. All was stillness and desolation. One reflected how many ages the plain had thus lasted, and how many more it was doomed thus to continue. Yet in passing over these scenes, without one bright object near, an ill-defined but strong sense of pleasure is vividly excited.

In the evening we sailed a few miles further up, and then pitched the tents for the night. By the middle of the next day the yawl was aground, and from the shoalness of the water could not proceed any higher. The water being found partly fresh, Mr. Chaffers took the dingey, and went up two or three miles further, where she also grounded, but in a fresh-water river. The water was muddy, and though the stream was most insignificant in size, it would be difficult to account for its origin, except from the melting snow on the Cordillera. At the spot where we bivouacked, we were surrounded by bold cliffs and steep pinnacles of porphyry. I do not think I ever saw a spot, which appeared more secluded from the rest of the world, than this rocky crevice in the wide plain.

The second day after our return to the anchorage, a party of officers and myself went to ransack an old Indian grave, which I had found on the summit of a neighbouring hill. Two immense stones, each probably weighing at least a couple of tons, had been placed in front of a ledge of rock, about six feet high. At the bottom of the grave on the hard rock, there was a layer of earth about a foot deep, which must have been brought up from the plain below. Above it a pavement of flat stones was placed, on which others were piled, so as to fill up the space between the ledge and the two great blocks. To complete the grave, the Indians had contrived to detach from the ledge a huge fragment, and to throw it over the pile so as to rest on the two blocks. We undermined the grave on both sides, but could not find any relics, or even bones. The latter probably had decayed long since (in which case the grave must have been of extreme antiquity), for I found in another place some smaller heaps, beneath which a very few crumbling fragments could yet be distinguished, as having belonged to a man. Falconer states, that where an Indian dies he is buried, but that subsequently his bones are carefully taken up and carried, let the distance be ever so great, to be deposited near the sea-coast. This custom, I think, may be accounted for, by recollecting, that before the introduction of horses, these Indians must have led nearly the same life as the Fuegians, and therefore generally have resided in the neighbourhood of the sea. The common prejudice of lying where one's ancestors have lain, would make the now roaming Indians bring the less perishable part of their dead to their ancient burial-grounds.

January 9th, 1834. Before it was dark the *Beagle* anchored in the fine spacious harbour of Port St. Julian, situated about one hundred and ten miles to the south of Port Desire. We remained here eight days. The country is nearly similar to that of Port Desire, but perhaps rather more sterile. One day a party accompanied Captain FitzRoy on a long walk round the head of the harbour. We were eleven hours without tasting any water, and some of the party were quite exhausted. From the summit of a hill (since well named Thirsty Hill) a fine lake was spied, and two of the party proceeded with concerted signals to show whether it was fresh water. What was our disappointment to find a snow-white expanse of salt, crystallized in great cubes!

We attributed our extreme thirst to the dryness of the atmosphere; but whatever the cause might be, we were exceedingly glad late in the evening to get back to the boats. Although we could nowhere find, during our whole visit, a single drop of fresh water, yet some must exist; for by an odd chance I found on the surface of the salt water, near the head of the bay, a Colymbetes not quite dead, which in all probability had lived in some not far distant pool. Three other kinds of insects, a Cincindela, like *hybrida*, a Cymindis, and a Harpalus, which all live on muddy flats occasionally overflowed by the sea, and one other beetle found dead on the plain, completes the list of the coleoptera. A good-sized fly (Tabanus) was extremely numerous, and tormented us by its painful bite. The common horsefly, which is so troublesome in the shady lanes of England, belongs to this same genus. We here have the puzzle, that so frequently occurs in the case of musquitoes, on the blood of what animals do these insects commonly feed? The guanaco is nearly the only warmblooded quadruped, and they are present in numbers quite inconsiderable, compared with the multitude of flies.

The foundation of porphyry is not here present, as it was at Port Desire, and in consequence the tertiary deposits are arranged with greater regularity. Five successive plains of different altitudes are very distinct. The lower one is a mere fringe nearly on a level with the sea, but the upper one is elevated 950 feet. This latter is represented in this neighbourhood only by a few truncate conical hills, of exactly the same height. It was very interesting to stand on one of these flat patches of gravel, and viewing the wide surrounding country, to speculate on the enormous quantity of matter which must have been removed, thus to leave these mere points, as measures of the former table-land.

I will now give a brief sketch of the geology of the grand tertiary formation of Patagonia, which extends from the Strait of Magellan to the Bay of S. Antonio. In Europe, deposits of the more recent eras have generally been accumulated in small basins or trough-shaped hollows. In South America, however, the entire plains of Patagonia extending seven hundred miles in length, and backed on the one hand by the chain of the Andes, and fronted on the other by the shores of the Atlantic, are thus constituted. Moreover the northern boundary is merely assumed in consequence of a mineralogical change in the strata: if organic remains were present, it probably would be found to be only an artificial limit. Again to the northward (1300 miles distant from the Strait of Magellan) we have the Pampas deposit, which though very different in composition, belongs to the same epoch with the superficial covering of the plains of Patagonia.

The cliffs on the coast give the following section: The lower part consists of a soft sandstone, containing large concretions of a harder nature. These strata contain many organic remains—immense oysters nearly a foot in diameter, curious pectens, echini, turritellae, and other shells, of which the greater portion are extinct, but a few resemble those now existing on the coast. Above these fossiliferous beds, a mass of friable stone or earth is superimposed, which, from its extreme whiteness, has been mistaken for chalk. It is, however, quite different; and closely resembles the less argillaceous varieties of decomposed felspar. This substance never contains organic remains. Lastly, the cliff is surmounted by a thick bed of gravel, almost exclusively derived from porphyritic rocks. For the sake of making the following description more easily intelligible, I have subjoined an imaginary section of the plains near the coast. It must be observed, that the width of each plain is in nature *very much* greater in proportion to the height, than here represented.

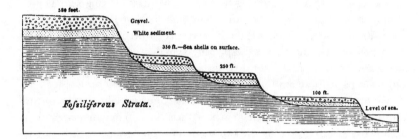

The whole series is horizontally stratified, and I do not recollect ever seeing signs of violence, not even such as a fault. The gravel covers the entire surface of the land, from the Rio Colorado to the Strait of Magellan, a space of 800 miles, and is one chief cause of the desert character of Patagonia. Judging from a section across the continent at the St. Cruz river, and from some other reasons, I believe the gravel beds gradually thickening as they ascend, every where reach the base of the Cordillera. It is to these mountains we must look for the parent rocks, of at least a portion of the well-rounded fragments. I apprehend so great an area covered by shingle, could scarcely be pointed out in any other part of the world.

Having said thus much of the constitution, let us look at the external configuration of the mass. The level plains are cut off along the whole line of coast by perpendicular cliffs, which are necessarily of different altitudes, because any one of the successive terraces, which, as I have already noticed, rise like steps one above the other, may form the sea cliff. These steps are often several miles broad; but from one point of view I have seen four very distinct lines of escarpment abutting one over the other. Having observed that the plains appeared to run for great distances along the coast at the same level, I measured barometrically the elevation of some of them, and compared these measurements, and took all those made by the officers employed in the survey. I was astonished to find at how great distances, even of 600 miles, plains occurred that had, within a few feet of difference, the same elevation. I believe I can distinguish seven or eight distinct terraces which occur along the line of the coast, and which include heights between 1200 feet and the level of the sea. It will be understood that they are not always present, for the lower ones have in some parts been removed by the action of the sea sooner than in others. When any broad valley enters the country, the terraces sweep round and run up on each side; in which case, the correspondence on the opposite sides is beautifully illustrated.

I have called these step-like plains level, because they appear to be absolutely so to the eye, but in truth they rise a little between the edge of one line of cliff and the base of the next above it. Their slope is about the same as that of the gradually shoaling bottom of the neighbouring sea. The elevation of 350 feet is gained by three steps; one of about 100 feet, the second 250, and the third 350. Over these three plains marine remains are frequently scattered, but they are especially abundant on the lower one. The shells are the same as the now existing littoral species, and the muscle and turbo yet partially retain their blue and purple colours.

We have now stated the problem, which is to be explained so as to connect together these various phenomena. At first I could only understand the grand covering of gravel, by the supposition of some epoch of extreme violence, and the successive lines of cliff, by as many great elevations, the precise action of which I could not however follow out. Guided by the *Principles of Geology*, and having under my view the vast changes going on in this continent, which at the present day seems the great workshop of nature, I came to another, and I hope more satisfactory conclusion. The importance of any view which may explain the agency by which such vast beds of shingle have been transported over the surface of the successive plains, cannot be doubted. Whatever the cause may have been, it has determined the condition of this desert country, with respect to its form, nature, and capabilities of supporting life.

There are proofs, that the whole coast has been elevated to a considerable height within the recent period; and on the shores of the Pacific, where successive terraces likewise occur, we know that these changes have latterly been very gradual. There is indeed reason for believing, that the uplifting of the ground during the earthquakes in Chile, although only to the height of two or three feet, has been a disturbance which may be considered as a great one, in comparison to the series of lesser and scarcely sensible movements which are likewise in progress. Let us then imagine the consequence of the shoaling bed of an ocean, elevated at a perfectly equable rate, so that the same number of feet should be converted into dry land in each succeeding century. Every part of the surface would then have been exposed for an equal length of time to the action of the beach-line, and the whole in consequence equally modified. The shoaling bed of the ocean would thus be changed into a sloping land, with no marked line on it. If, however, there should occur a long period of repose in the elevations, and the currents of the sea should tend to wear away the land (as happens along this whole coast), then there would be formed a line of cliff. Accordingly as the repose was long, so would be the quantity of land consumed, and the consequent height of such cliffs. Let the elevations recommence, and another sloping bank (of shingle, or sand, or mud, according to the nature of the successive beach-lines) must be formed, which again will be broken by as many lines of cliff, as there shall be periods of rest in the action of the subterranean forces. Now this is the structure of the plains of Patagonia; and such gradual changes harmonize well with the undisturbed strata, extending over so many hundred miles.

I must here observe, that I am far from supposing that the entire coast of this part of the continent has ever been lifted up, to the height of even a foot, at any one moment of time; but, drawing our analogies from the shores of the Pacific, that the whole may have been insensibly rising, with every now and then a paroxysmal or accelerated movement in certain spots. With respect to the alternation of the periods of such continued rise and those of quiescence, we may grant that they are probable, because such alternation agrees with what we see in the action, not only of a single volcano, but like-wise of the disturbances affecting whole regions of the earth. At the present day, to the north of the parallel 44°, the subterranean forces are constantly manifesting their power over a space of more than one thousand miles. But to the southward of that line, as far as Cape Horn, an earthquake is seldom or never experienced, and there is not a single point of eruption; yet in former ages, as we shall hereafter show, deluges of lava flowed from that very part. It is in conformity with our hypothesis that this southern region of repose, is at present suffering from the inroads of the ocean, as attested by the long line of cliff on the Patagonian coast. Such we believe to have been the causes of this singular configuration of the land. Nevertheless, we confess that it at first appears startling, that the most marked intervals between the heights of the successive plains should, instead of some great and sudden action of the subterranean forces, only indicate a longer period of repose.

In explaining the widely-spread bed of gravel, we must first suppose a great mass of shingle to be collected by the action of innumerable torrents, and the swell of an open ocean, at the submarine basis of the Andes, prior to the elevation of the plains of Patagonia. If such a mass should then be lifted up, and left exposed during one of the periods of subterranean repose; a certain breadth, for instance a mile, would be washed down, and spread out over the bottom of the invading waters. (That the sea near the coast can carry out pebbles, we may feel sure from the circumstance of their gradual decrease in size, according to the distance from the coast-line.)

If this part of the sea should now be elevated, we should have a bed of gravel, but it would be of less thickness than in the first mass, both because it is spread over a larger area, and because it has been much reduced by attrition. This process being repeated, we might carry beds of gravel, always decreasing in thickness (as happens in Patagonia) to a considerable distance from the line of parent rock. For instance, on the banks of the St. Cruz at the distance of one hundred miles above the mouth of the river, the bed of gravel is 212 feet thick, whereas, near the coast, it seldom exceeds 25 or 30 feet; the thickness being thus reduced to nearly one-eighth.

I have already stated that the gravel is separated from the fossiliferous strata by some white beds of a friable substance, singularly resembling chalk, but which cannot be compared, as far as I am aware, with any formation in Europe. With respect to its origin, I may observe that the well-rounded pebbles all consist of various felspathic porphyries; and that, from their prolonged attrition, during the successive remodellings of the whole mass, much sediment must have been produced. I have already remarked that the white earthy matter more closely resembles decomposed felspar, than any other substance. If such is its origin, it would always, from its lightness, be carried further to seaward than the pebbles. But as the land was elevated, the beds would be brought nearer the coast-line, and so become covered by the fresh masses of gravel which were travelling outwards. When these white beds were themselves elevated, they would hold a position intermediate between the gravel and the common foundation, or the fossiliferous strata. To explain my meaning more clearly, let us suppose the bottom of the present sea covered to a certain distance from the coast-line, with pebbles gradually decreasing in size, and beyond it the white sediment. Let the land rise, so that the beach-line, by the fall of the water, may be carried outwards; then likewise the gravel, by the same agency as before, will be transported so much further from the coast, and will cover the white sediment, and these beds again will invade the more distant parts of the bottom of the sea. By this outward progress, the order of supposition must always be gravel, white sediment, and the fossiliferous strata.

Such is the history of the changes by which the present condition of Patagonia has, I believe, been determined. These changes all result from the assumption of a steady but very gradual elevation, extending over a wide area, and interrupted at long intervals by periods of repose. But we must now return to Port St. Julian. On the south side of the harbour, a cliff of about ninety feet in height intersects a plain constituted of the formations above described; and its surface is strewed over with recent marine shells. The gravel, however, differently from that in every other locality, is covered by a very irregular and thin bed of a reddish loam, containing a few small calcareous concretions. The matter somewhat resembles that of the Pampas, and probably owes its origin either to a small stream having formerly entered the sea at that spot, or to a mud-bank similar to those now existing at the head of the harbour. In one spot this earthy matter filled up a hollow, or gully, worn quite through the gravel, and in this mass a group of large bones was embedded. The animal to which they belonged, must have lived, as in the case at Bahia Blanca, at a period long subsequent to the existence of the shells now inhabiting the coast. We may feel sure of this, because the formation of the lower terrace or plain, must necessarily have been posterior to those above it, and on the surface of the two higher ones, sea-shells of recent species are scattered. From the small physical change, which the last one hundred feet elevation of the continent could have produced, the climate, as well as the general condition of Patagonia, probably was nearly the same, at the time when the animal was embedded, as it now is. This conclusion is moreover supported by the identity of the shells belonging to the two ages. Then immediately occurred the difficulty, how could any large quadruped have subsisted on these wretched deserts in lat. 49° 15′? I had no idea at the time, to what kind of animal these remains belonged. The puzzle, however, was soon solved when Mr. Owen examined them: for he considers that they formed part of an animal allied to the guanaco or llama, but fully as large as the true camel. As all the existing members of the family of Camellidae are inhabitants of the most sterile countries, so may we suppose was this extinct kind. The structure of the cervical vertebrae, the transverse processes not being perforated for the vertebral artery, indicates its affinity: some other parts, however, of its structure, probably are anomalous.

The most important result of this discovery, is the confirmation of the law that existing animals have a close relation in form with extinct species. As the guanaco is the characteristic quadruped of Patagonia, and the vicuna of the snow-clad summits of the Cordillera, so in bygone days, this gigantic species of the same family must have been conspicuous on the southern plains. We see this same relation of type between the existing and fossil Ctenomys, between the capybara (but less plainly, as shown by Mr. Owen) and the gigantic Toxodon, and lastly, between the living and extinct Edentata. At the present day, in South America, there exist probably nineteen species of this order, distributed into several genera; while throughout the rest of the world there are but five. If, then, there is a relation between the living and the dead, we should expect that the Edentata would be numerous in the fossil state. I need only reply by enumerating the megatherium, and the three or four other great species, discovered at Bahia Blanca; the remains of some of which are also abundant over the whole immense territory of La Plata. I have already pointed out the singular relation between the armadilloes and

their great prototypes, even in a point apparently of so little importance as
their external covering.

Darwin's assertion of the "law" that animals now living are related to
extinct species with the same ranges—especially the living and fossil
Guanaco, the armadillo and the Megatherium—was an important key
to his early evolutionary thought. See *Notebook* entries below, on pp.
50, 54, 58, 61.

The order of rodents at the present day, is most conspicuous in South
America, on account of the vast number and size of the species, and the
multitude of individuals: according to the same law, we should expect to find
their representatives in a fossil state. Mr. Owen has shown how far the Toxodon is thus related; and it is moreover not improbable that another large
animal has likewise a similar affinity.

The teeth of the rodent nearly equalling in size those of the Capybara,
which were discovered near Bahia Blanca, must also be remembered.

The law of the succession of types, although subject to some remarkable
exceptions, must possess the highest interest to every philosophical naturalist, and was first clearly observed in regard to Australia, where fossil remains of a large and extinct species of Kangaroo and other marsupial animals were discovered buried in a cave. In America the most marked change
among the mammalia has been the loss of several species of Mastodon, of an
elephant, and of the horse. These Pachydermata appear formerly to have
had a range over the world, like that which deer and antelopes now hold. If
Buffon had known of these gigantic armadilloes, llamas, great rodents, and
lost pachydermata, he would have said with a greater semblance of truth,
that the creative force in America had lost its vigour, rather than that it had
never possessed such powers.

It is impossible to reflect without the deepest astonishment, on the changed
state of this continent. Formerly it must have swarmed with great monsters,
like the southern parts of Africa, but now we find only the tapir, guanaco,
armadillo, and capybara; mere pigmies compared to the antecedent races.
The greater number, if not all, of these extinct quadrupeds lived at a very
recent period; and many of them were contemporaries of the existing mol-
luscs. Since their loss, no very great physical changes can have taken place
in the nature of the country. What then has exterminated so many living
creatures? In the Pampas, the great sepulchre of such remains, there are no
signs of violence, but on the contrary, of the most quiet and scarcely sensible
changes. At Bahia Blanca I endeavoured to show the probability that the an-

cient Edentata, like the present species, lived in a dry and sterile country, such as now is found in that neighbourhood. With respect to the camel-like llama of Patagonia, the same grounds which, before knowing more than the size of the remains, perplexed me, by not allowing any great change of climate, now that we can guess the habits of the animal, are strangely confirmed. What shall we say of the death of the fossil horse? Did those plains fail in pasture, which afterwards were overrun by thousands and tens of thousands of the successors of the fresh stock introduced with the Spanish colonist? In some countries, we may believe, that a number of species subsequently introducd, by consuming the food of the antecedent races, may have caused their extermination; but we can scarcely credit that the armadillo has devoured the food of the immense Megatherium, the capybara of the Toxodon, or the guanaco of the camel-like kind. But granting that all such changes have been small, yet we are so profoundly ignorant concerning the physiological relations, on which the life, and even health (as shown by epidemics) of any existing species depends, that we argue with still less safety about either the life or death of any extinct kind.

One is tempted to believe in such simple relations, as variation of climate and food, or introduction of enemies, or the increased numbers of other species, as the cause of the succession of races. But it may be asked whether it is probable that any such cause should have been in action during the same epoch over the whole northern hemisphere, so as to destroy the *Elephas primigenus*, on the shores of Spain, on the plains of Siberia, and in Northern America; and in a like manner, the *Bos urus*, over a range of scarcely less extent? Did such changes put a period to the life of *Mastodon angustidens*, and of the fossil horse, both in Europe and on the Eastern slope of the Cordillera in Southern America? If they did, they must have been changes common to the whole world; such as gradual refrigeration, whether from modifications of physical geography, or from central cooling. But on this assumption, we have to struggle with the difficulty that these supposed changes, although scarcely sufficient to affect molluscous animals either in Europe or South America, yet destroyed many quadrupeds in regions now characterized by *frigid*, *temperate*, and *warm* climates! These cases of extinction forcibly recall the idea (I do not wish to draw any close analogy) of certain fruit-trees, which, it has been asserted, though grafted on young stems, planted in varied situations, and fertilized by the richest manures, yet at one period, have all withered away and perished. A fixed and determined length of life has in such cases been given to thousands and thousands of buds (or individual germs), although produced in long succession. Among the greater number of animals, each individual appears nearly independent of its kind; yet all of one kind may be bound together by common laws, as well as a certain number of individual buds in the tree, or polypi in the Zoophyte.

I will add one other remark. We see that whole series of animals, which have been created with peculiar kinds of organization, are confined to certain areas; and we can hardly suppose these structures are only adaptations to peculiarities of climate or country; for otherwise, animals belonging to a distinct type, and introduced by man, would not succeed so admirably, even to the extermination of the aborigines. On such grounds it does not seem a necessary conclusion, that the extinction of species, more than their creation, should exclusively depend on the nature (altered by physical changes) of their country. All that at present can be said with certainty, is that, as with the individual, so with the species, the hour of life has run its course, and is spent.

INTRODUCTION: SECOND EDITION (1845)

The *Journal of Researches*, now commonly referred to as *The Voyage of the Beagle*, was extensively revised for the 1845 second edition, and no chapter was more thoroughly reworked than that on the Galapagos Islands. The revisions reflect the fact that Darwin had accomplished much in his career between 1839 and 1845. He had completed the publication of the mammoth *Zoology of the Beagle*, with its beautifully illustrated scientific descriptions of his biological specimens from the voyage. Such rapid publication of a major voyage collection, which included Richard Owen's descriptions of the spectacular South American fossil mammalia, was important for Darwin's scientific reputation. No less important was Darwin's private science during this period. Between 1837 and 1844, he formulated the essence of his theory of natural selection first in notebooks and then in the essays that would form the structural core of *On the Origin of Species* (see chapters 4, 6 and 8). Yet these momentous documents—and the full-blown theory of evolution they contained—were indeed private; after 1844, Darwin stored them away for more than a decade.

It is the extensively revised 1845 Galapagos chapter of the *Journal of Researches* that provides the link between Darwin's public and private science. The new chapter is organized along evolutionary lines, and for those who knew how to read them, Darwin's covert messages were as plain as day. The major emphasis is on what Darwin calls "aboriginal creations." Today, we call them endemic species—that is, species that are found in only one place in the world. Darwin found that the Galapagos Archipelago, with its numerous islands hundreds of miles off the coast of South America, abounds in "aboriginal creations" in group after group of animals and plants. As Darwin puts it:

> The natural history of these islands is eminently curious, and well deserves attention. Most of the organic productions are aboriginal creations, found nowhere

else; there is even a difference between the inhabitants of the different islands; yet all show a marked relationship with those of America. . . . The archipelago is a little world within itself, or rather a satellite attached to America . . . (*Journal of Researches,* 27)

Darwin understood that the Galapagos Islands—those isolated volcanoes only recently emerged from the Pacific—were a living laboratory in the early stages of the origin of new species. Certain examples were outstanding: the distinct species of giant tortoises whose shell anatomy revealed their island of origin to the experienced observer (see Ornithological Notes, chapter 2, 49), the flowering plants with rates of endemism over 50 percent, and substantial numbers of new species found only on individual islands. The most famous example was "a most singular group of finches"—thirteen species, all endemic, with a "perfect gradation in the size of the beaks in the different species . . . from one as large as that of the hawfinch to that of a chaffinch . . . to that of a warbler" (*Journal of Researches,* 28).[1]

As he describes this differentiation of species within the limited confines of this "little world within itself," Darwin rather broadly drops various evolutionary hints. For example, in the following passage, well-read naturalists recognize the language of Sir John Herschel's appeal to Charles Lyell to solve natural history's "mystery of mysteries," the origin of species:

Hence, both in space and time, we seem to be brought somewhat near to that great fact—that mystery of mysteries—the first appearance of new beings on this earth. (*Journal of Researches,* 27)

The same language would appear in the very first paragraph of the *Origin* fifteen years later, in 1859. But for now, Darwin not only brings us near to "that great fact," he also brings us near to an evolutionary interpretation of that "fact":

Seeing this gradation and diversity of structure in one small, intimately-related group of birds, one might really fancy that from an original paucity of birds in this archipelago one species has been taken and modified for different ends. (*Journal of Researches,* 29)

Darwin even went beyond evolutionary hints in his revision of the Patagonia chapter for the 1845 edition. There he tipped his Malthusian hand, introducing Malthusian explanations that began by addressing the "perplexing" problem of mass extinctions but wandered toward questions of evolutionary ecology.

1. See Frank Sulloway's fascinating accounts (1982) of the complex history of Darwin's finches; see bibliography.

We do not steadily bear in mind, how profoundly ignorant we are of the conditions of existence of every animal; nor do we always remember that some check is constantly preventing the too rapid increase of every organized being left in a state of nature. The supply of food, on an average, remains constant; yet the tendency in every animal to increase by propagation is geometrical; and its surprising effects have nowhere been more astonishingly shown, than in the case of the European animals run wild during the last few centuries in America. Every animal in a state of nature regularly breeds; yet in a species long established, any *great* increase in numbers is obviously impossible, and must be checked by some means. We are, nevertheless, seldom able with certainty to tell in any given species, at what period of life, or at what period of the year, or whether only at long intervals, the check falls; or, again, what is the precise nature of the check. Hence probably it is, that we feel so little surprise at one, of two species closely allied in habits, being rare and the other abundant in the same district; or, again, that one should be abundant in a neighbouring district, differing very little in conditions. If asked how this is, one immediately replies that it is determined by some slight difference in climate, food, or the number of enemies: yet how rarely, if ever, we can point out the precise cause and manner of the check! We are, therefore, driven to the conclusion, that causes generally quite inappreciable by us, determine whether a given species shall be abundant or scanty in numbers. (*Journal of Researches*, Chap. 8)

For many of its original readers, the *Journal of Researches* was a wonderful account of a naturalist's voyage around the world. For some few, it provided insight into the grand theory of nature its author had already fully conceived. It is intriguing to speculate about what Alfred Russel Wallace made of the second edition of the *Journal of Researches*. In 1858, in a malarial dream, he "independently" discovered natural selection thirty years after Darwin (see appendix 2). Some scholars have suggested that Wallace influenced the final maturation of Darwin's conception of natural selection. It is perhaps more likely that Darwin influenced the very first steps of Wallace's discovery. Certainly, it is little wonder that when the openly evolutionist *Vestiges of the Natural History of Creation* was published anonymously in 1845, Darwin was suspected of being Mr. Vestiges.

GALAPAGOS ARCHIPELAGO

September 15th.—This archipelago consists of ten principal islands, of which five exceed the others in size. They are situated under the equator, and between five and six hundred miles westward of the coast of America. They are all formed of volcanic rocks; a few fragments of granite curiously glazed and altered by the heat can hardly be considered as an exception. Some of the craters surmounting the larger islands are of immense size, and they rise to a height of between three and four thousand feet. Their flanks are studded by innumerable smaller orifices. I scarcely hesitate to affirm that there must be in the whole archipelago at least two thousand craters. These consist either of lava or scoriae, or of finely-stratified, sandstone-like tuff. Most of the latter are beautifully symmetrical; they owe their origin to eruptions of volcanic mud without any lava: it is a remarkable circumstance that every one of the twenty-eight tuff-craters which were examined, had their southern sides either much lower than the other sides, or quite broken down and removed. As all these craters apparently have been formed when standing in the sea, and as the waves from the trade wind and the swell from the open Pacific here unite their forces on the southern coasts of all the islands, this singular uniformity in the broken state of the craters, composed of the soft and yielding tuff, is easily explained.

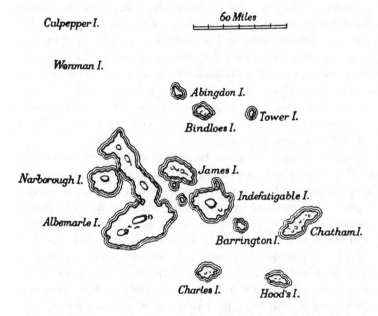

Considering that these islands are placed directly under the equator, the climate is far from being excessively hot; this seems chiefly caused by the singularly low temperature of the surrounding water, brought here by the great southern polar current. Excepting during one short season, very little rain falls, and even then it is irregular; but the clouds generally hang low. Hence, while the lower parts of the islands are very sterile, the upper parts, at a height of a thousand feet and upward, possess a damp climate and a tolerably luxuriant vegetation. This is especially the case on the windward sides of the islands, which first receive and condense the moisture from the atmosphere.

In the morning (17th) we landed on Chatham Island, which, like the others, rises with a tame and rounded outline, broken here and there by scattered hillocks, the remains of former craters. Nothing could be less inviting than the first appearance. A broken field of black basaltic lava, thrown into the most rugged waves, and crossed by great fissures, is everywhere covered by stunted, sunburned brushwood, which shows little signs of life. The dry and parched surface, being heated by the noonday sun, gave to the air a close and sultry feeling, like that from a stove: we fancied even that the bushes smelled unpleasantly. Although I diligently tried to collect as many plants as possible, I succeeded in getting very few; and such wretched-looking little weeds would have better become an arctic than an equatorial Flora. The brushwood appears, from a short distance, as leafless as our trees during winter; and it was some time before I discovered that not only almost every plant was now in full leaf, but that the greater number were in flower. The commonest bush is one of the Euphorbiaceae: an acacia and a great odd-looking cactus are the only trees which afford any shade. After the season of heavy rains, the islands are said to appear for a short time partially green. The volcanic island of Fernando Noronha, placed in many respects under nearly similar conditions, is the only other country where I have seen a vegetation at all like this of the Galapagos Islands.

The *Beagle* sailed round Chatham Island, and anchored in several bays. One night I slept on shore on a part of the island, where black truncated cones were extraordinarily numerous: from one small eminence I counted sixty of them, all surmounted by craters more or less perfect. The greater number consisted merely of a ring of red scoriae or slags, cemented together: and their height above the plain of lava was not more than from fifty to a hundred feet; none had been very lately active. The entire surface of this part of the island seems to have been permeated, like a sieve, by the subterranean vapors: here and there the lava, while soft, has been blown into great bubbles; and in other parts, the tops of caverns similarly formed have fallen in, leaving circular pits with steep sides. From the regular form of the many craters, they gave to the country an artificial appearance, which vividly reminded me of those parts of Staffordshire, where the great iron foundries are most numerous. The day was glowing hot, and the scrambling over the rough surface and through the intricate thickets, was very fatiguing; but I was well repaid by the strange Cyclopean scene. As I was walking along I met two large tortoises, each of which must have weighed at least two hundred pounds: one was eating a piece of cactus, and as I approached, it stared at me and slowly stalked away; the other gave a deep hiss, and drew in its head. These huge reptiles, surrounded by the black lava, the leafless shrubs, and large cacti, seemed to my fancy like some antediluvian animals. The few dull-coloured birds cared no more for me than they did for the great tortoises.

23rd.—The *Beagle* proceeded to Charles Island. This archipelago has long been frequented, first by the buccaneers, and latterly by whalers, but it is only within the last six years that a small colony has been established here. The inhabitants are between two and three hundred in number: they are nearly all people of color, who have been banished for political crimes from the Republic of the Equator, of which Quito is the capital. The settlement is placed about four and a half miles inland, and at a height probably of a thousand feet. In the first part of the road we passed through leafless thickets, as in Chatham Island. Higher up, the woods gradually became greener; and as soon as we crossed the ridge of the island, we were cooled by a fine southerly breeze, and our sight refreshed by a green and thriving vegetation. In this upper region coarse grasses and ferns abound; but there are no tree-ferns: I saw nowhere any member of the Palm family, which is the more singular, as 360 miles northward, Cocos Island takes its name from the number of cocoanuts. The houses are irregularly scattered over a flat space of ground, which is cultivated with sweet potatoes and bananas. It will not easily be imagined how pleasant the sight of black mud was to us, after having been so long accustomed to the parched soil of Peru and northern Chile. The inhabitants, although complaining of poverty, obtain, without much trouble, the means of subsistence. In the woods there are many wild pigs and goats; but the staple article of animal food is supplied by the tortoises. Their numbers have of course been greatly reduced in this island, but the people yet count on two days' hunting giving them food for the rest of the week. It is said that formerly single vessels have taken away as many as seven hundred, and that the ship's company of a frigate some years since brought down in one day two hundred tortoises to the beach.

September 29th.—We doubled the southwest extremity of Albemarle Island, and the next day were nearly becalmed between it and Narborough Island. Both are covered with immense deluges of black naked lava, which have flowed either over the rims of the great caldrons, like pitch over the rim of a pot in which it has been boiled, or have burst forth from smaller orifices on the flanks; in their descent they have spread over miles of the sea-coast. On both of these islands, eruptions are known to have taken place; and in Albemarle, we saw a small jet of smoke curling from the summit of one of the great craters. In the evening we anchored in Bank's Cove, in Albemarle Island. The next morning I went out walking. To the south of the broken tuff-crater, in which the *Beagle* was anchored, there was another beautifully symmetrical one of an elliptic form; its longer axis was a little less than a mile, and its depth about 500 feet. At its bottom there was a shallow lake, in

the middle of which a tiny crater formed an islet. The day was overpower-
ingly hot, and the lake looked clear and blue: I hurried down the cindery
slope, and, choked with dust, eagerly tasted the water—but, to my sorrow,
I found it salt as brine.

The rocks on the coast abounded with great black lizards, between three
and four feet long; and on the hills, an ugly yellowish-brown species was
equally common. We saw many of this latter kind, some clumsily running
out of our way, and others shuffling into their burrows. I shall presently describe in more detail the habits of both these reptiles. The whole of this
northern part of Albemarle Island is miserably sterile.

October 8th.—We arrived at James Island: this island, as well as Charles
Island, were long since thus named after our kings of the Stuart line. Mr.
Bynoe, myself, and our servants were left here for a week, with provisions
and a tent, while the *Beagle* went for water. We found here a party of
Spaniards, who had been sent from Charles Island to dry fish and to salt
tortoise-meat. About six miles inland, and at the height of nearly 2,000 feet,
a hovel had been built in which two men lived, who were employed in catching tortoises, while the others were fishing on the coast. I paid this party two
visits, and slept there one night. As in the other islands, the lower region
was covered by nearly leafless bushes, but the trees were here of a larger
growth than elsewhere, several being two feet and some even two feet nine
inches in diameter. The upper region being kept damp by the clouds supports a green and flourishing vegetation. So damp was the ground that there
were large beds of a coarse cyperus, in which great numbers of a very small
water-rail lived and bred. While staying in this upper region we lived entirely upon tortoise-meat: the breast-plate roasted (as the Gauchos do *carne
con cuero*), with the flesh on it, is very good; and the young tortoises make
excellent soup; but otherwise the meat to my taste is indifferent.

One day we accompanied a party of the Spaniards in their whale-boat to
a salina, or lake from which salt is procured. After landing, we had a very
rough walk over a rugged field of recent lava, which has almost surrounded
a tuff-crater, at the bottom of which the salt lake lies. The water is only three
or four inches deep, and rests on a layer of beautifully crystallized white salt.
The lake is quite circular, and is fringed with a border of bright green succulent plants; the almost precipitous walls of the crater are clothed with wood,
so that the scene was altogether both picturesque and curious. A few years
since, the sailors belonging to a sealing-vessel murdered their captain in this
quiet spot; and we saw his skull lying among the bushes.

During the greater part of our stay of a week the sky was cloudless, and
if the trade wind failed for an hour, the heat became very oppressive. On
two days, the thermometer within the tent stood for some hours at 93°; but
in the open air, in the wind and sun, at only 85°. The sand was extremely

hot; the thermometer placed in some of a brown colour immediately rose to 137°, and how much above that it would have risen I do not know, for it was not graduated any higher. The black sand felt much hotter, so that even in thick boots it was quite disagreeable to walk over it.

The natural history of these islands is eminently curious, and well deserves attention. Most of the organic productions are aboriginal creations, found nowhere else; there is even a difference between the inhabitants of the different islands; yet all show a marked relationship with those of America, though separated from that continent by an open space of ocean between 500 and 600 miles in width. The archipelago is a little world within itself, or rather a satellite attached to America, whence it has derived a few stray colonists and has received the general character of its indigenous productions. Considering the small size of the islands, we feel the more astonished at the number of their aboriginal beings, and at their confined range. Seeing every height crowned with its crater, and the boundaries of most of the lava-streams still distinct, we are led to believe that within a period geologically recent the unbroken ocean was here spread out. Hence, both in space and time, we seem to be brought somewhat near to that great fact—that mystery of mysteries—the first appearance of new beings on this earth.

Of terrestrial mammals, there is only one which must be considered as indigenous, namely, a mouse (Mus Galapagoensis), and this is confined, as far as I could ascertain, to Chatham Island, the most easterly island of the group. It belongs, as I am informed by Mr. Waterhouse, to a division of the family of mice characteristic of America. At James Island there is a rat sufficiently distinct from the common kind to have been named and described by Mr. Waterhouse; but as it belongs to the old-world division of the family, and as this island has been frequented by ships for the last hundred and fifty years, I can hardly doubt that this rat is merely a variety produced by the new and peculiar climate, food, and soil to which it has been subjected. Although no one has a right to speculate without distinct facts, yet, even with respect to the Chatham Island mouse, it should be borne in mind that it may possibly be an American species imported here; for I have seen, in a most unfrequented part of the Pampas, a native mouse living in the roof of a newly-built hovel, and therefore its transportation in a vessel is not improbable: analogous facts have been observed by Dr. Richardson in North America.

Of land-birds I obtained twenty-six kinds, all peculiar to the group and found nowhere else, with the exception of one lark-like finch from North America (Dolichonyx oryzivorus), which ranges on that continent as far north as 54°, and generally frequents marshes. The other twenty-five birds consist, firstly, of a hawk, curiously intermediate in structure between a buzzard and the American group of carrion-feeding Polybori; and with these latter birds it agrees most closely in every habit and even tone of voice. Secondly, there are two owls, representing the short-eared and white barn-owls of Europe. Thirdly, a wren, three tyrant-flycatchers (two of them species of Pyrocephalus, one or both of which would be ranked by some ornithologists as only varieties), and a dove—all analogous to, but distinct from, American species. Fourthly, a swallow, which though differing from the Progne purpurea of both Americas, only in being rather duller colored, smaller, and slenderer, is considered by Mr. Gould as specifically distinct. Fifthly, there are three species of mocking-thrush—a form highly characteristic of America. The remaining land-birds form a most singular group of finches, related to each other in the structure of their beaks, short tails, form of body and plumage: there are thirteen species, which Mr. Gould has divided into four sub-groups. All these species are peculiar to this archipelago; and so is the whole group, with the exception of one species of the sub-group Cactornis, lately brought from Bow Island, in the Low Archipelago. Of Cactornis, the two species may be often seen climbing about the flowers of the great cactus-trees; but all the other species of this group of finches, mingled together in flocks, feed on the dry and sterile ground of the lower districts. The males of all, or certainly of the greater number, are jet black; and the females (with perhaps one or two exceptions) are brown. The most curious fact is the perfect gradation in the size of the beaks in the different species of Geospiza, from one as large as that of a hawfinch to that of a chaffinch, and (if Mr. Gould is right in including his sub-group, Certhidea, in the main group) even to that of a warbler. The largest beak in the genus Geospiza is shown in Fig. 1, and the smallest in Fig. 3; but instead of there being only one intermediate species, with a beak of the size shown in Fig. 2, there are no

1. Geospiza magnirostris.   2. Geospiza fortis.
3. Geospiza parvula.   4. Certhidea olivasea.

less than six species with insensibly graduated beaks. The beak of the sub-group Certhidea is shown in Fig. 4. The beak of Cactornis is somewhat like that of a starling; and that of the fourth sub-group, Camarhynchus, is slightly parrot-shaped. Seeing this gradation and diversity of structure in one small, intimately-related group of birds, one might really fancy that from an original paucity of birds in this archipelago, one species had been taken and modified for different ends. In a like manner it might be fancied that a bird originally a buzzard had been induced here to undertake the office of the carrion-feeding Polybori of the American continent.

Of waders and water-birds I was able to get only eleven kinds, and of these only three (including a rail confined to the damp summits of the islands) are new species. Considering the wandering habits of the gulls, I was surprised to find that the species inhabiting these islands is peculiar, but allied to one from the southern parts of South America. The far greater peculiarity of the land-birds, namely, twenty-five out of twenty-six being new species, or at least new races, compared with the waders and web-footed birds, is in accordance with the greater range which these latter orders have in all parts of the world. We shall hereafter see this law of aquatic forms, whether marine or fresh water, being less peculiar at any given point of the earth's surface than the terrestrial forms of the same classes, strikingly illustrated in the shells, and in a lesser degree in the insects of this archipelago.

Two of the waders are rather smaller than the same species brought from other places: the swallow is also smaller, though it is doubtful whether or not it is distinct from its analogue. The two owls, the two tyrant-catchers (Pyrocephalus) and the dove, are also smaller than the analogous but distinct species to which they are most nearly related; on the other hand, the gull is rather larger. The two owls, the swallow, all three species of mocking-thrush, the dove in its separate colours, though not in its whole plumage, the Totanus, and the gull, are likewise duskier coloured than their analogous species; and in the case of the mocking-thrush and Totanus than any other species of the two genera. With the exception of a wren with a fine yellow breast, and of a tyrant-flycatcher with a scarlet tuft and breast, none of the birds are brilliantly coloured, as might have been expected in an equatorial district. Hence it would appear probable that the same causes which here make the immigrants of some species smaller, make most of the peculiar Galapageian species also smaller, as well as very generally more dusky coloured. All the plants have a wretched, weedy appearance, and I did not see one beautiful flower. The insects, again, are small sized and dull coloured, and, as Mr. Waterhouse informs me, there is nothing in their general appearance which would have led him to imagine that they had come from under the equator. The birds, plants, and insects have a desert character, and are not more brilliantly colored than those from southern Patagonia; we may, therefore, conclude that the usual gaudy coloring of the intertropical productions is not related either to the heat or light of those zones, but to some other cause, perhaps to the conditions of existence being generally favourable to life.

We will now turn to the order of reptiles, which gives the most striking character to the zoology of these islands. The species are not numerous, but the numbers of individuals of each species are extraordinarily great. There is one small lizard belonging to a South American genus, and two species (and probably more) of the Amblyrhynchus—a genus confined to the Galapagos Islands. There is one snake which is numerous; it is identical, as I am informed by M. Bibron, with the Psammophis Temminckii from Chile. Of sea-turtle I believe there are more than one species; and of tortoises there are, as we shall presently show, two or three species or races. Of toads and frogs there are none: I was surprised at this, considering how well suited for them the temperate and damp upper woods appeared to be. It recalled to my mind the remark made by Bory St. Vincent, namely, that none of this family are found on any of the volcanic islands in the great oceans. As far as I can ascertain from various works, this seems to hold good throughout the Pacific, and even in the large islands of the Sandwich archipelago. Mauritius offers an apparent exception, where I saw the Rana Mascariensis in abundance: this frog is said now to inhabit the Seychelles, Madagascar, and Bourbon; but on the other hand, Du Bois, in his voyage in 1669, states that there were no reptiles in Bourbon except tortoises; and the Officier du Roi asserts that before 1768 it had been attempted, without success, to introduce frogs into Mauritius—I presume, for the purpose of eating: hence it may be well doubted whether this frog is an aboriginal of these islands. The absence of the frog family in the oceanic islands is the more remarkable when contrasted with the case of lizards, which swarm on most of the smallest islands. May this difference not be caused by the greater facility with which the eggs of lizards, protected by calcareous shells, might be transported through salt water than could the slimy spawn of frogs?

I will first describe the habits of the tortoise (Testudo nigra, formerly called Indica), which has been so frequently alluded to. These animals are found, I believe, on all the islands of the archipelago; certainly on the greater number. They frequent in preference the high damp parts, but they likewise live in the lower and arid districts. I have already shown, from the numbers which have been caught in a single day, how very numerous they must be. Some grow to an immense size: Mr. Lawson, an Englishman, and vice-governor of the colony, told us that he had seen several so large that it required six or eight men to lift them from the ground; and that some had afforded as much as two hundred pounds of meat. The old males are the largest, the females rarely growing to so great a size: the male can readily be distinguished from the female by the greater length of its tail. The tortoises which live on those islands where there is no water, or in the lower and arid parts of the others, feed chiefly on the succulent cactus. Those which frequent the higher and damp regions eat the leaves of various trees, a kind of berry (called guayavita) which is acid and austere, and likewise a pale green filamentous lichen (Usnera plicata), that hangs in tresses from the boughs of the trees.

The tortoise is very fond of water, drinking large quantities and wallowing in the mud. The larger islands alone possess springs, and these are always situated towards the central parts, and at a considerable height. The tortoises, therefore, which frequent the lower districts, when thirsty, are obliged to travel from a long distance. Hence broad and well-beaten paths branch off in every direction from the wells down to the sea-coast; and the Spaniards, by following them up, first discovered the watering-places. When I landed at Chatham Island, I could not imagine what animal travelled so methodically along well-chosen tracks. Near the springs it was a curious spectacle to behold many of these huge creatures, one set eagerly travelling onwards with outstretched necks, and another set returning, after having drunk their fill. When the tortoise arrives at the spring, quite regardless of any spectator, he buries his head in the water above his eyes, and greedily swallows great mouthfuls, at the rate of about ten in a minute. The inhabitants say each animal stays three or four days in the neighbourhood of the water, and then returns to the lower country; but they differed respecting the frequency of these visits. The animal probably regulates them according to the nature of the food on which it has lived. It is, however, certain that tortoises can subsist even on these islands where there is no other water than what falls during a few rainy days in the year.

I believe it is well ascertained that the bladder of the frog acts as a reservoir for the moisture necessary to its existence: such seems to be the case with the tortoise. For some time after a visit to the springs, their urinary bladders are distended with fluid, which is said gradually to decrease in volume, and to become less pure. The inhabitants, when walking in the lower district, and overcome with thirst, often take advantage of this circumstance, and drink the contents of the bladder if full: in one I saw killed, the fluid was quite limpid, and had only a very slightly bitter taste. The inhabitants, however, always first drink the water in the pericardium, which is described as being best.

The tortoises, when purposely moving towards any point, travel by night and day, and arrive at their journey's end much sooner than would be expected. The inhabitants, from observing marked individuals, consider that they travel a distance of about eight miles in two or three days. One large tortoise, which I watched, walked at the rate of sixty yards in ten minutes, that is 360 yards in the hour, or four miles a day—allowing a little time for it to eat on the road. During the breeding season, when the male and female are together, the male utters a hoarse roar or bellowing, which, it is said, can be heard at the distance of more than a hundred yards. The female never uses her voice, and the male only at these times; so that when the people hear this noise, they know that the two are together. They were at this time (October) laying their eggs. The female, where the soil is sandy, deposits them together, and covers them up with sand; but where the ground is rocky she drops them indiscriminately in any hole: Mr. Bynoe found seven placed in a fissure. The egg is white and spherical; one which I measured was seven inches and three-eighths in circumference, and therefore larger than a hen's egg. The young tortoises, as soon as they are hatched, fall a prey in great numbers to the carrion-feeding buzzard. The old ones seem generally to die from accidents, as from falling down precipices: at least several of the inhabitants told me that they never found one dead without some evident cause.

The inhabitants believe that these animals are absolutely deaf; certainly they do not overhear a person walking close behind them. I was always amused when overtaking one of these great monsters, as it was quietly pacing along, to see how suddenly, the instant I passed, it would draw in its head and legs, and uttering a deep hiss fall to the ground with a heavy sound, as if struck dead. I frequently got on their backs, and then giving a few raps on the hinder part of their shells, they would rise up and walk away—but I found it very difficult to keep my balance. The flesh of this animal is largely employed, both fresh and salted; and a beautifully clear oil is prepared from the fat. When a tortoise is caught, the man makes a slit in the skin near its tail, so as to see inside its body, whether the fat under the dorsal plate is thick. If it is not, the animal is liberated and it is said to recover soon from this strange operation. In order to secure the tortoise, it is not sufficient to turn them like turtle, for they are often able to get on their legs again.

There can be little doubt that this tortoise is an aboriginal inhabitant of the Galapagos; for it is found on all, or nearly all, the islands, even on some of the smaller ones where there is no water; had it been an imported species, this would hardly have been the case in a group which has been so little frequented. Moreover, the old buccaneers found this tortoise in greater numbers even than at present: Wood and Rogers also, in 1708, say that it is the opinion of the Spaniards that it is found nowhere else in this quarter of the world. It is now widely distributed; but it may be questioned whether it is in any other place an aboriginal. The bones of a tortoise at Mauritius, associated with those of the extinct Dodo, have generally been considered as belonging to this tortoise: if this had been so, undoubtedly it must have been there indigenous; but M. Bibron informs me that he believes that it was distinct, as the species now living there certainly is.

The Amblyrhynchus, a remarkable genus of lizards, is confined to this archipelago; there are two species, resembling each other in general form, one being terrestrial and the other aquatic. This latter species (A. cristatus) was first characterized by Mr. Bell, who well foresaw, from its short, broad head, and strong claws of equal length, that its habits of life would turn out very peculiar, and different from those of its nearest ally, the Iguana. It is

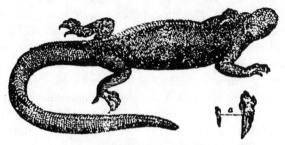

AMBLYRHYNCHUS CRISTATUS.
a, Tooth of natural size, and likewise magnified.

extremely common on all the islands throughout the group, and lives exclusively on the rocky sea-beaches, being never found, at least I never saw one, even ten yards in shore. It is a hideous-looking creature, of a dirty black colour, stupid, and sluggish in its movements. The usual length of a full-grown one is about a yard, but there are some even four feet long; a large one weighed twenty pounds: on the island of Albemarle they seem to grow to a greater size than elsewhere. Their tails are flattened sideways, and all four feet partially webbed. They are occasionally seen some hundred yards from the shore, swimming about; and Captain Collnett, in his "Voyage," says, "They go to sea in herds a-fishing, and sun themselves on the rocks; and may be called alligators in miniature." It must not, however, be supposed that they live on fish. When in the water this lizard swims with perfect ease and quickness, by a serpentine movement of its body and flattened tail—the legs being motionless and closely collapsed on its sides. A seaman on board sank one, with a heavy weight attached to it, thinking thus to kill it directly; but when, an hour afterwards, he drew up the line, it was quite active. Their limbs and strong claws are admirably adapted for crawling over the rugged and fissured masses of lava which everywhere form the coast. In such situations, a group of six or seven of these hideous reptiles may oftentimes be seen on the black rocks, a few feet above the surf, basking in the sun with outstretched legs.

I opened the stomachs of several, and found them largely distended with minced sea-weed (Ulvae), which grows in thin foliaceous expansions of a bright green or a dull red colour. I do not recollect having observed this sea-weed in any quantity on the tidal rocks, and I have reason to believe it grows at the bottom of the sea, at some little distance from the coast. If such be the case, the object of these animals occasionally going out to sea is explained. The stomach contained nothing but the sea-weed. Mr. Bynoe, however, found a piece of crab in one; but this might have got in accidentally, in the same manner as I have seen a caterpillar, in the midst of some lichen, in the paunch of a tortoise. The intestines were large, as in other herbivorous animals. The nature of this lizard's food, as well as the structure of its tail and feet, and the fact of its having been seen voluntarily swimming out at sea, absolutely prove its aquatic habits; yet there is in this respect one strange anomaly, namely, that when frightened it will not enter the water. Hence it is easy to drive these lizards down to any little point overhanging the sea, where they will sooner allow a person to catch hold of their tails than jump into the water. They do not seem to have any notion of biting; but when much frightened they squirt a drop of fluid from each nostril. I threw one several times as far as I could into a deep pool left by the retiring tide; but it invariably returned in a direct line to the spot where I stood. It swam near the bottom, with a very graceful and rapid movement, and occasionally aided itself over the uneven ground with its feet. As soon as it arrived near the edge, but still being under water, it tried to conceal itself in the tufts of sea-weed, or it entered some crevice. As soon as it thought the danger was past, it crawled out on the dry rocks, and shuffled away as quickly as it could. I several times caught this same lizard, by driving it down to a point, and though possessed of such perfect powers of diving and swimming, nothing would induce it to enter the water; and as often as I threw it in, it returned in the manner above described. Perhaps this singular piece of apparent stupidity may be accounted for by the circumstance that this reptile has no enemy whatever on shore, whereas at sea it must often fall a prey to the numerous sharks. Hence, probably, urged by a fixed and hereditary instinct that the shore is its place of safety, whatever the emergency may be, it there takes refuge.

During our visit (in October), I saw extremely few small individuals of this species, and none I should think under a year old. From this circumstance it seems probable that the breeding season had not then commenced. I asked several of the inhabitants if they knew where it laid its eggs: they said that they knew nothing of its propagation, although well acquainted with the eggs of the land kind—a fact, considering how very common this lizard is, not a little extraordinary.

We will now turn to the terrestrial species (A. Demarlii), with a round tail, and toes without webs. This lizard, instead of being found like the other on all the islands, is confined to the central part of the archipelago, namely to Albemarle, James, Barrington, and Indefatigable Islands. To the southward, in Charles, Hood, and Chatham islands, and to the northward, in Towers, Bindloes, and Abingdon, I neither saw nor heard of any. It would appear as if it had been created in the centre of the archipelago, and thence had been dispersed only to a certain distance. Some of these lizards inhabit the high and damp parts of the islands, but they are much more numerous in the lower and sterile districts near the coast. I cannot give a more forcible proof of their numbers than by stating that, when we were left at James Island, we could not for some time find a spot free from their burrows on which to pitch our single tent. Like their brothers the sea-kind, they are ugly animals, of a yellowish orange beneath, and of a brownish red colour above: from their low facial angle they have a singularly stupid appearance. They are, perhaps, of a rather less size than the marine species; but several of them weighed between ten and fifteen pounds. In their movements they are lazy and half torpid. When not frightened, they slowly crawl along with their tails and bellies dragging on the ground. They often stop, and doze for a minute or two, with hind legs spread out on the parched soil.

They inhabit burrows, which they sometimes make between fragments of lava, but more generally on level patches of the soft sandstone-like tuff. The holes do not appear to be very deep, and they enter the ground at a small angle; so that when walking over these lizard-warrens, the soil is constantly giving way, much to the annoyance of the tired walker. This animal, when making its burrow, works alternately the opposite sides of its body. One front leg for a short time scratches up the soil, and throws it towards the hind foot, which is well placed so as to heave it beyond the mouth of the hole. That side of the body being tired, the other takes up the task, and so on alternately. I watched one for a long time, till half its body was buried; I then walked up and pulled it by the tail, at this it was greatly astonished, and soon shuffled up to see what was the matter; and then stared me in the face, as much as to say, "What made you pull my tail?"

They feed by day, and do not wander far from their burrows; if frightened, they rush to them with a most awkward gait. Except when running down hill, they cannot move very fast, apparently from the lateral position of their legs. They are not at all timorous: when attentively watching any one, they curl their tails, and, raising themselves on their front legs, nod their heads vertically, with a quick movement, and try to look very fierce; but in reality they are not at all so: if one just stamps on the ground, down go their tails, and off they shuffle as quickly as they can. I have frequently observed small fly-eating lizards, when watching anything, nod their heads in precisely the same manner; but I do not at all know for what purpose. If this Amblyrhynchus is held and plagued with a stick, it will bite it very severely; but I caught many by the tail, and they never tried to bite me. If two are placed on the ground and held together, they will fight and bite each other till blood is drawn.

The individuals, and they are the greater number, which inhabit the lower country, can scarcely taste a drop of water throughout the year; but they consume much of the succulent cactus, the branches of which are occasionally broken off by the wind. I several times threw a piece to two or three of them when together; and it was amusing enough to see them trying to seize and carry it away in their mouths, like so many hungry dogs with a bone. They eat very deliberately, but do not chew their food. The little birds are aware how harmless these creatures are: I have seen one of the thick-billed finches picking at one end of a piece of cactus (which is much relished by all the animals of the lower region), while a lizard was eating at the other end; and afterwards the little bird with the utmost indifference hopped on the back of the reptile.

I opened the stomachs of several, and found them full of vegetable fibres and leaves of different trees, especially of an acacia. In the upper region they live chiefly on the acid and astringent berries of the guayavita, under which trees I have seen these lizards and the huge tortoises feeding together. To obtain the acacia-leaves they crawl up the low stunted trees; and it is not uncommon to see a pair quietly browsing, while seated on a branch several feet above the ground. These lizards, when cooked, yield a white meat, which is liked by those whose stomachs soar above all prejudices. Humboldt has remarked that in intertropical South America, all lizards which inhabit dry regions are esteemed delicacies for the table. The inhabitants state that those which inhabit the upper damp parts drink water, but that the others do not, like the tortoises, travel up for it from the lower sterile country. At the time of our visit, the females had within their bodies numerous, large, elongated eggs, which they lay in their burrows: the inhabitants seek them for food.

These two species of Amblyrhynchus agree, as I have already stated, in their general structure, and in many of their habits. Neither have that rapid movement so characteristic of the genera Lacerta and Iguana. They are both herbivorous, although the kind of vegetation on which they feed is so very different. Mr. Bell has given the name to the genus from the shortness of the snout: indeed, the form of the mouth may almost be compared to that of the tortoise: one is led to suppose that this is an adaptation to their herbivorous appetites. It is very interesting thus to find a well-characterized genus, having its marine and terrestrial species, belonging to so confined a portion of the world. The aquatic species is by far the most remarkable, because it is the only existing lizard which lives on marine vegetable productions. As I at first observed, these islands are not so remarkable for the number of the species of reptiles, as for that of the individuals, when we remember the well-beaten paths made by the thousands of huge tortoises—the many turtles—the great warrens of the terrestrial Amblyrhynchus—and the groups of the marine species basking on the coast-rocks of every island—we must admit that there is no other quarter of the world where this Order replaces the herbivorous mammalia in so extraordinary a manner. The geologist on hearing this will probably refer back in his mind to the Secondary epochs, when lizards, some herbivorous, some carnivorous, and of dimensions comparable only with our existing whales, swarmed on the land and in the sea. It is, therefore, worthy of his observation that this archipelago, instead of possessing a humid climate and rank vegetation, cannot be considered otherwise than extremely arid, and, for an equatorial region, remarkably temperate.

To finish with the zoology: the fifteen kinds of sea-fish which I procured here are all new species; they belong to twelve genera, all widely distributed, with the exception of Prionotus, of which the four previously known species live on the eastern side of America. Of land-shells I collected sixteen kinds (and two marked varieties, of which, with the exception of one Helix found at Tahiti, all are peculiar to this archipelago: a single fresh-water shell (Paludina) is common to Tahiti and Van Diemen's Land. Mr. Cuming, before our voyage, procured here ninety species of sea-shells, and this does not include several species not yet specifically examined of Trochus, Turbo, Monodonta, and Nassa. He has been kind enough to give me the following interesting results: of the ninety shells, no less than forty-seven are unknown elsewhere—a wonderful fact, considering how widely distributed sea-shells generally are. Of the forty-three shells found in other parts of the world, twenty-five inhabit the western coast of America, and of these eight are distinguishable as varieties; the remaining eighteen (including one variety) were found by Mr. Cuming in the Low Archipelago, and some of them also at the Philippines. This fact of shells from islands in the central parts of the Pacific occurring here deserves notice, for not one single sea-shell is known to be common to the islands of that ocean and to the west coast of America. The space of open sea running north and south off the west coast separates two quite distinct conchological provinces; but at the Galapagos Archipelago we have a halting-place, where many new forms have been created, and whither these two great conchological provinces have each sent up several colonists. The American province has also sent here representative species; for there is a Galapageian species of Monoceros, a genus only found on the west coast of America; and there are Galapageian species of Fissurella and Cancellaria, genera common on the west coast, but not found (as I am informed by Mr. Cuming) in the central islands of the Pacific. On the other hand, there are

Galapageian species of Oniscia and Stylifer, genera common to the West Indies and to the Chinese and Indian Seas, but not found either on the west coast of America or in the central Pacific. I may here add that, after the comparison by Messrs. Cuming and Hinds of about 2,000 shells from the eastern and western coasts of America, only one single shell was found in common, namely, the Purpura patula, which inhabits the West Indies, the coast of Panama, and the Galapagos. We have, therefore, in this quarter of the world, three great conchological sea-provinces, quite distinct, though surprisingly near each other, being separated by long north and south spaces either of land or of open sea.

I took great pains in collecting the insects, but excepting Tierra del Fuego, I never saw in this respect so poor a country. Even in the upper and damp region I procured very few, excepting some minute Diptera and Hymenoptera, mostly of common mundane forms. As before remarked, the insects, for a tropical region, are of very small size and dull colours. Of beetles I collected twenty-five species (excluding a Dermestes and Corynetes imported, wherever a ship touches); of these, two belong to the Harpalidæ, two to the Hydrophilidæ, nine to three families of the Heteromera, and the remaining twelve to as many different families. This circumstance of insects (and I may add plants), where few in number, belonging to many different families, is, I believe, very general. Mr. Waterhouse, who has published an account of the insects of this archipelago, and to whom I am indebted for the above details, informs me that there are several new genera: and that of the genera not new one or two are American, and the rest of mundane distribution. With the exception of a wood-feeding Apate, and of one or probably two water-beetles from the American continent, all the species appear to be new.

The botany of this group is fully as interesting as the zoology. Dr. J. Hooker will soon publish in the "Linnean Transactions" a full account of the Flora, and I am much indebted to him for the following details. Of flowering plants there are, as far as at present is known, 185 species, and 40 cryptogamic species, making together 225; of this number I was fortunate enough to bring home 193. Of the flowering plants, 100 are new species, and are probably confined to this archipelago. Dr. Hooker conceives that, of the plants not so confined, at least 10 species found near the cultivated ground at Charles Island have been imported. It is, I think, surprising that more American species have not been introduced naturally, considering that the distance is only between 500 and 600 miles from the continent; and that (according to Collnet, p. 58) drift-wood, bamboos, canes, and the nuts of a palm, are often washed on the southeastern shores. The proportion of 100 flowering plants out of 185 (or 175, excluding the imported weeds) being new, is sufficient, I conceive, to make the Galapagos Archipelago a distinct botanical province;

but this Flora is not nearly so peculiar as that of St. Helena, nor, as I am informed by Dr. Hooker, of Juan Fernandez. The peculiarity of the Galapageian Flora is best shown in certain families—thus there are 21 species of Compositae, of which 20 are peculiar to this archipelago; these belong to twelve genera, and of these genera no less than ten are confined to the archipelago! Dr. Hooker informs me that the Flora has an undoubted Western American character; nor can he detect in it any affinity with that of the Pacific. If, therefore, we except the eighteen marine, the one fresh-water, and one land-shell, which have apparently come here as colonists from the central islands of the Pacific, and likewise the one distinct Pacific species of the Galapageian group of finches, we see that this archipelago, though standing in the Pacific Ocean, is zoologically part of America.

If this character were owing merely to immigrants from America, there would be little remarkable in it; but we see that a vast majority of all the land animals, and that more than half of the flowering plants, are aboriginal productions. It was most striking to be surrounded by new birds, new reptiles, new shells, new insects, new plants, and yet by innumerable trifling details of structure, and even by the tones of voice and plumage of the birds, to have the temperate plains of Patagonia, or the hot dry deserts of northern Chile, vividly brought before my eyes. Why, on these small points of land, which within a late geological period must have been covered by the ocean, which are formed by basaltic lava, and therefore differ in geological character from the American continent, and which are placed under a peculiar climate—why were their aboriginal inhabitants, associated, I may add, in different proportions both in kind and number from those on the continent, and therefore acting on each other in a different manner—why were they created on American types of organization? It is probable that the islands of the Cape de Verd group resemble, in all their physical conditions, far more closely the Galapagos Islands than these latter physically resemble the coast of America; yet the aboriginal inhabitants of the two groups are totally unlike; those of the Cape de Verd Islands bearing the impress of Africa, as the inhabitants of the Galapagos Archipelago are stamped with that of America.

I have not as yet noticed by far the most remarkable feature in the natural history of this archipelago; it is, that the different islands to a considerable extent are inhabited by a different set of beings. My attention was first called to this fact by the Vice-Governor, Mr. Lawson, declaring that the tortoises differed from the different islands, and that he could with certainty tell from which island any one was brought. I did not for some time pay sufficient attention to this statement, and I had already partially mingled together the collections from two of the islands. I never dreamed that islands, about fifty or sixty miles apart, and most of them in sight of each other, formed of precisely the same rocks, placed under a quite similar climate, rising to a nearly equal height, would have been differently tenanted; but we shall soon see that this is the case. It is the fate of most voyagers, no sooner to discover what is most interesting in any locality than they are hurried from it; but I ought, perhaps, to be thankful that I obtained sufficient materials to establish this most remarkable fact in the distribution of organic beings.

The inhabitants, as I have said, state that they can distinguish the tortoises from the different islands; and that they differ not only in size, but in other characters. Captain Porter has described those from Charles and from the nearest island to it, namely, Hood Island, as having their shells in front thick and turned up like a Spanish saddle, while the tortoises from James Island are rounder, blacker, and have a better taste when cooked. M. Bibron, moreover, informs me that he has seen what he considers two distinct species of tortoise from the Galapagos, but he does not know from which islands. The specimens that I brought from three islands were young ones; and probably owing to this cause neither Mr. Gray nor myself could find in them any specific differences. I have remarked that the marine Amblyrhynchus was larger at Albemarle Island than elsewhere; and M. Bibron informs me that he has seen two distinct aquatic species of this genus; so that the different islands probably have their representative species or races of the Amblyrhynchus, as well as of the tortoise. My attention was first thoroughly aroused by comparing together the numerous specimens shot by myself and several other parties on board, of the mocking-thrushes, when, to my astonishment, I discovered that all those from Charles Island belonged to one species (Mimus trifasciatus); all from Albemarle Island to M. parvulus; and all from James and Chatham Islands (between which two other islands are situated, as connecting links) belonged to M. melanotis. These two latter species are closely allied, and would by some ornithologists be considered as only well-marked races or varieties; but the Mimus trifasciatus is very distinct. Unfortunately most of the specimens of the finch tribe were mingled together; but I have strong reasons to suspect that some of the species of the sub-group Geospiza are confined to separate islands. If the different islands have their representatives of Geospiza, it may help to explain the singularly large number of the species of this sub-group in this one small archipelago, and as a probable consequence of their numbers, the perfectly graduated series in the size of their beaks. Two species of the sub-group Cactornis, and two of the Camarhynchus, were procured in the archipelago; and of the numerous specimens of these two sub-groups shot by four collectors at James Island, all were found to belong to one species of each; whereas the numerous specimens shot either on Chatham or Charles Island (for the two sets were mingled together) all belonged to the two other species: hence we may feel almost sure that these islands possess their representative species of these two sub-groups. In land-shells this law of distribution does not appear to hold good. In my very small collection of insects, Mr. Waterhouse remarks that, of those which were ticketed with their locality, not one was common to any two of the islands.

If we now turn to the Flora, we shall find the aboriginal plants of the different islands wonderfully different. I give all the following results on the high authority of my friend Dr. J. Hooker. I may premise that I indiscriminately collected everything in flower on the different islands, and fortunately kept my collections separate. Too much confidence, however, must not be placed in the proportional results, as the small collections brought home by some other naturalists, though in some respects confirming the results, plainly show that much remains to be done in the botany of this group: the Leguminosae, moreover, have as yet been only approximately worked out:

Name of Island.	Total No. of Species.	No. of Species found in other parts of the world.	No. of Species confined to the Galapagos Archipelago.	No. confined to the one Island.	No. of Species confined to the Galapagos Archipelago, but found on more than the one Island.
James Island	71	33	38	30	8
Albemarle Island	46	18	26	22	4
Chatham Island	32	16	16	12	4
Charles Island	68	39 (or 29, if the probably imported plants be subtracted)	29	21	8

Hence we have the truly wonderful fact that, in James Island, of the thirty-eight Galapageian plants, or those found in no other part of the world, thirty are exclusively confined to this one island; and in Albemarle Island, of the twenty-six aboriginal Galapageian plants, twenty-two are confined to this one island, that is, only four are at present known to grow in the other islands of the archipelago; and so on, as shown in the above table, with the plants from Chatham and Charles Islands. This fact will, perhaps, be rendered even more striking, by giving a few illustrations: thus, Scalesia, a remarkable arborescent genus of the Compositae, is confined to the archipelago: it has six species: one from Chatham, one from Albemarle, one from Charles Island, two from James Island, and the sixth from one of the three latter islands, but it is not known from which: not one of these six species grows on any two islands. Again, Euphorbia, a mundane or widely distributed genus, has here eight species, of which seven are confined to the archipelago, and not one found on any two islands: Acalypha and Borreria, both mundane genera, have respectively six and seven species, none of which have the same species on two islands, with the exception of one Borreria, which does occur on two islands. The species of the Compositae are particularly local; and Dr. Hooker has furnished me with several other most striking illustrations of the difference of the species on the different islands. He remarks that this law of distribution holds good both with those genera confined to the archipelago, and those distributed in other quarters of the world: in like manner we have seen that the different islands have their proper species of the mundane genus of tortoise, and of the widely distributed American genus of the mocking-thrush, as well as of two of the Galapageian sub-groups of finches, and almost certainly of the Galapageian genus Amblyrhynchus.

The distribution of the tenants of this archipelago would not be nearly so wonderful, if, for instance, one island had a mocking-thrush, and a second island some other quite distinct genus; if one island had its genus of lizard, and a second island another distinct genus, or none whatever; or if the different islands were inhabited, not by representative species of the same genera of plants, but by totally different genera, as does to a certain extent hold good; for, to give one instance, a large berry-bearing tree at James Island has no representative species in Charles Island. But it is the circumstance that several of the islands possess their own species of the tortoise, mocking-thrush, finches, and numerous plants, these species having the same general habits, occupying analogous situations, and obviously filling the same place in the natural economy of this archipelago, that strikes me with wonder. It may be suspected that some of these representative species, at least in the case of the tortoise and of some of the birds, may hereafter prove to be only well-marked races; but this would be of equally great interest to the philosophical naturalist. I have said that most of the islands are in sight of each other: I may specify that Charles Island is fifty miles from the nearest part of Chatham Island, and thirty-three miles from the nearest part of Albemarle Island. Chatham Island is sixty miles from the nearest part of James Island, but there are two intermediate islands between them which were not visited by me. James Island is only ten miles from the nearest part of Albemarle Island, but the two points where the collections were made are thirty-two miles apart. I must repeat that neither the nature of the soil, nor height of the land, nor the climate, nor the general character of the associated beings, and therefore their action one on another, can differ much in the different islands. If there be any sensible difference in their climates, it must be between the windward group (namely, Charles and Chatham Islands), and that to leeward; but there seems to be no corresponding difference in the productions of these two halves of the archipelago.

The only light which I can throw on this remarkable difference in the inhabitants of the different islands is that very strong currents of the sea, running in a westerly and W.N.W. direction, must separate, as far as transportal by the sea is concerned, the southern islands from the northern ones; and between these northern islands a strong N.W. current was observed, which must effectually separate James and Albemarle Islands. As the archipelago is free to a most remarkable degree from gales of wind, neither the birds, insects, nor lighter seeds, would be blown from island to island. And lastly, the profound depth of the ocean between the islands, and their apparently recent (in a geological sense) volcanic origin, render it highly unlikely that they were ever united; and this, probably, is a far more important consideration than any other, with respect to the geographical distribution of their inhabitants. Reviewing the facts here given, one is astonished at the amount of creative force, if such an expression may be used, displayed on these small, barren, and rocky islands; and still more so at its diverse yet analogous action on points so near each other. I have said that the Galapagos Archipelago might be called a satellite attached to America, but it should rather be called a group of satellites, physically similar, organically distinct, yet intimately related to each other, and all related in a marked, though much lesser degree, to the great American continent.

I will conclude my description of the natural history of these islands by giving an account of the extreme tameness of the birds.

This disposition is common to all the terrestrial species; namely, to the mocking-thrushes, the finches, wrens, tyrant-flycatchers, the dove, and carrion-buzzard. All of them often approached sufficiently near to be killed with a switch, and sometimes, as I myself tried, with a cap or hat. A gun is here almost superfluous; for with the muzzle I pushed a hawk off the branch of a tree. One day, while lying down, a mocking-thrush alighted on the edge of a pitcher, made of the shell of a tortoise, which I held in my hand, and began very quietly to sip the water; it allowed me to lift it from the ground while seated on the vessel: I often tried, and very nearly succeeded, in catching these birds by their legs. Formerly the birds appear to have been even tamer than at present. Cowley (in the year 1684) says that the "Turtledoves were so tame that they would often alight on our hats and arms, so as that we could take them alive: they not fearing man, until such time as some of our company did fire at them, whereby they were rendered more shy." Dampier also, in the same year, says that a man in a morning's walk might kill six or seven dozen of these doves. At present, although certainly very tame, they do not alight on people's arms, nor do they suffer themselves to be killed in such large numbers. It is surprising that they have not become wilder; for these islands during the last hundred and fifty years have been frequently visited by buccaneers and whalers: and the sailors, wandering through the wood in search of tortoises, always take cruel delight in knocking down the little birds.

These birds, although now still more persecuted, do not readily become wild: in Charles Island, which had then been colonized about six years, I saw a boy sitting by a well with a switch in his hand, with which he killed the doves and finches as they came to drink. He had already procured a little heap of them for his dinner; and he said that he had constantly been in the habit of waiting by this well for the same purpose. It would appear that the birds of this archipelago, not having as yet learned that man is a more dangerous animal than the tortoise or the Amblyrhynchus, disregard him, in the same manner as in England shy birds, such as magpies, disregard the cows and horses grazing in our fields.

The Falkland Islands offer a second instance of birds with a similar disposition. The extraordinary tameness of the little Opetiorhynchus has been remarked by Pernety, Lesson, and other voyagers. It is not, however, peculiar to that bird: the Polyborus, snipe, upland and lowland goose, thrush, bunting, and even some true hawks, are all more or less tame. As the birds are so tame there, where foxes, hawks, and owls occur, we may infer that the absence of all rapacious animals at the Galapagos is not the cause of their tameness here. The upland geese at the Falklands show, by the precaution they take in building on the islets, that they are aware of their danger from the foxes; but they are not by this rendered wild toward man. This tameness of the birds, especially of the waterfowl, is strongly contrasted with the habits of the same species in Tierra del Fuego, where for ages past they have been persecuted by the wild inhabitants. In the Falklands, the sportsman may sometimes kill more of the upland geese in one day than he can carry home; whereas in Tierra del Fuego it is nearly as difficult to kill one as it is in England to shoot the common wild goose.

In the time of Pernety (1763), all the birds there appear to have been much tamer than at present; he states that the Opetiorhynchus would almost perch on his finger; and that with a wand he killed ten in half an hour. At that period the birds must have been about as tame as they now are at the Galapagos. They appear to have learned caution more slowly at these latter islands than at the Falklands, where they have had proportionate means of experience; for besides frequent visits from vessels, those islands have been at intervals colonized during the entire period. Even formerly, when all the birds were so tame, it was impossible by Pernety's account to kill the black-necked swan—a bird of passage, which probably brought with it the wisdom learned in foreign countries.

I may add that, according to Du Bois, all the birds at Bourbon in 1571–72, with the exception of the flamingoes and geese, were so extremely tame, that they could be caught by the hand, or killed in any number with a stick. Again, at Tristan d'Acunha in the Atlantic, Carmichael states that the only two land-birds, a thrush and a bunting, were "so tame as to suffer themselves to be caught with a hand-net." From these several facts we may, I think, conclude, first, that the wildness of birds with regard to man, is a particular instinct directed against *him*, and not dependent on any general degree of caution arising from other sources of danger; secondly, that it is not acquired by individual birds in a short time, even when much persecuted; but that in the course of successive generations it becomes hereditary. With domesticated animals we are accustomed to see new mental habits or instincts acquired and rendered hereditary; but with animals in a state of nature, it must always be most difficult to discover instances of acquired hereditary knowledge. In regard to the wildness of birds towards man, there is no way of accounting for it, except as an inherited habit: comparatively few young birds, in any one year, have been injured by man in England, yet almost all, even nestlings, are afraid of him; many individuals, on the other hand, both at the Galapagos and at the Falklands, have been pursued and injured by man, but yet have not learned a salutary dread of him. We may infer from these facts, what havoc the introduction of any new beast of prey must cause in a country, before the instincts of the indigenous inhabitants have become adapted to the stranger's craft or power.

Chapter 2

DARWIN'S NOTEBOOKS

INTRODUCTION

From his student days, Darwin was aware of the idea of evolution as conceived by his grandfather Erasmus Darwin and his Lamarckian mentor Robert Grant. But he did not begin to think seriously about the origin of species until near the end of the five-year *Beagle* voyage. It remains an open question what Darwin meant when he wrote in his Ornithological Notes, "If there is the slightest foundation for these remarks the zoology of Archipelagoes—will be well worth examining," and added the tantalizing comment, "for such facts would undermine the stability of Species." While on board the *Beagle*, however, Darwin did begin the practice of recording his field observations in small notebooks. And in the spring of 1837, shortly after returning to England, he used one of these field books, the Red Notebook, to record his first halting attempts to formulate an evolutionary theory.[1] For example, he proposed that new species are formed "at one blow." One is left to wonder whether this was evolution by saltation or creation by natural means.

By June 1837, Darwin had completed the *Journal of Researches*, which contains cryptic evolutionary ideas akin to those sketched more explicitly in the private Red Notebook (see, for example, the discussion of the "law of the succession of types" in chapter 1, 17–20). One month later, Darwin began Notebook B and moved from tentative notes to a bold and systematic search for the laws of the transmutation of species. This search would cul-

1. See Sandra Herbert, "The Place of Man in the Development of Darwin's Theory of Transmutation." Pts. I & II. *Journal of the History of Biology* 7:217–58, 10:155–227, 1974–77.

minate between September 1838 and December 1839 with Darwin's understanding of the basic principles of natural selection. He began the search with several key framing questions: How do species produce adaptations to a changing world? How do new species form? How does the hierarchy of relationships in the natural system of classification form? These are the signature questions that define evolutionary theory: adaptation, speciation, and phylogeny. They reflect Darwin's synthesis of contemporary and traditional debates in natural theology, transformist theory, comparative anatomy, and biogeography. From the first, Darwin focused on the crucial question of the origin of adaptation, which he treated as a problem in finding the evolutionary meaning of reproduction, or generation, as it was known. He concluded that "living beings become permanently changed or subject to variety, according <<to>> circumstance . . . we see generation here seems a means to vary or adaptation" (Notebook B:3). As to how new species may be formed, in Notebook B, geographic isolation and especially island endemism dominated Darwin's conception of speciation. Darwin's final question early in Notebook B concerned the relationship between transmutation and systematics. Using striking diagrams of the "coral of life" and the "tree of life," Darwin explained taxonomic similarities as the result of descent from "ancient types" and the gaps between groups as the consequences of extinction. These diagrams may be seen as the starting point for Darwin's later formulation of the principle of divergence (see "*Natural Selection: The Big Species Book* [1856–1858]," chapter 6, 140).

The ideas Darwin first sketched early on in Notebook B established the structure and direction of the Darwinian research program. More broadly, all the books excerpted in the present collection have their roots in the notebooks. In Notebook C, Darwin carried forward this coherent framework of ideas, which led him to postulate empirical laws and clarifying definitions in a variety of fields: the hereditary transmission of form, the distribution of local and wide-ranging species, the distinction between systematic affinity and analogy, and the relation between habit (behavior) and structure.

Always, Darwin explored the various subfields of natural history more with an eye to formulating evolutionary explanations than to solving the internal problems of the field. We see a characteristic example of this evolutionary bias in Darwin's study of heredity in Notebook C, where, rather than pursuing the laws of hereditary transmission *per se*, Darwin focused on the notion that hereditary characters tend to become indelible with the passage of time. A long-lasting pattern of thought seems to have been established here. See chapter 11, "The Provisional Hypothesis of Pangenesis," where we find that, some thirty years after Notebook C, Darwin continued to view the laws of heredity as subordinate to the theory of evolution.

Darwin's interest in heredity was paralleled by an interest in the laws of animal breeding (Notebook C:133-134). Initially, this interest did not undermine his conclusion that "picking varieties" (that is, artificial selection) is an "unnatural circumstance" (C:120). Rather, behavioral adaptations became a major concern in Notebook C, as Darwin's thinking was shaped by the idea that behavior—or habit, as he calls it—is hereditary and that adaptive changes in habit therefore precede changes in anatomical structure (C:63). This Lamarckian law, which was an important extension of the adaptive mechanism developed in Notebook B, became the dominant expression of Darwin's search for a transformist explanation in the latter part of Notebook C. The related questions of behavior and materialism became so important that upon completing Notebook C, Darwin established Notebooks M and N as a separate series of "metaphysical enquiries." Therein are to be found the foundational ideas of the *Descent of Man* (see chapter 10) and the *Expression of Emotions*.

On 15 July 1838, Darwin began Notebook D, and by the time he had filled it on 1 October, Darwin had seized upon the embryonic theory of natural selection. Thus, insofar as the notebooks on transmutation are a record of the conceptual growth of Darwin's theory, Notebook D is the climactic document of the series. In evocative language, Darwin expressed a metaphorical yet decisive grasp of the adaptive role of the competition that attends "the warring of species as inference from" the Malthusian law of population: "One may say there is a force like a hundred thousand wedges trying force <into> every kind of adapted structure into the gaps <of> in the oeconomy of Nature, or rather forming gaps by thrusting out weaker ones" (D:134–135). The full articulation of the theory of natural selection from this first formulation took place over the subsequent months and is to be found in Notebook E, but the image of natural selection as a wedging force persisted through the first edition of the *Origin of Species* (see "The Struggle for Existence," chapter 8, 169). Relying on the classical political economy of Malthus as a resource for his science, Darwin made competition in man the model for his understanding of nature. Moreover, Darwin conceived of his wedging force in the language of teleology. "The final cause of all this wedging, must be to sort out proper structure & adapt it to change" (D:135).

EDITORIAL NOTE

The British Museum (Natural History)/Cornell University Press edition of *Charles Darwin's Notebooks, 1836–1844* is edited to present the most precise transcription and accurate reconstruction of Darwin's condensed, telegraphic, and often heavily revised notes. Here we have simplified the editing in the interests of readability. Darwin's interlined revisions (marked << >>)

have all been included. We have deleted most of his deletions; a few that add to the understanding of the entry have been left (marked < >). Notes added at a later date, often in a different writing medium, have been printed in bold type. Darwin was a frenetic punctuator. This we have simplified. He was also an erratic capitalizer. This we have left alone. We have also left his spelling peculiarities intact. In a few cases, we have supplied missing letters and punctuation in brackets.

Finally, we must comment about dates. The Ornithological Notes passage was written in June 1836, while the *Beagle* was on the last leg of its homeward journey. In it, Darwin reflects on his visit to the Galapagos Islands in 1835. The Red Notebook passages were written in London in the spring of 1837 in a notebook Darwin began on the *Beagle*. The remainder of the notebooks were all written in London: Notebook B was begun in July 1837, Notebook C was begun in February 1838, Notebook D was begun July 1838, and Notebook E was begun in October 1839 and completed in July 1839. Notebooks M and N parallel Notebooks D and E: Notebook M was begun in July 1838, whereas Notebook N was begun in October 1838 and was virtually complete by July 1839.

THE NOTEBOOKS

ORNITHOLOGICAL NOTES

[Galapagos, 1835] When I recollect, the fact that the form of the body, shape of scales & general size, the Spaniards can at once pronounce, from which Island any Tortoise may have been brought. When I see these islands in sight of each other, & possessed of but a scanty stock of animals, tenanted by these birds, but slightly differing in structure & filling the same place in Nature, I must suspect they are only varieties. The only fact of a similar kind of which I am aware, is the constant asserted difference—between the wolf-like Fox of East & West Falkland Islds.— If there is the slightest foundation for these remarks the zoology of Archipelagoes—will be well worth examining; for such facts <<would>> undermine the stability of Species.

RED NOTEBOOK

[127] Speculate on neutral ground of 2. ostriches; bigger one encroaches on smaller.— change not progressif:[1] produced at one blow. if one species altered: Mem:[2] my idea of Volc: islands. elevated. then peculiar plants created. if for such mere points; then any mountain, one is falsely less surprised at

1. progressif: Note that Darwin uses a French term to express this Lamarckian idea. See n. 6.
2. The abbreviation "Mem" means memorandum.

new creation for large.—Australia's = if for volc. isl^d. then for any spot of land. = Yet new creation affected by Halo of neighbouring continent: = as if any [128] creation <<taking place>> over certain area must have peculiar character: plants, Cacti: & with limits of no vegetation at S. Shetland =

Go steadily through all the limits of birds & animals in S. America. Zorilla:

[129] wide limits of Waders: Ascension. Keeling: at sea so commonly seen. at long distances; generally first arrives:—

New Zealand rats offering in the history of rats, in the antipodes a parallel case.—

Should urge that extinct Llama.[3a] owed its death not to change of circumstances; reversed argument. knowing it to be a desert.— Tempted to believe animals created for a definite time: not extinguished by change of circumstances:

[130] ∴ The same kind of relation that common ostrich bears to Petisse & diff kinds of Fourmillier; extinct Guanaco[3b] to recent: in former case position, in latter time. (or changes consequent on lapse) being the relation.— As in first cases distinct species inosculate,[4] so must we believe ancient ones: not *gradual* change or degeneration. from circumstances: if one species does change into another it must be per saltum—or species may perish. = This <inosculation> <<representation>> of species important, each its own limit & represented.— Chiloe creeper: Furnarius. <Caracara> Calandria: inosculation alone shows not gradation.—

[132] Propagation. whether ordinary, hermaphrodite, or by cutting an animal in two. (gemmiparous, by nature or accident). we see an individual divided either at one moment or through lapse of ages.—Therefore we are not so much surprised at seeing Zoophite producing distinct animals, still partly united, & eggs which become quite separate.—Considering all individuals of all species, as <<each>> one individual <<divided>> by different methods, associated life only adds one other method where the division is not perfect.

[133] Dogs, Cats, Horses, Cattle, Goat, Asses have all run wild & bred, no doubt with perfect success—showing non Creation does not bear upon solely

3a−b. "extinct Llama" and "extinct Guanaco" refer to the large fossil mammal Darwin collected in South America. The animal was described by Richard Owen as *Macrauchenia* (long-necked) on the basis of its long cervical vertabrae.

4. inosculate: anastomosing. Refers here to the "neutral ground" of Red Notebook 127, that is, the area of overlap in the contiguous ranges of the two South American species of ostrich.

adaptation of animals.— extinction in same manner may not depend. There is no more wonder in extinction of species than of individual.

[153] When we see Avestruz two species. certainly different. not insensible change.— Yet one is urged to look to common parent? why should two of the most closely allied species occur in the same country? In botany instances diametrically opposite have been instanced: it is

NOTEBOOK B
Zoonomia[5]

[1] Two kinds of generation

the coeval kind, all individuals absolutely similar; for instance fruit trees, probably polypi, gemmiparous propagation. bisection of Planariae. &c &c.— The ordinary kind which is a longer process, the new individual passing throug[h] several stages (typical, or shortened repetition of what the original molecule has done).— This [2] appears highest office in organization (especially in lower animals, where mind & therefore relations to other life has not come into play)—See Zoonomia arguements, fails in hybrids where every thing else is perfect; mothes apparently only born to breed —annuals rendered perennial. &c &c.—

Yet Eunuchs nor <<cut>> stallions nor nuns are longer lived

Why is life short, Why such high object generation.— We *know* world subject to cycle of change, temperature & all circumstances which [3] influence living beings.—

We see <living beings>. the young of living beings, become permanently changed or subject to variety, according <<to>> circumstance—seeds of plants sown in rich soil, many kinds, are produced, though new individuals produced by buds are constant, hence we see generation here seems a means to vary, or adaptation—Again we <believe> <<know>> in course of generations even mind & instinct becomes influenced.

[4] child of savage not civilized man.—birds rendered wild <through> generations, acquire ideas ditto. V. Zoonomia.

There may be unknown difficulty with *full grown* individual <<with fixed organization>> thus being modified—therefore generation to adapt & alter the race to *changing* world.—

On other hand, generation destroys the effect of accidental injuries, which if animals lived for ever would be endless

5. Erasmus Darwin (1731–1802). Physician, poet, scientific generalist. Darwin's grandfather. Author of *Zoonomia* (1794–1796.)

[5] (that is with our present system of body & universe) therefore final cause[6] of life

With this tendency to vary by generation, why are species constant over whole country; beautiful law of intermarriages <separating> partaking of characters of both parents & these infinite in Number

[9] <Granting> Species according to Lamarck[7] disappear as collection made perfect.—truer even than in Lamarck's time. Gray's[8] remark, best known species. (as some common land shells) Most difficult to separate

Every character continues to vanish, bones instinct &c &c &c

[10] non fertility of hybridity &c &c

<assuming all> if species (1) may be derived from form (2). &c. Then (remembering Lyells arguments of transportal) island near continents might have some species same as nearest land, which were late arrivals

[11] others old ones, (of which none of same kind had in interval arrived) might have grown altered

Hence the type would be of the continent though species all different. In cases as Galapagos & Juan Fernandez. When contine[n]t of Pacific existed might have been Monsoons. . when they ceased importation ceased & [12] changes commenced.— or intermediate land existed.— or they may represent some large country long separated.—

On this idea of propagation of species we can see why, a form peculiar to continents; all bred in from one parent why. Myothera several [13] species in S. America why, 2 of ostriches in. S. America—This is answer to Decandoelle.[9] **(his argument applies only to hybridity** —genera being usually peculiar to same country, different genera different countries.

6. Final cause: purpose, function, or significance. One of Aristotle's four causes.

7. Jean-Baptiste Monet de Lamarck (1744–1829). Preeminent French naturalist and biological theorist who produced works in botany, invertebrate zoology, geology, and meteorology. Formulated a comprehensive, controversial, and influential system of evolution, which depends principally on adaptation through the inheritance of acquired characters, the importance of behavior in evolution, the assumption of an inherent—but material—principle of progressive change, spontaneous generation, and the denial of extinction. Author of *Philosophie Zoologique* (1809). See Darwin's annotations of Lamarck in chapter 3.

8. John Edward Gray (1800–1875). Zoologist.

9. Augustin-Pyramus de Candolle (1778–1841). Swiss botanist. Author of *Géographie Botanique Raisonnée* (1820) : 412.

[14] Propagation explains why modern animals same type as extinct which is law almost proved. We can see <<why>> structure is common in certain countries when we can hardly believe necessary, but if it was necessary to one forefather, the result, would be as it is. Hence Antelopes at C. of Good Hope— [15] Marsupials at Australia—

Will this apply to whole organic kingdom, when our planet first cooled.—

Countries longest separated greatest differences— if separated from im-mens[e] ages possibly two distinct type, but each having its representatives— <<as in Australia>>

This presupposes time when no Mammalia existed; Australian; Mamm were produced from propagation from different set, as the rest of the world.—

[16] This view supposes that in course of ages, & therefore changes, every animal has tendency to change.—

this difficult to prove cats &c from Egypt no answer because time short & no great change has happened

I look at two ostriches as strong argument of possibility of such [17] change— as we see them in space, so might they in time

As I have before said *isolate* species especially with some change probably vary quicker

Unknown causes of change. Volcanic isl^d.— Electricity

[18] Each species changes. does it progress.

Man gains ideas.

the simplest cannot help becoming more complicated; & if we look to first origin there must be progress.

if we suppose monads[10] are <<constantly>> formed ꞩwould they not be pretty similar over whole world under [19] similar climates & as far as world has been uniform, at former epoch— How is this Ehrenberg[11] every suc-cessive animal is branching upwards different types of organization improv-

---

10. Monad: hypothetical first life form. Equivalent to Darwin's "original molecule" (B:1) and "Monucule" (B:35).

11. Christian Gottfried Ehrenberg (1795–1876). German natural historian and early student of microbial paleontology.

ing as Owen[12] says simplest coming in & most perfect <<& others>> occasionally dying out— for instance secondary terebratula may [20] have propagated recent terebratula, but Megatherium[13] nothing.

We may look at Megatheria, armadillos & sloths as all offsprings of some still older type[.] some of the branches dying out.— with this tendency to change, (& to multiplications when isolated, requires deaths of species to keep numbers of forms equable:— **but is there any reason for supposing number of forms equable: this being due to subdivisions & amount of differences, so forms would be about equally numerous.**

changes not result of will of animal, but law of adaptation as much as acid & alkali

organized beings represent a tree. *irregularly branched* some branches far more branched.— Hence Genera.— <<as many terminal-buds dying, as new ones generated>>

[22] There is nothing stranger in death of species, than individuals

If we suppose monad definite existence, as we may suppose is the case. their creation being dependent on definite laws, then those which have changed most. <<owing to the accident of positions>> must in each state of existence have shortest [23] life. Hence shortness of life of Mammalia.—

Would there not be a triple branching in the tree of life owing to three elements air, land & water, & the endeavor of each typical class to extend his domain into the other domains. & subdivision three more, double arrangement.—

[24] if each Main stem of the tree is adapted for these three elements, there will be certainly points of affinity in each branch

A species as soon as once formed by separation of change in part of country. repugnance to intermarriage <increases it—> settles it

[25] ⸠We need not think that fish & penguins really pass into each other.—

The tree of life should perhaps be called the coral of life, base of branches dead; so that passages cannot be seen.— this again offers [26] contradiction

---

12. Richard Owen (1804–1892). Leading British comparative anatomist. Described Darwin's *Beagle* fossil mammalia; entertained evolutionary speculations in the 1830s and 1840s, but became a harsh critic of the *Origin of Species* and natural selection.

13. Darwin recounts his discovery of megatherium fossils near Bahia Blanca in Argentina in the *Journal of Researches* (chapter 1, 16–20).

to constant succession of germs in progress.— <<no only makes it excessively complicated.>>

Is it thus fish can be traced right down to simple organization.—

birds—not

[27] We may fancy, according to shortness of life of species that in perfection, the bottom of branches deaden.— *so that* in Mammalia <<birds>> it would only appear like circles;— & insects amongst articulata.— but in lower classes, perhaps as more linear arrangement.—

[28] ʿHow is it that there come aberant species in each genus <<(with well characterized parts belonging to each)>> approaching another.

<Petrels have divided themselves into many species, so have the awks, there is particular circumstances, to which> is it an index of the point whence, two favourable points of organization commenced branching.—

[29] As all the species of some genera have died; have they all one determinate life dependent on genus, that genus upon another, whole class would die out, therefore not.—

Monad has not definite existence.

There does appear some connection shortness of existence, <<in>> perfect <<species from many>> changes and base of branches being dead from which they bifurcated.—

[32] Dr. Smith[14] says he is certain that when White Men & Hottentots or Negros cross at C. of Good. Hope the children cannot be made intermediate, the first children partake more of the mother, the later ones of the father; is not this owing to each copulation **producing its effect; as when bitches puppies are less purely bred owing to having once borne Mon-**

14. Andrew Smith (1797–1872). Medical doctor and zoologist stationed in South Africa (1821–1837). Darwin met Smith when the *Beagle* landed in Cape Town in 1836.

grels he has thus seen the black blood come out from the grandfather, (when the mother was nearly quite white) in the two first children

[33] No doubt wild men do not cross readily, distinctness of tribes in T. del Fuego. the existence of whiter tribes in centre of S. America shows this. Is there a tendency in plants hybrids to go back? If so Men & plants together would establish law. =as above stated: no one can doubt that lesser trifling differences are blended [34] by intermarriages, then the black & white is so far gone, that the species (for species they certainly are according to all common language) will keep to their type: in animals so far removed, with instinct in lieu of reason, there would probably be repugnance & art required to make marriage.— as Dr Smith remarked Man & **wild** animals in this respect are differently circumstanced.—

[35] ⸮Is this shortness of life of *species* in certain orders connected with gaps in the *series of connection?* <<if sta[r]ting from same epoch *certainly*>>

The absolute end of certain form from considering S. America (*independent of external causes*) does appear very probable:— Mem: Horse, Llama. &c &c—

If we <<grant>> similarity of animals in one country owing to springing from one branch, & the monucle has definite life, then all die at one period, which is not case. MONUCULE NOT DEFINITE LIFE

[36] I think

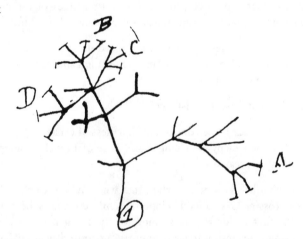

Case must be that one generation then should be as many as living now. **To do this & to have many species in the same genus (as is). REQUIRES extinction.**

Thus between A. & B. immens[e] gap of relation. C & B. the finest gradation, B & D rather greater distinction

Thus genera would be formed —bearing relation [37] to ancient types.—
with several extinct forms, for if each species <<an ancient (1)>> is capable
of making, 13 recent forms. Twelve of the contemporarys must have left no
offspring at all, so as to keep number of species constant.— With respect to
extinction we can easily see that variety of ostrich, Petise may not be well
adapted, & thus perish out, or on other hand like Orpheus, being favourable
[38] many might be produced.— This requires principle that the permanent
varieties produced by <inter> confined breeding & changing circumstances
are continued & produce according to the adaptation of such circumstances
& therefore that death of species is a consequence (contrary to what would
appear from America) [39] of non adaptation of circumstances.

Vide two. pages back. Diagram

The largeness of present genera renders it probable that <<many>> contemporary, would have left scarcely any type of their existence in the present
world.— or we may suppose only each species in each generation only breeds;
like individuals in a country not rapidly increasing.—

[40] If we thus go very far back to look to the source of the Mammalian type
of organization; it is extremely improbable that any of the successors of his
relatives shall now exist.— In same manner, if we take <<a man from>> any
large family of 12 brothers & sisters <<in a state which does not increase>>
[41] it will be chances against any one <<of them>> having progeny living
ten thousand years hence; because at present day many are relatives, so that
by tracing back. the fathers would be reduced to small percentage.— & therefore the chances are excessively great against, any two of the 12. having progeny. after that distant period.—

[42] Hence if this is true, that the *greater the groups* the *greater the gaps* (or
solutions of continuous structure) <<between them>>.— for instance there
would be great gap between birds & mammalia, Still greater between [43]
Vertebrate and Articulata. still greater between animals & Plants

But yet besides affinities from three elements, from the <<infinite>> variations, & all coming from one stock & obeying one law, they may approach—
some birds may approach animals, & some of the vertebrata invertebrates—

Such on few on each side will yet present some anomaly & bearing [44]
stamp of great main type, & the gradation will be sudden—

Heaven know whether this agrees with Nature: *Cuidado*

The above speculations are applicable to non progressive development

which certainly is the case at least during subsequent ages.

[45] The Creator has made tribes of animals adapted preeminently for each element, but it seems law that such tribes, as far as compatible with such structure are in minor degrees adapted for other elements.

every part would probably be not complete, if birds were fitted solely for air & fishes for water.

The condition of every animal is partly due to direct adaptation & partly to hereditary taint; hence the [47] resemblances & differences for instance of finches of Europe & America.

[49] Progressive development gives final cause for enormous periods anterior to Man. difficult for man to be unprejudiced about self, but considering power, extending range, reason & futurity. it does as yet appear clim[bing]

[50] In Mr Gould[15] Australian work some most curious cases. of close but certainly distinct species between Australia & Van Diemen's land. & Austral & New Zealand

Mr Gould says in sub-genera, they undoubtedly come from same countries. In mundine genera, [51] the nearest species often come very remote quarters. (NB. if Plata Partridge <<or Orpheus>> was introduced into Chili. in present states. it might continue & thus two species be created) & live in same country.

[53] ¶Whether extinction of great S. American quadrupeds, part of some great system acting over the whole world, the period of great quadrupeds declining as great reptiles must once have declined.—

Read his theory of the Earth attentively
Cuvier[16] objects to propagation of species, by saying, why not have some intermediate forms been discovered. between palaeotherium, megalonyx, mastodon, & the species now living.— Now according [54] to my view, in S. America parent of all armadilloes might be brother to Megatherium.— uncle now dead.

[55] ¶There certainly appears attempt in each dominant structure to accommodate itself to as many situations as possible.— Why should we have in

15. John Gould (1804–1881). Ornithologist. Described the birds collected on Darwin's *Beagle* voyage, including the Galapagos birds whose marked endemism contributed to Darwin's adoption of transmutation.

16. Georges Cuvier (1769–1865). Leading French comparative anatomist. Staunch opponent of Lamarck and advocate of geological catastrophism and special creation. Credited with first demonstrating that fossils are the remains of extinct organisms. Author of *Essay on the Theory of the Earth* (1827).

open country a ground woodpecker— <<ditto parrot>> a desert King-fisher.— mountain tringas.— Upland goose.— water chionis water rat with land structures; **law of chance would cause this to happen in all, but less in water birds**— carrion eagles — This is but carrying on. attempt at adaptation of each element.

May this not be explained on principle, of animal having come to island. where it could live — but there were causes to induce great change. like the Buzzard which has changed into Cara cara at the Galapagos.

[61] Law: existence *de*finite without change, superinduced, or new species; therefore animals would perish, if there was nothing in country to super-induce a change?

[62] Seeing animal die out in S. America; with no change, agrees with belief. that Siberian animals lived in cold countries & therefore not killed by cold countries.—

Seeing how horse & Elephant reached S. America.— explains how Zebras reached South Africa—

It is a wonderful fact Horse, Elephant & Mastodon dying out [63] about same time in such different quarters.—

Will Mr Lyell[17] say that some circumstance killed it over a tract from Spain to S. America. Never

They die; without they change; like Golden Pippens, it is a *generation of species* like generation of *individuals.*—

[78] An originality is given (& power of adaptation) is given by *true* generation, through means of every step of progressive increase of organization being imitated in the womb, which had been passed through to form that species. <Man is derived from Monad, each fresh—>

[80] The motion of the earth must be excessive up & down.— Elephants in Ceylon— East Indian archipelago.— West Indies = Opossum & Agouti same as on continent— **3 Paradupasi in common to Van Diemen's Land & Australia** From the consideration of these archipelagos ups & downs in full conformity with European formation— **England & Europe Ireland common animals—+++** for instance tertiary deposits between East Indian islets—

17. Charles Lyell (1797–1875). Leading British geologist. Exponent of uniformitarian geology and Darwin's scientific mentor.

(+++ Ireland no longer separated. Hare of two countries different.—
Ireland & Isle of Man possessed Elk not England. Did Ireland possess Mastodons?? Negative facts tell for little)

[81] Geographic distribution of Mammalia more valuable than any other,
because less easily transported— Mem plants on Coral islets. Next to animals land birds— & life shorter or change greater—

[82] Species formed by subsidence. Java & Sumatra. Rhinoceros. Elevate &
join keep distinct. two species made elevation & subsidence continually forming species.—

[85] In some of the lower orders a perfect gradation can be found from forms
marking good genera— by steps so insensible, that each is not more change
than we know *varieties* can produce. Therefore all genera MAY have had intermediate steps.—

But it is other question, whether there [86] have existed *all* those intermediate steps especially in those classes where species not numerous. (NB in
those classes with few species greatest jumps strongest marked genera? Reptiles?)

For instance there never may have been grade between pig & tapir, yet from
some [87] common progenitor.— Now if the intermediate ranks had produced infinite species probably the series would have been more perfect, because in each there is possibility of such organization.

As we have one Marsupial animal in Stonefied slate, the father of all [88]
Mammalia in ages long gone past, & still more so <<known>> with fishes
& reptiles.—

In mere eocine rocks. we can only expect some steps. I may ask whether the
series is not more perfect by the discovery of fossil Mammalia than before
[89] & that is all that can be expected— This answers Cuvier[18]—

Perhaps the father of Mammalia as Heterodox as ornithorhyncus.[19] If this
last animal bred— might ʃnot new classes be brought into play.—

[90] The father being climatized, climatizes the child. ʃ—whether every animal produces in course of ages ten thousand varieties, (influenced itself perhaps by circumstances) & those along preserved which are well adapted. This
would account for each tribe <<acting as>> in vacuum to each other

[96] ʃDo not plants, which have male & female organs together, yet receive
influence from other plants.— Does not Lyell give some argument about

18. See Notebook B:53.
19. Ornithorhyncus: duck-billed platypus.

varieties being difficult to keep on account of pollen from other plants because this may be applied to show all plants do receive intermixture. But
how with <<hermaphrodite>> shells!!!?

[97] We have not the slightest right to say there never was a common progenitor to Mammalia & fish. when there now exist such strange forms as
ornithorhyncus

[98] The question if creative power acted at Galapagos it so acted that bi[r]ds
with plumage <<&>> tone of voice partly American North & South— (&
**geographical division are arbitrary & not permanent. this might be
made very strong.— if we believe the Creator creates by any laws.
which, I think is shown by the very facts of the Zoological character
of these islands** so permanent a breath cannot reside in space before island
existed.— Such an influence Must exist in such spots. We know birds do
arrive & seeds.— [99] The same remarks applicable to fossil animals same
type, armadillo like covering created.— passage for vertebrae in neck same
cause, such beautiful adaptations yet other animals live so well. This view of
propagation gives hiding place for many unintelligible structures. it might
have been of use in progenitor— or it may be of use — like Mammae on
mens' breasts.

[100] How does it come wandering birds. such sandpipers. not new at
Galapagos.— did the creative force know that <<these>> species could,
arrive— did it only create those kinds not so likely to wander. Did it create
two species closely allied to Mus[cicapa] coronata, but not coronata.— We
know that domestic animals vary in countries, without any assignable reason.—

[101] Astronomers might formerly have said God ordered, each planet to
move in its particular destiny.— In same manner God orders each animal
created with certain form in certain country, but how much more simple, &
sublime power let attraction act according to certain laws such are inevitable
consequen[ce] let animal be created, then by the fixed laws of generation,
such will be their successors.— [102] let the powers of transportal be such
& so will be the form of one country to another.— let geological changes go
at such a rate, so will be the numbers & distribution of the species!!

[104] Absolute knowledge that species die & others replace them— two hypotheses fresh creations is mere assumption, it explains nothing further, points
gained if any facts are connected.—

[108] There must be progressive development; for instance none—? of the
vertebrates could exist without plants & insects had been created; but on
other hand creation of small animals must have gone on since from para-
sitical nature of insects & worms.— In abstract we may say that vegetables

& mass of insects could live without animals [109] but not vice versa. ¿could plants live without carbonic acid gaz.)— Yet unquestionably animals most dependent on vegetables. of the two great Kingdoms.—

[120] In intermarriages; smallest differences blended, rather stronger tendency to imitate one of the parents; repugnance <<generally>> to marriage <if offspring not fertile> <but producing> <<before domestication, afterwards none or little with>> fertile offspring; marriage never probably excepting from <<strict>> domestication offspring not fertile. or at least most rarely & perhaps never female.—no offspring: physical impossibility to marriage.—

[123] Race permanent, because every trifle heredetary, without some cause of change, yet such causes are most obscure. without doubt:— Vide cattle:

[130] A Race of domestic animals made from influences in one country is permanent in another.— **Good argument for species not being so closely adapted.**

[146] If population of place be constant <<say 2000>> and at present day, every ten living souls on average are related to the (200^{dth} year) degree. Then 200 years ago, there were 200 people living who now have successors.— Then the chance of 200 people being related within 200 years backward might be calculated & this number elimanated say 150 people four hundred years since were progenitors of present people, and so on backwards to one progenitor, who might have continued breeding from eternity <<backwards.—>>

[147] If population was increasing between each lustrum, the number related at the first start must be greater, & this number would vary at each lustrum, & the calculation of chance of the relationship of the progenitors would have different formula for each lustrum.—

We may conclude that there will be a period though long distant, when of the present men (of all races) not more than a few will have successors. at the present day. in looking at two fine families one with [148] successors <<for>> centuries, the other will become extinct. Who can analyze causes, dislike to marriage, heredetary disease, effects of contagions & accidents: yet some causes are evident, as for instance one man killing another. So is it with *varying* races of man: these races may be overlooked mere variations consequent on climate &c— the whole races act towards each other, and are acted on, just like the two fine families <<no doubt a different set of causes must act in the two case>> May this not be extended to all animals first consider species of cats.— **Exclude mothers & then try this as simile**

[149] In a decreasing population at any one moment fewer closely related; ∴(few species of genera) ultimately few genera (for otherwise the relationship would converge sooner) & lastly perhaps some one single one.—

Will not this account for the odd genera with few species which stand between great groups, which we are bound to consider the increasing ones.—
NB As Illustrations are there many anomalous lizards living; or of the tribe
fish extinct, or of Pachydermata, or of coniferous trees; or in certain shell
cephalopoda. <<Read Buckland[20]>>

[154] We need not expect to find varieties, intermediate between every
species.— Who can find trace or history of species between [155] Indian
cow with hump & Common — between Esquimaux & European dog? Yet
man has had no interest in perpetuating these particular varieties.

If species made by isolation; then their distribution (after physical changes)
would be in rays— from certain spots.— **Agrees with old Linnaean doctrine & Lyells. to certain extent**[21]

[161] Mr Owen suggested to me, that the production <<of monsters>>
(which, Hunter says owe their origin to very early stage) & which, follow
certain laws according to species. present an analogy to production of
species.—

Animals have no notions of beauty, therefore instinctive feelings against each
other <<for sexual ends>> species, whereas Man has such instincts very
little.

[169] Man in *savage* state may be called, **species**. in domesticated **races**.—
If all men were dead then monkeys make men.— Men makes angels—

[172] I think it is certain strata could not now accumulate without seal-
bones & cetaceans.— both found in every sea, from Equatorial to extreme
poles.— Oh Wealden.— Wealden.

[196] Before Attract of Gravity discovered. it might have been said it was as
gre[a]t a difficulty to account for movement of all, by one law. as to account
for each separate one, so to say that all Mammalia, were born from one stock,
& since distributed by such means as we can recognize, may be thought to
explain nothing.— it being as easy to produce <<for the creator>> two quadrupeds at S. America Jaguar & Tiger [197] & Europe, as to produce same one.
Although in plants, you cannot say that instinct perverted, yet organization
<<especially>> connected with generation certainly is.= The dislike of two
species to each other is evidently an instinct— & this prevents breeding.

20. William Buckland (1784–1856). Oxford geologist. Darwin wrote an abstract of
the paleontological section of Buckland's Bridgewater Treatise: *Geology and Miner-
alogy Considered with Reference to Natural Theology* (1836).

21. Lyell discusses Linnaeus's views on centers of creation in *Principles of Geology*.
Darwin copiously annotated his copy of the fifth edition (1837).

now domestication depends on perversion of instinct (in plants domestication on perversion of structures especially reproductive organs) & therefore the one distinction of species would fail. But this applies only to coition & not production. But who can say, whether offspring does not depend on mind or instinct of parent. Mem Lord Moreton's Mare.[22] The fact of plants going back

[198] hybrid plants; analogous to Men & dogs. Now if we take structure as criterion of species Hogs different species, dogs not, but if we take character of offspring. Hogs not different. some dogs different.—

[204] My idea of propagation almost infers, what we call improvement, —All Mammalia from one stock, & now that one stock cannot be supposed to be <<most>> perfect (according to our ideas of perfection); but intermediate [205] in character, the same reasoning will allow of decrease in character. (which perhaps is) Case with fish— as some of the most perfect kinds the shark. lived in remotest epochs. ʃlizards of secondary period in same predicament. It is another question, whether whole scale of Zoology may not be perfecting by change of Mammalia for Reptiles, which can only be adaptation to changing world:— I cannot for a [206] moment doubt, but what cetaceae & Phocae now replace Saurians of Secondary epoch: it is impossible to suppose such an accumulation at present day & not include Mammalian remains. The Father of all insects gives same argument as father of Mammalia; but have improvement in system of articulation. ʃwhether type of each order may not be supposed that form. which has wandered least from ancestral form. If so are present typical [207] species most near in form to ancient; in shells alone can this comparison be instituted.

People often talk of the wonderful event of intellectual Man appearing.— the appearance of insects with other sense is more wonderful. its mind more different probably & introduction of Man. Nothing compared to the first thinking being. although hard to draw line.— [208] not so great as between perfect insect & former hard to tell whether articulate or intestinal, or even a mite.— a bee <<compared with cheese mite>> with its wonderful instincts. The difference is that there is wide gap between Man & next animals in mind, more than in structures.—

[209] The reason why there is not perfect *gradation* of change in species, as physical changes are *gradual*, is this if after isolation (seed blown into desert)

22. Lord Moreton's [*sic*] mare: reputed case of telegony (hereditary influence transmitted by a prior mating). Morton's chestnut mare was first mated with a striped stallion. She bore a striped foal when later mated with a chestnut stallion. George Morton, "A Communication of a Singular Fact in Natural History," *Phil. Trans. Roy. Soc.* 110 (1821): 20–22.

or separation by mountain chains &c the species have not been *much* altered
they will cross (perhaps more fertility & so make that sudden step. species
or not.

[210] A plant submits to more individual change, (as some animals do more
than others, & cut off limbs & new ones are formed) but yet propagates
varieties according to same law with animals??

Why are species not formed, during ascent of mountain or approach of
desert?— because the crossing of species less altered prevents the complete
adaptation which would ensue

[211] A. B. C. D.— (A) crossing with (B) (& B having crossed with (C)
prevents offspring of A. becoming a good species well adapted to locality A.
but it is instead a stunted & diseased form a plant, adapted to A. B. C. D.—
Destroy plants B. C. D. & A will soon form good species!

The increased fertility of slightly different species & intermediate character
of offsprings accounts for *uniformity* of species & we Must confess that we
can[n]ot tell, what is the amount [212] of difference, which improves & checks
it.— It does not bear any precise relation to structures

The passage in last page explains that between Species from <<moder-
ately>> distant countries. there is not test but generation, <<(but experience according to each group)>> whether good species, & hence the importance Naturalists attach to Geographical range of species.—

[213] Definition of Species: one that remains <<at large>> with constant
characters, together with other beings of very near structure.— Hence species may be good ones & differ scarcely in any external character:— For
instance two wrens forced to haunt two islands one with one kind of herbage
& one with other, might change organization of stomach & hence remain
distinct.

[214] Where country changes rapidly, we should expect most species.

The difference intellect of Man & animals not so great as between living
thing without thought (plants) & living things with thoughts (animal).

<<∴ my theory very distinct from Lamarcks>>
Without *two* species will generate common kind, which is not probable; then
monkeys will never produce man, but [215] both monkeys & man may produce other species. man already has produced marked varieties & may some-
day produce something else, but not probable owing to mixture of races.—
When all mixed & physical changes (ʃintellectual being acquired alters case)
other species or angels produced

[216] Has the Creator since the Cambrian formations gone on creating animals with same general structure.— miserable limited view. With respect to how species are. Lama[r]cks "willing" doctrine absurd. (as equally are arguments against it— namely how did otter live before being made otter— why to be sure there were a thousand intermediate [217] forms.— Opponent will say. show me them, I will answer yes, if you will show me every step between bull Dog & Greyhound). I should say the changes were effects of external causes, of which we are as ignorant. as why millet seed turns a Bullfinch black, or **iodine on glands of throat** (or colour of plumage altered during passage of birds (where is this statement I remember. L. Jenyns[23] talking of it) or how to make Indian Cow with bump & pigs foot with cloven hoof)

[224] If my theory true, we get (1) a *horizontal* history of earth <<within recent times>> & many curious points of speculation; for having ascertained means of transport, we should then know whether former lands intervened.— (2^d) By character of any <<two>> ancient fauna, we may form some idea of connection of those two countries. **Hence India, Mexico & Europe one gre[a]t sea (Coral reefs ∴ shallow water at Melville Is^d.(3^d)** We know that structure of every organ in A. B. C. three species <<of one genus>> can pass into each other <<any steps we see>>: but this cannot be predicated of structures in two genera.— although D. E. F. follow close to [225] A. B. C. we cannot be sure that structure (C) could pass into (D).— We may foretell species. limits of good species being known.— It explains the blending of two genera— It explains typical structure.— Every species is due to adaptation + heredetary structure. **latter far chief element. ∴ little service habits in classification. or rather the fact that they are not far the most serviceable.** We may speculate of durability of succession from what we have seen. in old world, & on amount changes which may happen— It leads you to believe the world older than *geologists* think. it agrees with excessive inequality of numbers of species in divisions. look at Articulata!!! [226] It leads to Nature of physical change between one group of animals & a successive one. It leads to knowledge of what kinds of structure may pass into each other: now on this view no one need look for intermediate structures say in brain. between lowest Mammal & Reptile (or between extremities of any great divisions) thus a knowledge of possible changes is discovered, for speculating on future.

[227] With belief of transmutation & geographical grouping we are led to endeavour to discover causes of change — the manner of adaptation (wish

23. Leonard Jenyns (1800–1893). Naturalist and clergyman. A friend of Darwin's at Cambridge. Described Darwin's fish specimens from the *Beagle* voyage.

of parents??) instinct & structure becomes full of speculation & line of observation.—

View of generation. being condensation, test of highest organization intelligible.— may look to first germ— [228] —led to comprehend affinities. My theory would give zest to recent & Fossil Comparative Anatomy & it would lead to study of instincts. heredetary. & mind heredetary, whole metaphysics.— it would lead to closest examination of hybridity <<to what circumstances favour crossing & what prevents it—>> & generation, causes of change <<in order>> to know what we have come from & to what we tend.— this & <<direct>> examination of direct passages of structure in species, might lead to laws of change, which would then be main object of study, to guide our speculations [229] with respect to past & future. **The Grand Question, which every naturalist ought to have before him, when dissecting a whale, or classifying a mite, a fungus, or an infusorian. is "What are the laws of life".—**

[230] It really would be worthy trying to isolate some plants, under glass bells & see what offspring would come from them. Ask Henslow for some plant, whose seeds go back again, not a monstrous plant, but any marked variety. Strawberry produced by seeds??

Universality of generation strongly shown by hybridity of ferns.— hybridity showing connexion of two plants.

[231] Animals— whom we have made our slaves we do not like to consider our equals.— <<Do not slaveholders wish to make the black man other kind?>> Animals with affections, imitation, fear <of death>, pain, sorrow for the dead.—respect

We have no more reason to expect the father of mankind. than Macrauchenia[24] yet he may be found: We must not compare <<chances of embedment in>> man in present state, with what he is as former species. His arts would not then have taken him over whole world.

[232] <<—the soul by consent of all is superadded, animals not got it, not look forward>> if we choose to let conjecture run wild then animals our fellow brethren in pain, disease, death & suffering <<& famine>>; our slaves in the most laborious work, our companion in our amusements. they may partake, from our origin in one common ancestor we may be all netted together.—

[235] It may be argued against theory of changes that if so in approaching desert country or ascending mountain you ought to have a gradation of spe-

24. See n. 3a—b and Red Notebook 129—130.

cies, now this notoriously is [236] not the case, you have stunted species, but not such as would make species (except perhaps in some plants & then a chain of steps is found in same mountain).—

How is this explained by law of small differences producing more fertile offspring.— 1st. All variation of animal is either effect or adaptation, ∴ animal best fitted to that country when change has taken place, Nature

[239] Any change suddenly acquired is with difficulty permanently transmitted.— a plant will admit of a certain quantity of change at once, but afterwards will not alter. This need not apply to very slow changes, without crossing.— Now a gradual change can only be traced geologically (& then monuments imperfect) or horizontally & then cross breeding prevents perfect change.—

[248] Those will not object to my theory, those the philosophers who soar above the pride of the savage, they perceive the superiority of man over animals, without such resorts

[252] When we talk of higher orders, we should always say, intellectually higher.— But who with the face of the earth covered with the most beautiful savannahs & forests dare to say that intellectuality is only aim in this world

NOTEBOOK C

[17] The changes in species must be very slow, owing to physical change, slow & offspring not picked[25]—as men do when making varieties.—

[30] ∴ The most hypoth: part of my theory, that <<two>> varieties of many ages standing, will not readily breed together: The argument must thus be taken, as <<in>> wild state (where instinct not interfered with, or generative organs affected as with plants) no animals VERY different will breed together, so when we grant <<(which can be shown probable)>> varieties may be made in wild state, there will be presumption that they would not breed together.— We see even in domesticated varieties a tendency to go back to oldest race, which evidently is tending to same end, as the law of hybridity, namely the [31-32 missing]

[34] If varieties <<produced by slow causes, without picking>> become more & more impressed in blood with time, then generation will <<only>> produce an offspring capable of producing such as itself.— therefore two different varieties will produce hybrids but not varieties, which are not deeply impressed on blood. will cross & produce fertile offspring

25. Picked: selected.

[51] Instinct goes before structure (habits of ducklings and chickens) **Young water ouzels** hence aversion to generation, before great difficulty in propagation.—

[57] The circumstance of ground woodpeckers.— birds that cannot fly &c &c. seem clearly to indicate those very changes which at first it might be doubted were possible.— it has been asked how did the otter live before it had its web-feet— all Nature answers to the possibility.—

[60] Descent, or true relationship, tends to keep to species to one form, (but is modified), the relationship of Analogy is a divellent power[26] & tends to make forms remote antagonist powers.— Every animal in cold country has some analogy in hot gaudy colours so all changes may be considered in this light.— [61]. . . Hence relation of analogy may chiefly be looked for in the aberrant groups.[27]— It is having walking fly catcher, woodpecker &c & which causes the confusion in this system of nature— Whether species may not be made by a little more vigour being given to the chance offspring who have any slight peculiarity of structure. <<hence seals take victorious seals, hence deer victorious deer, hence males armed & pugnacious (all order; cocks all warlike)>> <<this wars against in any class: those points which are different from each other, & resemble some other class analogy>>

[62] All the discussion about affinity & how one order first becomes developed & then another— (according as parent types are present) must follow after there is proof of the non creation of animals.— then argumen[t] May be.— subterranean lakes, hot springs &c &c inhabited therefore mud wood be inhabited, then how is this effected by— for instance fish, being excessively abundant [63] & tempting the Jaguar to use its feet much in swimming, & every developement giving greater vigour to the parent so tending to produce effect on offspring— but WHOLE race of that species must take to that particular habit.— All structures either direct effect of habit, or heredetary <<& combined>> effect of habit.— perhaps in process of change.— Are any men born with any peculiarity. or any race of plants— Lamark's willing absurd, ∴ not applicable to plant,

[73] Study the wars of organic being.— the fact of guavas having overrun— Tahiti. thistle, Pampas. show how nicely things adapted.—

[74] The believing that monkey would breed (if mankind destroyed) some intellectual being though not MAN.— is as difficult to understand as Lyells

26. Divellent power: drawing apart or decomposing.

27. Aberrant groups: taxonomic groups with unusual characters. In the 1850s, Darwin will see aberrants as small groups with isolated geographic distributions, that is, relics or living fossils.

doctrine of slow movements &c &c. [75] this multiplication of little means & bringing the mind to grapple with great effect produced, is a most laborious, & painful effort of the mind (although this may appear an absurd saying) & will never be conquered by anyone (if has any kind of prejudices) who just takes up & lay down the subject without long meditation— His best chance is to have profoundly over the enormous difficulty of reproductions of species & certainty of destruction; then he will choose & firmly believe in his new faith of the lesser of the difficulties [76] Once grant that <<species>> one genus may pass into other.— grant that one instinct to be acquired (if the medullary point in ovum, has such organization as to force in one man the developement of a brain capable of producing more glowing imagining or more profound reasoning than other— if this be granted!!) & whole fabric totters & falls.— look abroad, study gradation. study unity of type— Study geographical distribution [77] study relation of fossil with recent. the fabric falls! But Man—wonderful Man. "divino ore versus coelum attentus"[28] is an exception.— He is Mammalian.— his origin has not been indefinite— he is not a deity, his end <<under present form>> will come, (or how dredfully we are deceived) then he is no exception.— he possesses some of the same general instincts & <moral> feelings as animals.— they on other hand can reason— but Man has reasoning powers in excess. instead of [78] definite instincts.— this is a replacements in mental machinery— so analogous to what we see in bodily, that it does not stagger me.— What circumstances may have been necessary to have made man! Seclusion want &c & perhaps a train of animals of hundred generations of species to produce contingents proper.— Present monkeys might not,— but probably would.— the world [79] now being fit, for such an animal.— man, (rude, uncivilized man) might not have lived when certain other animals were alive, which have perished.—

[84] With respect to question what is adaptation.— Ermine, ptarmigan hare becoming white in Arctic countries few will say it is direct effect, according to Physical laws, as sulphuric acid disorganizes [85] wood, but adaptation.— albino however is monster, yet albino may so far be considered an adaptation as best attempt of nature, colouring matter being absent.— again dwarf plant in alpine district & dwarf plants from seed, one adaptation, other monster.—

[102] The intimate relation of Life with laws of Chemical combination, & the universality of latter render— spontaneous generation not improbable.—

[106] Two grand classes of varieties; one where offspring picked, one where not.— the latter made by man & Nature; but cannot be counteracted by Man.— effect of external contingencies & long bred in—

28. Latin: "with divine face turned toward heaven."

[123] Mention persecution of early Astronomers.— then add chief good of individual scientific men is to push science a few years in advance only of their age. (differently from literary men.—) must remember that if they *believe* & do not openly avow their belief, they do as much to retard, as those, whose opinion they believe have endeavoured to advance cause of truth

[124] It is of the utmost importance to show that habits sometimes go before structures.— the only argument can be, a bird practising imperfectly some habit, which the whole rest of other family practise with a peculiar structure, then Milvulus forficatus **Tyrannus Sulphureus** if compelled solely to fish, structure would alter.—

[133] Sir J. Sebright—pamphlet[29]—most important, showing effects of peculiarities being long in blood.—++ thinks difficulty in crossing race.— bad effects of incestuous intercourse.— excellent observations of sickly offspring being cut off—so that not propagated by nature.— Whole art of making varieties may be inferred from facts stated.—

[147] The *quantity of life* on planet at different periods, depends,— on relations of desert, open ocean, &c this probably on long average, equal quantity, 2^d on relations of heat & cold. therefore probably fewer now than formerly.— *The number of forms depends* on the external relations (a fixed quantity) & on subdivision of stations & diversity— this perhaps on long average equal.—

[166] Thought (or desires more properly) being heredetary.— it is difficult to imagine it anything but structure of brain heredetary. analogy points out to this.— love of the deity effect of organization. oh you Materialist!— Read Barclay[30] on organization!!
Avitism[31] in mental structure or disposition, & avitism in corporeal structure are facts full of meaning.—
Why is thought being a secretion of brain, more wonderful than gravity a property of matter? It is our arrogance, it our admiration of ourselves.—

[171] Reflect much over my view of particular instinct being memory transmitted without consciousness <<a most possible thing. see men walking in

29. John Saunders Sebright (1767–1846). Author of *The Art of Improving the Breeds of Domestic Animals* (1809).

30. John Barclay, *An Inquiry into the Opinions, Ancient and Modern, Concerning Life and Organization* (1822).

31. Avitism: tendency in heredity for characters to revert to ancestral form; also atavism, reversion. Darwin saw this tendency as one of the basic laws of inheritance. See chapter 10 on pangenesis.

sleep>>.— an action becomes habitual is probably first stage, & an habitual action implies want of consciousness & will & therefore may be called instinctive.— But why do some actions become heredetary & instinctive & not others.— We even see they must be done often <<to be habitual>> or of great importance to cause long memory.— structure is only gained slowly.— therefore it can only be those actions, which *Many* successive generations are impelled to do in same way— The improvement of reason implies diversity & therefore would banish individual, but general ones might yet be transmitted.= [172] Memory springing up after long intervals of forgetfulness.— after sleep, <> analogies with memory in offspring.— Some association in such cases recall the idea. <<or simple structure in brain people in fevers recollecting things utterly forgotten>>—it is scarcely more wonderful, that it should be remembered in next generation. ...

[173] My view of instinct explains its loss ¶if it explains its acquirement.— Analogy. a bird can swim without being web footed yet with much practice & led on by circumstanc[e] it becomes web footed, now Man by effort of Memory can remember how to swim after having once learnt, & if that was a regular contingency the brain would become webfooted & there would be no act of memory.—

[174] Wax of Ear, bitter perhaps to prevent insects lodging there. Now these exquisite adaptations can hardly be accounted for by My method of breeding there must be some corelation, but, the whole Mechanism is so beautiful.— The corelations are not, however, perfect, else one animal would not cause misery to other.— else smell of Man would be disagreeable to Musquitoes.

[175] We may never be able to trace the steps by which the organization of the eye, passed from simpler stage to more perfect, preserving its relations.— the wonderful power of adaptation given to organization.— This really perhaps greatest difficulty to whole theory.—

[220] Educate all classes— avoid the contamination of castes. improve the women. (double influence) & mankind must improve—

NOTEBOOK D

[36] 16th Aug.[32]— What a magnificent view one can take of the world Astronomical <& unknown> causes, modified by unknown ones. cause changes in geography & changes of climate superadded to change of climate from physical causes.— these superinduce changes of form in the organic world, as adaptation. . . . How far grander than idea from cramped [37] imagina-

32. 1838.

tion that God created, (warring against those very laws he established in all organic nature) the Rhinoceros of Java & Sumatra, that since the time of the Silurian, he has made a long succession of vile Molluscous animals— How beneath the dignity of him, who <<is supposed to have>> said let there be light & there was light.— <<bad taste {whom it has been declared "he said let there be light & there was light".—>>

[44] The very many breeds of animals in Britain shows, with the aid of *seclusion* in breeding, how easy races or varieties are made.—
The Highland Shepherd dogs, coloured like Magellanic Fox.[33]— peculiar hair & appearance— good case of Provincial Breed— Highland Sheep jet black legs, & face & tail, just like species.— high active breedin[g]

[60] Lyell's excellent view of geology, of each formation being merely a page torn out of a history, & the geologist being obliged to fill up the gaps.— is possibly the same with the <Zoologist> <<philosopher>>, who has trace the structure of animals & plants.— he get merely a few pages.—

[99] September 13th The passion of the doe to the victorious stag, who rubs the skin off horns to fight— is analogous to the love of woman (as Mitchell remarks seen in savages) to brave men.—

[107] One can perceive that Natural varieties or species, all the structure of which is adaptation to habits (& habit second nature) may be more in constitutional.— more conformable to the structure which has been adapted to former changes, than a mere monstrosity propagated by art.

[134] 28th [Sept]. <<I do not doubt, every one till he thinks deeply has assumed that increase of animals exactly proportiona[l] to the number that can live.—>> We ought to be far from wondering of changes in number of species, from small changes in nature of locality. Even the energetic language of <Malthus> <<Decandoelle>> does not convey the warring of the species as inference from Malthus.[34]— <<increase of brutes, must be prevented solely by positive checks, excepting that famine may stop desire.—>> in Nature production does not increase, whilst no checks prevail, but the positive check of famine & consequently death. . . [135] population in increase at geometrical ratio in FAR SHORTER time than 25 years— yet until the one sentence of Malthus no one clearly perceived the great check amongst men.— <<Even a *few* years plenty, makes population in Men increase, & an *ordi-*

---

33. Magellanic Fox: encountered by Darwin in the Falklands during the *Beagle* voyage. See Ornithological Notes, 49.

34. Thomas Robert Malthus (1766–1834). Political economist and Anglican clergyman. Author of *An Essay on the Principle of Population* (1798). Darwin owned the sixth edition (1826).

nary crop, causes a dearth then in Spring, like food used for other purposes as wheat for making brandy.—>> take Europe on an average, every species must have same number killed, year with year, by hawks, by, cold &c— . . even one species of hawk decreasing in number must effect instantaneously all the rest.— One may say there is a force like a hundred thousand wedges trying force <into> every kind of adapted structure into the gaps <of> in the oeconomy of Nature, or rather forming gaps by thrusting out weaker ones. <<The final cause of all this wedgings, must be to sort out proper structure & adapt it to change.— to do that, for form, which Malthus shows, is the final effect, (by means however of volition) of this populousness, on the energy of Man>>

[167] One <<invisible>> animalcule in four days could form 2. cubic stone. like that of Billin.—

NOTEBOOK E

[3] Epidemics— seem intimately related to famines, yet very inexplicable.— d[itt]o p.529. "It accords with the most *liberal*! spirit of philosophy to believe that no stone can fall, or plant rise, without the immediate agency of the deity. But we know from *experience*! that these operations of what we call nature, have been conducted *almost*! invariably according to fixed laws: And since the world began, the causes of population & depopulation have been probably as constant as any of the laws of nature with which we are acquainted."— this applies to one species— I would apply it not only to population & depopulation, but extermination & production of new forms.— their number & corelations

[4] Octob. 4th. It cannot be objected to my theory, that the amount of change within historical times has been small— because change in forms is solely adaptation of whole of one race to some change of circumstances; now we know how slowly & insensibly such changes are in progress.— we feel interest in discovering a change of level of a few feet during last two thousand years in Italy, but what <<changes>> would such a change produce in climate vegetation &c.— . . .

[5] The difficulty of multiplying effects & to conceive the result with that clearness of conviction, absolutely necessary as the <<basal>> foundation stone of further inductive reasoning is immense.
It is curious that geology, by giving proper ideas of these subjects, should be *absolutely* necessary to arrive at right conclusion about species

[9] No structure will last, without it is adaptation to *whole* life of animal, & not if it be solely to womb, as in monster, or solely to childhood, or solely

to manhood.— it will decrease & be driven outwards in the grand crush of population.—

[48] My theory gives great final cause <<I do not wish to say only cause, but one great final cause.— nothing probably exists for one cause>> of *sexes* <<in separate>> <<*animals*>>: for otherwise, there would be as many species, as individuals, & though we may not trace out all the ill effects.— we see it is not the order in this perfect <uni> world, either [49] at the present, or many anterior epochs.— but we can see if all species, there would not be social animals. <<this is stated too strongly. for there would be innumerable species, & hence few only social[.] there could not be one body of animals, living with certainty on other>> hence no social instincts, which as I hope to show is <<probably>> the foundation of all that is most beautiful in the moral sentiments of the animated beings.— &c

[50] Without sexual crossing, there would be endless changes, & hence no feature would be deeply impressed on it, & hence there could not be *improvement*. <<& hence not be higher animals>>—it was absolutely necessary that Physical changes should act not on individuals, but on masses of individuals.— so that the changes should be slow & bear relation to the whole changes of country, & not to the local [51] changes. == this could only be effected by sexes:

[58] Three principles, will account for all
    (1) Grandchildren. like. grandfathers
    (2) Tendency to small change . . <<especially with physical change>>
    (3) Great fertility in proportion to support of parents

[71] It is a beautiful part of my theory, that <<domesticated>> races, of organics, are made by percisely same means as species— but latter far more perfectly & infinitely slower.— No domesticated animal is perfectly adapted to external conditions.— (hence great variation in each birth) from man arbitrarily destroying certain forms & not others.—

[109] In early stages of transmutations, the relations of animals & plants to each other would rapidly increase, & hence number of forms, once formed, would remain stationary, hence all present types are ancient.

[114] March 12th—[35] It is difficult to believe in the dreadful <<but quiet>> war of organic beings, going on the peaceful woods, & smiling fields.— we must recollect the multitudes of plants introduced into our gardens (opportunities of escape for foreign birds & insects) which are propagated with very little care.— & which might spread themselves, as well as our wild

35. 1839.

plants, we see how full nature. how firmly each holds its place.— When we hear from authors [. . .] that in the Pyrenees, that the [115] Rhododendron ferrugineum, begins at 1600 metres precisely & stops at 2600. & yet know that plant can be cultivated with ease near London.— what makes the line, as trees in Beagle Channel.— it is not elements.— we cannot believe in such a line, it is other plants.— a broad border of Killed trees would form fringe.— but there is a contest, & a grain of sand turns the balance.—

NOTEBOOK M

[26] The common remark that fat men are good natured, & vice versa Walter Scotts remark how odious an illtempered fat man looks, shows same connection between organization & [27] mind.— thinking over these things, one doubts existence of free will every action determined by heredetary constitution, example of others or teaching of others.—

[31] I verily believe free-will & chance are synonymous.— Shake ten thousand grains of sand together & one will be uppermost:— so in thoughts, one will rise according to law.

[32] Beauty is instinctive feeling, & thus cuts the Knot:— Sir J. Reynolds[36] explanation may perhaps account for our acquiring <<the *instinct*>> our notion of beauty & negroes another; but it does not explain the *feeling* in any one man.—

[34] Now that I have a test of hardness of thought, from weakness of my stomach I observe a long castle in the air, is as hard work (abstracting it being done in open air, with exercise &c no organs of sense being required) as the closest train of geological thought.— the capability of such trains of thought makes a discoverer, & therefore (independent of improving powers of invention) such castles in the air are highly dangerous, before real train of inventive [35] thoughts are brought into play & then perhaps the sooner castles in the air are banished the better.—

[36] Analysis of pleasures of scenery.—
There is absolute pleasure independent of imagination, (as in *hearing* music), this probably arises from (1) harmony of colours, & their *absolute* beauty. (which is as real a cause as in music) from the splendour of light, especially when coloured.— that light is a beautiful object one knows from seeing artificial lights in the night.— from the mere exercise of the [37] organ of sight, which is common to every kind of view— as likewise is novelty of

36. Joshua Reynolds (1723–1792). English Painter. See *The Works of Sir Joshua Reynolds* (1798), 1:219–220.

view even old one, every time one looks at it.— these two causes very weak.—
(2^d) form. some forms seem instinctively beautiful <<as round, ovals>>;—
then there the pleasure of perspective, which cannot be doubted if we look
at buildings, even ugly ones.— the pleasure from perspective is derived in
a river from seeing how the serpentine lines narrow in the distance.— &
even on paper two waving *perfectly parallel* lines are elegant.— [38] Again
there is beauty in rhythm & symmetry, of forms— the beauty of some as
Norfolk I^{sd}. fir shows this, or sea weed, &c &c— this gives beauty to a single
tree.— & the leaves of the foreground either owe their beauty to absolute
forms or to the repetition of similar forms as in angular leaves.— (this Rhythmical beauty is shown by Humboldt from occurrence in Mexican & Grae-
cian to be single cause) this symmetry & rhythm applies [39] to the view as
a whole.— Colour <<& light>> has very much to do, as may be known by
autumn, on *clear* day.— 3^d pleasure association *warmth, exercise*, birds
singings.—
4th. Pleasure of imagination, which correspond to those awakened during
music.— connection with poetry, abundance, fertility, rustic life, virtuous
forgotten.—

[57] To avoid stating how far, I believe, in Materialism, say only that emotions, instincts degrees of talent, which are heredetary are so because
brain of child resemble, parent stock.— (& phrenologists state that brain
alters)

[61] Hensleigh[37] says to say. *Brain* per se thinks is nonsense; yet who will
venture to say germ within egg cannot think— as well as animal born with
instinctive knowledge.— but if so, yet this knowledge acquired by senses.—
then thinking consists of sensation of images before your eyes, or ears (language mere means [62] of exciting association.)— or of memory of such sensations, & memory is repetition of whatever takes place in brain, when sensation is perceived.==

[72] With respect to free will, seeing a puppy playing cannot doubt that
they have free will, if so all animals, then an oyster has & a polype (& a
plant in some senses, perhaps, though from not having pain or pleasure
actions unavoidable & only to be changed by habits). now free will of oyster, one can fancy to be direct effect of organisation, by the capacities its

37. Hensleigh Wedgwood (1803–1891). Philologist and barrister. Darwin's cousin
and fellow student at Christ College, Cambridge. Married Fanny Mackintosh, daugher
of Sir James Mackintosh (see Notebook M:151). The Hensleigh Wedgwoods were a
principal part of Darwin's social set, along with Darwin's brother Erasmus and Harriet Martineau, during the London years when he was using the notebooks to formulate his theory of evolution.

senses give it of pain or pleasure, if so free will is to mind, what chance is to matter <<(M. Le Compte[38])>>— the free will (if so called) makes change [73] in bodily organization of oyster, so may free will make change in man.— the real argument fixes on heredetary disposition & instincts—. Put it so.— Probably some error in argument, should be grateful if it were pointed out.— . . .

[74] The above views would make a man a predestinarian of a new kind, because he would tend to be an atheist. Man thus believing, would more earnestly pray "deliver us from temptation," he would be most humble, he would strive <to do good> <<to improve his organization>> for his children's sake & for the effect of his example on others.

[75] Martineau.[39] How to observe, p. 21–26, argues <<with examples>> very justly there is no universal moral sense.— <<from difference of action of approved>> Yet as, I think, the opposite side has been shown— see Mackintosh.— Must grant, that the conscience varies in different races.— no more wonderful than dogs should have different instincts.— Fact most opposed to this view, where [76] the moral sense seems to have changed suddenly— but are not such <<sudden>> changes rare,— as when Polynesian mothers ceased to destroy their offspring— ʃyet perhaps if they had murdered their children, this moral sense, would have been so much, as in other races of mankind.—

[81] Aug. 12th. 38. At the Athenaeum Club. was very much struck with an intense headache <<after good days work>> which came on from reading <<review of>> M. Comte Phil.[40] which made me <<endeavour to>> remember, & to think deeply, & the immediate manner in which my head got well when reading article by Boz.— now in this I was interested as was I in the other, & read so intently as to be unconscious of all around, yet there was no strain on the intellectual powers— the difference is of a man wagging his foot & working with his toe to perform some difficult task.—

[84] . . . Origin of man now proved.— Metaphysic must flourish.— He who understands baboon <will> would do more towards metaphysics than Locke

38. Auguste Comte (1798–1857). French philosopher and mathematician; founder of positivism. Author of *Cours de philosophie positive* (1830–1835), 2 vols.

39. Harriet Martineau (1802–1876). Author and reformer. *How to Observe. Morals and Manners* (1838).

40. Darwin read the review of Comte by the Scottish physicist David Brewster in *Edinburgh Review* 67:271–308.

[88] Fine poetry, or a strain of music, when the mind is rendered ductile by grief, or by bodily weakness, melts into tears, with sensations of sorrowful delight, very like best feeling of sympathy.— Mem: Burke's[41] idea of Sympathy, being real pleasure at pain of others, with rational [89] desire to assist them.— otherwise as he remarks sympathy could be barren, & lead people from scenes of distress.— see how a crowd collects at an accident.— children with other children naughty.— Why does person cry for joy?

[108] Adam Smith (D. Stewart life of, p. 27), says we can only know what others think by putting ourselves in their situation, & then we feel like them.— hence sympathy very unsatisfactory because does not like Burke explain pleasure.

[128] Plato <<Erasmus says>> in Phaedo that our *"necessary ideas"* arise from the preexistence of the soul, are not derivable from experience.— read monkeys for preexistence—

[132] Sept. 8th. I am tempted to say that those actions which have been found necessary for long generation, (as friendship to fellow animals in social animals) are those which are good & consequently give pleasure, & not as Paleys rule is those that on long run *will* do good.— alter *will* in all cases to *have* & *origin* as well as *rule* will be given.— **Descent of Man Moral Sense**

[139] <<I saw>> Jenny untying a very difficult knot— the sailor on board the ship could not puzzle her— with aid of teeth & hands.— **Descent 1838** It was very curious to see her take bread from a visitor, & before eating <<everytime>>, look up to <<keeper>> see whether, this was permitted & eat it.— good case of association.—

[150] May not moral sense arise from our enlarged capacity <acting> <<yet being obscurely guided>> or strong instinctive sexual, parental & social instincts, giving rise "do unto others as yourself"."love thy neighbour as thyself". Analyse this out.— bearing [151] in mind many new relations from language.— the social instinct more than mere love.— fear for others acting in unison.— active assistance. &c &c. it comes to Miss Martineaus one principle of charity. ¶May not idea of God arise from our confused idea of "ought." joined with necessary notion of "causation", in reference to this "ought," as well as the works of the whole world.— Read Mackintosh on Moral sense & emotions.[42]—

41. Edmund Burke (1729–1797). English philosopher. Author of *Philosophical Inquiry into the Origin of Our Ideas of the Sublime and Beautiful* (1823).

42. James Mackintosh (1765–1832). Philosopher and historian. Author of *Dissertation on the Progress of Ethical Philosophy* (1837).

[154] This unwillingness to consider Creator as governing by laws is probably that as long as we consider each object an act of separate creation, we admire it more, because we can compare it to the standard of our own minds, which ceases to be the case when we consider the formation of laws invoking laws, & giving rise at last even to the perception of a final cause.—

NOTEBOOK N

[5] To study Metaphysic, as they have always been studied appears to me to be like puzzling at Astronomy without Mechanics.— Experience shows the problem of the mind cannot be solved by attacking the citadel itself.— the mind is function of body.— we must bring some *stable* foundation to argue from.—

[12] M. Le Comte's idea of theological state of science, grand *idea*: as before having analogy to guide one to conclusion that any one fact was connected with law.— as soon as any enquiry commenced, for instance probably such a thing as thunder, would be placed to the will of God. Zoology itself is now purely theological.—

[26] Consult the VII discourse by Sir J. Reynolds.— Is our idea of beauty, that which we have been most generally accustomed to:— analogous case to my idea of conscience.— deduction from this would be that a mountaineer born out of country yet would love mountains, & a negro, similarly treated would think [27] negress beautiful.— (male glow worm doubtless admires female, showing, no connection with male figure)— As forms change, so must idea of beauty.— (Old Graecians living amongst naked figures, & observing powers common to savages???].— The existence in human mind, is to me clear evidence, of the general ideas of our ancestors being impressed upon us.— Surely we have taste naturally all has [28] not been acquired by education, else why do some children acquire it soon, & why do all men, agree ultimately?— We acquire many notions unconsciously, without abstracting them & reasoning on them (as *justice*?? as ancients did high forehead sign of exalted character???) Why may not our heredetary nature thus acquire some general notions, which are taste?

[36] We can allow <<satellites>>, planets, suns, universe, nay whole systems of universe <of man> to be governed by laws, but the smallest insect, we wish to be created at once by special act, provided with its instincts its place in nature, its range, its— &c &c:— *must be a special act, or result of laws, yet we placidly* believe the Astronomer, when he tells us satellites &c &c <<*The Savage admires not a steam engine, but a piece*>> of coloured glass <&

admires> is lost in astonishment at the artificer.—>> Our faculties are more fitted to recognize the wonderful structure of a beetle than a Universe.—

Source Note: Darwin's Ornithological Notes, edited by Nora Barlow, *Bulletin of the British Museum (Natural History) Historical Series* 1963, 2:201–78. *Charles Darwin's Notebooks 1836–1844: Geology, Transmutation of Species, Metaphysical Enquiries*, transcribed and edited by Paul H. Barrett, Peter J. Gautrey, Sandra Herbert, David Kohn, and Sydney Smith. British Museum (Natural History), Cornell University Press, and Cambridge University Press 1987. The Red Notebook was edited by Sandra Herbert and previously published as *The Red Notebook of Charles Darwin*, British Museum (Natural History), 1980. Notebooks B–E were edited by David Kohn and previously published as *Darwin's Notebooks on Transmutation of Species*, edited by Gavin de Beer, M. J. Rowlands, and B. M. Skramovsky, *Bulletin of the British Museum (Natural History) Historical Series* 2 (2–6):185–200, 3 (5):129–176, 1960, 1961, 1967. Notebooks M and N were edited by Paul H. Barrett and previously published with Howard E. Gruber in *Darwin on Man*, 1974, Chicago University Press, and as *Metaphysics, Materialism, & the Evolution of Mind*, 1980, Chicago University Press.

Chapter 3

DARWIN AND LAMARCK:

MARGINALIA

If Darwin was uninhibited in his notebooks, he was even more so in annotations made in the margins of the many books he read. Here we examine Darwin's annotations of Jean-Baptiste Lamarck's *Zoological Philosophy* in the French edition of 1809.[1] Lamarck (1744–1829) was curator of invertebrates at the Paris Museum of Natural History. His theory, which was not quite a theory of evolution in the strictest sense of descent with modification from ancestors, was that plants and animals progressed upward along the great chain of being. He did not accept the notion of extinction: when a species disappeared from the fossil record, he reckoned it was because it had changed into something else. A corollary to this view was that he denied the reality of species; for him, the chain of being was continuous. We have seen in a notebook entry (p. 52) that Darwin saw that for Lamarck the notion of a "hard" species was simply an artifact of a deficient taxonomy. The more individuals that are described, the more do species blend into a continuous series.

Not only was there an innate progressive tendency, but changes from one form to another also occurred in response to environmental change. In Lamarck's view, environmental change created certain needs to which the animal responded according to a kind of inner drive which decided what organs

---

1. *Philosophie Zoologique* (Paris, Duminil-Leseur, 1809), cited hereinafter from the English translation, *Zoological Philosophy* (Chicago, University of Chicago Press, 1984), which we have compared to the French original. The annotations have been published by Mario A. Di Gregorio, *Charles Darwin's Marginalia*, vol. I (New York, Garland, 1990), cols. 477–480. In the interest of readability, we have simplified Darwin's punctuation somewhat.

would be affected by the change. Such responses (acquired characteristics) were then inherited by the individual's progeny. Paradigmatic examples are that of the short-necked giraffe, which stretched its neck in order to be able to feed from trees, and those of long-necked birds like the heron, which, according to Lamarck, had stretched their necks to avoid getting wet while catching fish.

Darwin made numerous short marginal annotations in the first volume of Lamarck's *Zoological Philosophy*, as well as longer, more comprehensive notes at the ends of chapters. By page 62, he had already concluded that it was a "Very poor & useless Book." He did not like Lamarck's non-approach to the issue of species. At the end of chapter 1, Darwin notes that Lamarck "argues that *all divisions* or *gaps* [in the chain of being] are artificial or that the series is either now perfect or has been so—Fallacy." Not surprisingly, therefore, he took a dim view of Lamarck's views on extinction: "Lamarck argues, species of shells, not killed by man, no apparent cause of death, but causes of change are present . . . therefore fossil same species with modern." In a similar passage, Darwin wrote, "Therefore every fossil species direct father of existing analogies & no extinction except through man! Hence cause of innumerable errors in Lamarck." He also wrote, "definition of species/ doubts any extinct animals! (hence theory must be false)."

Also in connection with extinction, Darwin, in a long annotation, comments on Lamarck's views of environmental change by invoking Lyell's uniformitarian concepts (though without alluding to them explicitly): "rate of change not uniform in world, except on great scale. Geologists judge of time by change of species, these changes effect of *physical* changes (dynamical changes). These we can only judge by present day. Therefore measure of past ages is reduced to observation of changes at present day."

Darwin wraps up Lamarck's chapter 3, on species, with an extended, rapid-fire critique: Lamarck, he writes, "argues against permanence, when conditions changing—series branching now once perfect—no genera—conditions change species & these changes by time become fixed—assumes some more species made by hybridity and fixed by time—wants produce habitudes, the source of actions, faculties & instincts."

Classification was one of two primary foci of Darwin's critique of Lamarck: "Classification . . . few animals at the limits of classes . . . animals in *series* & *not ramified* . . . (quite different from my own view)." Darwin, of course, conceptualized phylogeny as a branching tree whereon living species constituted the ends of smaller branches. Darwin considered that most of the plant and animal species that had lived in the course of geological time were already extinct. For Lamarck, however, there was no extinction, and hence no branching tree.

The other focus was the issue of the mechanism of evolution. In Lamarck's view, the environment acts directly upon plants and animals to modify their organs:

> Hence forth we have to distinguish the degradation of organisation which arises from the influence of environment and acquired habits, from that which results from the smaller progress in the perfected or complexity of organisation. We must be careful therefore about going into too much detail in this respect; because as I shall show the environment in which animals habitually live, their special habitats, the habits which circumstances have forced upon them, their manner of life, etc., have a great power to modify organs; so that the shapes of parts might be attributed to degradation when they are really due to other causes.[2]

Here Darwin's terse comment is: "Here is the difference between Lamarck & Me." Darwin may well have already formed the view that the environment's influence on species was indirect and worked through natural selection.

Lamarck's notions of "degradation" also caused problems for Darwin; at the end of Lamarck's discussion of the issue in chapter 6, Darwin observes: "On this scheme of organization lower down it would not be expected to find organs, smell more perfect. But in others as articulates[3] it is much more perfect. And further on, at the end of Lamarck's chapter on abortive organs: "This chapter must be looked over again: L. distinguishes between *degraded* or abortive organs, such as extremities of Cetacea, & less developed forms: discussion of this point fills much of this Chapt[er]."

Darwin was also bothered by the fact that Lamarck seemingly had formulated different sets of rules for the plant and animal kingdoms. Since for Lamarck, plants had no habits to attune, through volition, to changing environmental conditions, some other mechanism had to be found to account for their evolution: "The want of progression in the vegetable world serious fact Lamarck has rather overlooked," Darwin remarks. Lamarck specifies his views in the following passage:

> In plants, where there are no activities and consequently no habits, properly so-called, great changes of environment none the less lead to great differences in the development of their parts; so that these differences cause the origin and development of some, and the shrinkage and disappearance of others. But all this is here brought about by the changes sustained in the nutrition of the plant, in its absorption and transpiration, in the quantity of caloric, light, air and moisture

2. *Zoological Philosophy*, 74.

3. In Cuvier's classification scheme, segmented invertebrates.

that it habitually receives; lastly, in the dominance that some of the various vital movements acquire over others.[4]

Darwin here comments: "Therefore not same theory to plants & animals." And later, at the end of Lamarck's chapter 7, on the influence of the environment, Darwin continues: "Explain *how* animals & plants change. Lamarck's theory differs for plants & animals—it is absurd this way, he assumes the want of habit causes animals annihilation of organ and vice versa."

Darwin was keenly aware that his own theory would be judged, in part at least, on the basis of his success in convincing English readers that his theory was not simply that of Lamarck warmed over, as Charles Lyell had suggested to him. Such statements made Darwin defensive. As late as October 1859, he wrote to Lyell that Lamarck's book "appeared to me extremely poor; I got not a single fact or idea from it."[5] This disclaimer was a prelude to a famous letter of March 1863 that Darwin addressed to Lyell in which he complained that the geologist referred

> repeatedly to my view as a modification of Lamarck's doctrine of development and progression. If this is your deliberate opinion there is nothing to be said, but it does not seem so to me. Plato, Buffon, my grandfather before Lamarck and others, propounded the *obvious* views that if species were not created separately, they must have descended from other species, and I can see nothing else in common between the 'Origin' and Lamarck.[6]

Yet Darwin's relationship with Lamarck was self-contradictory. The theory as it appears in Notebooks B and C owes a great deal to Lamarck indeed. In Notebook B, both the inheritance of acquired characteristics and direct adaptation to the environment are plainly derived from Lamarck, and so also, in Notebook C, is the notion that change in habit precedes change in structure.

The relationship of Lamarck, Lyell, and Darwin was a complex one. Lyell dissented strongly from Lamarck's views on extinction, arguing that extinctions were natural and could be explained by observable processes. In order to explain the appearance of new species, Lyell could not accept Lamarck's version, and in rejecting it, he explained the species problem in

4. *Zoological Philosophy*, 108.
5. *Correspondence*, 7: 348.
6. *Life and Letters*, 3: 13–14.

a new way. Darwin would eventually solve that problem by adopting an approach to phylogeny that was extremely different from that of Lamarck.[7]

7. On the relationship of Lamarck, Lyell, and Darwin, see David Hull, "Lamarck among the Anglos," in *Zoological Philosophy*, xl–lxvi, especially xliii– xlv.

Chapter 4

1842 SKETCH AND
1844 ESSAY

INTRODUCTION

Darwin seems to have experienced considerable difficulty in making the transition between keeping notebooks—where speculative probings lead to short bursts of brilliant insight—and writing a disciplined exposition of his theory. After recording the Malthusian insights of September–December 1838, he completed Notebooks E and N by July 1839, and what followed was a tenuous trail of notebook scraps that all but petered out in 1841. Then, in May 1842, Darwin attempted to write up his theory. The result, known as the 1842 Sketch, is perhaps the most unstable of all the Darwinian manuscript texts. Ostensibly an essay, with topic headings, it is full of digressions, incomplete examples, vaguely positioned inserts, and illegible handwriting. Students should bear in mind that they are reading a rough first draft of ideas that have gone four years without being written down. The work is incomplete and the mark of the fluid and telegraphic notebook style is still evident, except that the notebook passages were often written with more mastery and closure.

It is, in part, precisely the textual difficulty of the Sketch that makes it an interesting historical and literary document. Yet what also makes the Sketch interesting is that the logic of Darwin's argument is tightly constructed. Indeed, in the Sketch, Darwin established the expository structure he would follow in the two subsequent drafts (1844 Essay and *Natural Selection*, known as the 1856–1858 "big species book") and the *Origin* (1859).

This logical structure has two parts. In the first part, Darwin presents the mechanisms and submechanisms that explain the broad conceptual region that constitutes "the origin of species." In the second part, he attempts to demonstrate that evolution is a fact by applying his mechanism to the solution of the core problems in morphology, and systematics, biogeography, paleontology, and behavior—in other words, all of natural history. The first part could be titled "The Argument for Natural Selection," the second part, "Evolution: The Argument from Natural Selection."

"The Argument for Natural Selection" begins with the origin of domesticated varieties. Plants and animals under domestication show tremendous variation in all hereditary characters. Breeders have selected from this individual variation and by inbreeding—or artificial selection—have produced highly specialized and distinctive varieties, everything from racehorses to drayhorses. Darwin then proposes an analogy between the human procedures that produce domesticated varieties by means of artificial selection and the natural processes that originate new species in the wild. Next Darwin personifies nature, proposing, as a rhetorical crutch, a hypothetical "being infinitely more sagacious than man (not an omniscient creator)." Finally, he proposes that there is "a natural means of selecting," which is the "war of nature" attendant upon the "enormous geometrical power of increase in every organism." As Darwin puts it, "In the course of a thousand generations infinitesimally small differences must inevitably tell."

In the summer of 1844, beginning from this base, Darwin wrote a full-blown essay of 230 pages, which he had copied out by his amanuensis. (See pp. 116–17 for Darwin's view of the importance of the 1844 Essay.)

Besides marking the transition to essays and serving Darwin as important way stations in the writing of *On the Origin of Species*, the 1842 Sketch and the 1844 Essay also mark a distinctive stage in the development of Darwin's theory. As Darwin's son Francis pointed out in 1909, both documents leave unresolved a key question of evolution: How is the branching taxonomic tree—the so-called natural system of taxonomy—actually formed? It would be a decade before Darwin solved this problem (see "On the Principle of Divergence," chapter 6, 130). For the time being, Darwin had solutions for the origin of adaptations and the origin of new species, but not for the origin of divergence—or the ordered relationship between higher taxa. In 1980, Dov Ospovat pointed out that the root cause of this apparent gap was Darwin's view of the nature and extent of variation in nature. In the 1842–1844 versions, Darwin emphasizes the limited extent of variation in wild species. Domestication seems to induce variability in organisms, making them plastic; but in nature, Darwin assumes, selection is so tight that little variation is present in natural populations. Therefore, natural selection, at this point in Darwin's thinking, is something like the eighteenth-century concept of "nature's broom," scrutinizing and adapting the organism to its environment and thereby maintaining the stability of species, rather than as a dynamically creative force.

Ultimately, in the decade of the 1850s, Darwin's attitude toward the amount of variation in nature would change, in part owing to his work on the highly complex taxonomy of the barnacles (chapter 5, below).

1842 SKETCH

ON SELECTION UNDER DOMESTICATION, NATURAL SELECTION, AND ORGANIC BEINGS IN THE WILD STATE

PART I

SECTION I

ON VARIATION UNDER DOMESTICATION, AND ON THE PRINCIPLES OF SELECTION

An individual organism placed under new conditions <often> sometimes varies in a small degree and in very trifling respects such as stature, fatness, sometimes colour, health, habits in animals and probably disposition. Also habits of life develop certain parts. Disuse atrophies. <Most of these slight variations tend to become hereditary.>

When the individual is multiplied for long periods by buds the variation is yet small, though greater and occasionally a single bud or individual departs widely from its type (example) and continues steadily to propagate, by buds, such new kind.

When the organism is bred for several generations under new or varying conditions, the variation is greater in amount and endless in kind <especially holds good when individuals have long been exposed to new conditions>. The nature of the external conditions tends to effect some definite change in all or greater part of offspring — little food, small size — certain foods harmless, etc. organs affected and diseases — extent unknown. A certain degree of variation (Müller's twins) seems inevitable effect of process of reproduction. But more important is that simple <<?>> generation, especially under new conditions <when no crossing> <<causes>> infinite variation and not direct effect of external conditions, but only in as much as it affects the reproductive functions. There seems to be no part (*beau ideal* of liver) of body, internal or external, or mind or habits, or instincts which does not vary in some small degree and <often> some <<?>> to a great amount.

<All such> variations <being congenital> or those very slowly acquired of all kinds <decidedly evince a tendency to become hereditary>, when not so become simple variety, when it does a race. Each parent transmits its peculiarities, therefore if varieties allowed freely to cross, except by the *chance*

of two characterized by same peculiarity happening to marry, such varieties will be constantly demolished. All bisexual animals must cross, hermaphrodite plants do cross, it seems very possible that hermaphrodite animals do cross — conclusion strengthened: ill effects of breeding in and in, good effects of crossing possibly analogous to good effects of change in condition <<?>>.

Therefore if in any country or district all animals of one species be allowed freely to cross, any small tendency in them to vary will be constantly counteracted. Secondly reversion to parent form — analogue of *vis medicatrix*. But if man selects, then new races rapidly formed — of late years systematically followed — in most ancient times often practically followed. By such selection make race-horse, dray-horse — one cow good for tallow, another for eating, etc. — one plant's good lay <<illegible>> in leaves another in fruit, etc.: the same plant to supply his wants at different times of year. By former means animals become adapted, as a direct effect to a cause, to external conditions, as size of body to amount of food. By this latter means they may also be so adapted, but further they may be adapted to ends and pursuits, which by no possibility can affect growth, as existence of tallow-chandler cannot tend to make fat. In such selected races, if not removed to new conditions, and <<if>> preserved from all cross, after several generations become very true, like each other and not varying. But man selects only <<?>> what is useful and curious — has bad judgement, is capricious — grudges to destroy those that do not come up to his pattern — has no <knowledge> power of selecting according to internal variations — can hardly keep his conditions uniform — <cannot> does not select those best adapted to the conditions under which <<the>> form <<?>> lives, but those most useful to him. This might all be otherwise.

SECTION II

ON VARIATION IN A STATE OF NATURE AND ON THE NATURAL MEANS OF SELECTION

Let us see how far above principles of variation apply to wild animals. Wild animals vary exceedingly little — yet they are known as individuals. British Plants, in many genera number quite uncertain of varieties and species: in shells chiefly external conditions. Primrose and cowslip. Wild animals from different <countries can be recognized>. Specific character gives some organs as varying. Variations analogous in kind, but less in degree with domesticated animals — chiefly external and less important parts.

Our experience would lead us to expect that any and every one of these organisms would vary if <<the organism were>> taken away <<?>> and placed under new conditions. Geology proclaims a constant round of change, bringing into play, by every possible <<?>> change of climate and the death of pre-existing inhabitants, endless variations of new conditions. These <<?>> generally very slow, doubtful though <<illegible>> how far the slowness <<?>> would produce tendency to vary. But Geolog<<ists>> show change in configuration which, together with the accidents of air and water and the means of transportal which every being possesses, must occasionally bring, rather suddenly, organisms to new conditions and <<?>> expose it for several generations. Hence <<?>> we should expect every now and then a wild form to vary; possibly this may be cause of some species varying more than others.

According to nature of new conditions, so we might expect all or majority of organisms born under them to vary in some definite way. Further we might expect that the mould in which they are cast would likewise vary in some small degree. But is there any means of selecting those offspring which vary in the same manner, crossing them and keeping their offspring separate and thus producing selected races: otherwise as the wild animals freely cross, so must such small heterogeneous varieties be constantly counterbalanced and lost, and a uniformity of character <kept up> preserved. The former variation as the direct and necessary effects of causes, which we can see can act on them, as size of body from amount of food, effect of certain kinds of food on certain parts of bodies, etc.; such new varieties may then become adapted to those external <natural> agencies which act on them. But can varieties be produced adapted to end, which cannot possibly influence their structure and which it is absurd to look <<at>> as effects of chance. Can varieties like some vars of domesticated animals, like almost all wild species be produced adapted by exquisite means to prey on one animal or to escape from another — or rather, as it puts out of question effects of intelligence and habits, can a plant become adapted to animals, as a plant which cannot be impregnated without agency of insect; or hooked seeds depending on animal's existence: woolly animals cannot have any direct effect on seeds of plant. This point which all theories about climate adapting woodpecker to crawl <<?>> up trees, <<illegible>> miseltoe, <<sentence incomplete>>. But if every part of a plant or animal was to vary <<illegible>>, and if a being infinitely more sagacious than man (not an omniscient creator) during thousands and thousands of years were to select all the variations which tended towards certain ends (<or were to produce causes <<?>> which tended to the same end>), for instance, if he foresaw a canine animal would be better off, owing to the country producing more hares, if he were longer legged and keener sight — greyhound produced. If he saw that aquatic <<animal would need>> skinned toes. If for some unknown cause he found it would advantage a plant, which <<?>> like most plants is occasionally visited by bees, etc.: if that plant's seed were occasionally eaten by birds and were then carried on to rotten trees, he might select trees with fruit more agreeable to such birds as perched, to ensure their being carried to trees; if he perceived those birds more often dropped the seeds, he might well have selected a bird who would <<illegible>> rotten trees or <gradually select plants which <<he>> had proved to live on less and less rotten trees>. Who, seeing how plants vary in garden, what blind foolish man has done in a few years, will deny an all-seeing being in thousands of years could effect (if the Creator chose to do so), either by his own direct foresight or by intermediate means — which will represent <<?>> the creator of this universe. Seems usual means. Be it remembered I have nothing to say about life and mind and *all* forms descending from one common type. I speak of the variation of the existing great divisions of the organized kingdom, how far I would go, hereafter to be seen.

Before considering whether <<there>> be any natural means of selection, and secondly (which forms 2nd Part of this sketch) the far more important point whether the characters and relations of animated <<things>> are such as favour the idea of wild species being races <<?>> descended from a common stock, as the varieties of potato or dahlia or cattle having so descended, let us consider probable character of <selected races> wild varieties.

Natural Selection. De Candolle's war of nature — seeing contented face of nature — may be well at first doubted; we see it on borders of perpetual cold. But considering the enormous geometrical power of increase in every organism and as <<?>> every country, in ordinary cases <<countries>> must be stocked to full extent, reflection will show that this is the case. Malthus on man — in animals no moral <check> restraint <<?>> — they breed in time of year when provision most abundant, or season most favourable, every country has its seasons — calculate robins — oscillating from years of destruction. If proof were wanted let any singular change of climate <<occur>> here <<?>>, how astoundingly some tribe <<?>> increase, also introduced animals, the pressure is always ready — capacity of alpine plants to endure other climates — think of endless seeds scattered abroad — forests regaining their percentage — a thousand wedges are being forced into the economy of nature. This requires much reflection; study Malthus and calculate rates of increase and remember the resistance — only periodical.

The unavoidable effect of this <<is>> that many of every species are destroyed either in egg or <young or mature (the former state the more common)>. In the course of a thousand generations infinitesimally small differences must inevitably tell; when unusually cold winter, or hot or dry summer comes, then out of the whole body of individuals of any species, if there be the smallest differences in their structure, habits, instincts <senses>, health, etc., <<it>> will on an average tell; as conditions change a rather larger

proportion will be preserved: so if the chief check to increase falls on seeds
or eggs, so will, in the course of 1,000 generations or ten thousand, those
seeds (like one with down to fly) which fly furthest and get scattered most
ultimately rear most plants, and such small differences tend to be hereditary
like shades of expression in human countenance. So if one parent <<?>>
fish deposits its egg in infinitesimally different circumstances, as in rather
shallower or deeper water, etc., it will then <<?>> tell.

Let hares increase very slowly from change of climate affecting peculiar
plants, and some other <<illegible>> rabbit decrease in same proportion
<let this unsettle organisation of>, a canine animal, who formerly derived
its chief sustenance by springing on rabbits or running them by scent, must
decrease too and might thus readily become exterminated. But if its form
varied very slightly, the long legged fleet ones, during a thousand years being selected, and the less fleet rigidly destroyed must, if no law of nature be
opposed to it, alter forms.

Remember how soon Bakewell on the same principle altered cattle and
Western, sheep — carefully avoiding a cross (pigeons) with any breed. We
cannot suppose that one plant tends to vary in fruit and another in flower,
and another in flower and foliage — some have been selected for both fruit
and flower: that one animal varies in its covering and another not — another
in its milk. Take any organism and ask what is it useful for and on that point
it will be found to vary — cabbages in their leaf — corn in size <<and>>
quality of grain, both in times of year — kidney beans for young pod and
cotton for envelope of seeds, etc.: dogs in intellect, courage, fleetness and
smell <<?>>: pigeons in peculiarities approaching to monsters. This requires consideration — should be introduced in first chapter if it holds, I
believe it does. It is hypothetical at best.

Nature's variation far less, but such selection far more rigid and scrutinizing. Man's races not <even so well> only not better adapted to conditions than other races, but often not <<?>> one race adapted to its conditions, as man keeps and propagates some alpine plants in garden. Nature
lets <<an>> animal live, till on actual proof it is found less able to do the
required work to serve the desired end, man judges solely by his eye, and
knows not whether nerves, muscles, arteries, are developed in proportion to
the change of external form.

Besides selection by death, in bisexual animals <<illegible>> the selection in time of fullest vigour, namely struggle of males; even in animals which
pair there seems a surplus <<?>> and a battle, possibly as in man more
males produced than females, struggle of war or charms. Hence that male
which at that time is in fullest vigour, or best armed with arms or ornaments
of its species, will gain in hundreds of generations some small advantage and
transmit such characters to its offspring. So in female rearing its young, the

most vigorous and skilful and industrious, <<whose>> instincts <<are>> best developed, will rear more young, probably possessing her good qualities, and a greater number will thus <<be>> prepared for the struggle of nature. Compared to man using a male alone of good breed. This latter section only of limited application, applies to variation of <specific> sexual characters. Introduce here contrast with Lamarck — absurdity of habit, or chance ?? or external conditions, making a woodpecker adapted to tree.

Before considering difficulties of theory of selection let us consider character of the races produced, as now explained, by nature. Conditions have varied slowly and the organisms best adapted in their whole course of life to the changed conditions have always been selected — man selects small dog and afterwards gives it profusion of food — selects a long-backed and short-legged breed and gives it no particular exercise to suit this function, etc. In ordinary cases nature has not allowed her race to be contaminated with a cross of another race, and agriculturists know how difficult they find always to prevent this — effect would be trueness. This character and sterility when crossed, and generally a greater amount of difference, are two main features, which distinguish domestic races from species.

<Sterility not universal admitted by all. *Gladiolus, Crinum, Calceolaria* must be species if there be such a thing. Races of dogs and oxen: but certainly very general; indeed a gradation of sterility most perfect very general. Some nearest species will not cross (crocus, some heath <<?>>), some genera cross readily (fowls and grouse, peacock, etc.). Hybrids no ways monstrous quite perfect except secretions hence even the mule has bred — character of sterility, especially a few years ago <<?>> thought very much more universal than it now is, has been thought the distinguishing character; indeed it is obvious if all forms freely crossed, nature would be a chaos. But the very gradation of the character, even if it always existed in some degree which it does not, renders it impossible as marks <<?>> those <<?>> suppose distinct as species.> Will analogy throw any light on the fact of the supposed races of nature being sterile, though none of the domestic ones are? Mr Herbert <<and>> Kölreuter have shown external differences will not guide one in knowing whether hybrids will be fertile or not, but the chief circumstance is constitutional differences, such as being adapted to different climate or soil, differences which <must> probably affect the whole body of the organism and not any one part. Now wild animals, taken out of their natural conditions, seldom breed. I do not refer to shows or to Zoological Societies where many animals unite, but <<do not?>> breed, and others will never unite, but to wild animals caught and kept *quite tame* left loose and well fed about houses and living many years. Hybrids produced almost as readily as pure breds. St Hilaire great distinction of tame and domestic — elephants — ferrets. Reproductive organs not subject to disease in

Zoological Garden. Dissection and microscope show that hybrid is in exactly same condition as another animal in the intervals of breeding season, or those animals which taken wild and *not bred* in domesticity, remain without breeding their whole lives. It should be observed that so far from domesticity being unfavourable in itself <<it>> makes more fertile: <when animal is domesticated and breeds, productive power increased from more food and selection of fertile races>. As far as animals go might be thought <<an>> effect on their mind and a special case.

But turning to plants we find same class of facts. I do not refer to seeds not ripening, perhaps the commonest cause, but to plants not setting, which either is owing to some imperfection of ovule or pollen. Lindley says sterility is the <curse> bane of all propagators — Linnaeus about alpine plants. American bog plants — pollen in exactly same state as in hybrids — same in geraniums. Persian and Chinese lilac will not seed in Italy and England. Probably double plants and all fruits owe their developed parts primarily <<?>> to sterility and extra food thus <<?>> applied. There is here gradation <<in>> sterility and then parts, like diseases, are transmitted hereditarily. We cannot assign any cause why the Pontic Azalea produces plenty of pollen and not American, why common lilac seeds and not Persian, we see no difference in healthiness. We known not on what circumstances these facts depend, why ferret breeds, and cheetah, elephant and pig in India will not.

Now in crossing it is certain every peculiarity in form and constitution is transmitted: an alpine plant transmits its alpine tendency to its offspring, an American plant its American-bog constitution, and <<with>> animals, those peculiarities, on which when placed out of their natural conditions they are incapable of breeding; and moreover they transmit every part of their constitution, their respiration, their pulse, their instinct, which are all suddenly modified, can it be wondered at that they are incapable of breeding? I think it may be truly said it would be more wonderful if they did. But it may be asked why have not the recognized varieties, supposed to have been produced through the means of man, <not refused to breed> have all bred. Variation depends on change of condition and selection, as far as man's systematic or unsystematic selection <<has>> gone; he takes external form, has little power from ignorance over internal invisible constitutional differences. Races which have long been domesticated, and have much varied, are precisely those which were capable of bearing great changes, whose constitutions were adapted to a diversity of climates. Nature changes slowly and by degrees. According to many authors probably breeds of dogs are another case of modified species freely crossing. There is no variety which <<illegible>> has been <<illegible>> adapted to peculiar soil or situation for a thousand years and another rigorously adapted to another, till such can be produced, the question is not tried. Man in past ages, could transport into different climates, animals and plants which would freely propagate in such new climates. Nature could effect, with selection, such changes slowly, so that precisely those animals which are adapted to submit to great changes have given rise to diverse races — and indeed great doubt on this head.

Before leaving this subject well to observe that it was shown that a certain amount of variation is consequent on mere act of reproduction, both by buds and sexually — is vastly increased when parents exposed for some generations to new conditions, and we find that many animals when exposed for first time to very new conditions, are (as) incapable of breeding as hybrids. It <probably> bears also on supposed fact of crossed animals when not infertile, as in mongrels, tending to vary much, as likewise seems to be the case, when true hybrids possess just sufficient fertility to propagate with the parent breeds and *inter se* for some generations. This is Kölreuter's belief. These facts throw light on each other and support the truth of each other, we see throughout a connection between the reproductive faculties and exposure to changed conditions of life whether by crossing or exposure of the individuals.

Difficulties on theory of selection. It may be objected such perfect organs as eye and ear, could never be formed, in latter less difficulty as gradations more perfect; at first appears monstrous and to <<the>> end appears difficulty. But think of gradation, even now manifest (Tibia and Fibula). Everyone will allow if every fossil preserved, gradation infinitely more perfect; for possibility of selection a perfect <<?>> gradation is required. Different groups of structure, slight gradation in each group — every analogy renders it probable that intermediate forms have existed. Be it remembered what strange metamorphoses; part of eye, not directly connected with vision, might come to be <thus used> gradually worked in for this end — swimming bladder by gradation of structure is admitted to belong to the ear system — rattlesnake. <Woodpecker best adapted to climb.> In some cases gradation not possible — as vertebrae — actually vary in domestic animals — less difficult if growth followed. Looking to whole animals, a bat formed not for flight. Suppose we had flying fish and not one of our now called flying fish preserved, who would have guessed intermediate habits. Woodpeckers and tree-frogs both live in countries where no trees.

The gradations by which each individual organ has arrived at its present state, and each individual animal with its aggregate of organs has arrived, probably never could be known, and all present great difficulties. I merely wish to show that the proposition is not so monstrous as it at first appears, and that if good reason can be advanced for believing the species have descended from common parents, the difficulty of imagining intermediate forms of structure not sufficient to make one at once reject the theory.

SECTION III

ON VARIATION IN INSTINCTS AND OTHER
MENTAL ATTRIBUTES

The mental powers of different animals in wild and tame state <present still greater difficulties> require a separate section. Be it remembered I have nothing to do with origin of memory, attention, and the different faculties of the mind, but merely with their differences in each of the great divisions of nature. Disposition, courage, pertinacity <<?>>, suspicion, restlessness, illtemper, sagacity and <<the>> reverse unquestionably vary in animals and are inherited (Cuba wildness dogs, rabbits, fear against particular object as man Galapagos). Habits purely corporeal, breeding season, etc., time of going to rest, etc., vary and are hereditary, like the analogous habits of plants which vary and are inherited. Habits of body, as manner of movement ditto and ditto. Habits, as pointing and setting on certain occasions ditto. Taste for hunting certain objects and manner of doing so — sheep-dog. These are shown clearly by crossing and their analogy with true instinct thus shown — retriever. Do not know objects for which they do it. Lord Brougham's definition. Origin partly habit, but the amount necessarily unknown, partly selection. Young pointers pointing stones and sheep — tumbling pigeons — sheep going back to place where born. Instinct aided by reason, as in the taylor-bird. Taught by parents, cows choosing foods, birds singing. Instincts vary in wild state (birds get wilder) often lost; more perfect — nest without roof. These facts <only clear way> show how incomprehensibly brain has power of transmitting intellectual operations.

Faculties distinct from true instincts — finding <way>. It must I think be admitted that habits whether congenital or acquired by practice <sometimes> often become inherited; instincts, influence, equally with structure, the preservation of animals; therefore selection must, with changing conditions tend to modify the inherited habits of animals. If this be admitted it will be found *possible* that many of the strangest instincts may be thus acquired. I may observe, without attempting definition, that an inherited habit or trick (trick because may be born) fulfils closely what we mean by instinct. A habit is often performed unconsciously, the strangest habits become associated, ditto tricks, going in certain spots, etc., even against will, is excited by external agencies, and looks not to the end — a person playing a pianoforte. If such a habit were transmitted it would make a marvellous instinct. Let us consider some of the most difficult cases of instincts, whether they could be *possibly* acquired. I do not say *probably*, for that belongs to our 3rd Part. I beg this may be remembered, nor do I mean to attempt to show exact method. I want only to show that whole theory ought not at once to be rejected on this score.

Every instinct must, by my theory, have been acquired gradually by slight changes <<illegible>> of former instinct, each change being useful to its then species. Shamming death struck me at first as remarkable objection. I found none really sham death, and that there is gradation; now no one doubts that those insects which do it either more or less, do it for some good, if then any species was led to do it more, and then <<?>> escaped, etc.

Take migratory instincts, faculty distinct from instinct, animals have notion of time — like savages. Ordinary finding way by memory, but how does savage find way across country — as incomprehensible to us, as animal to them — geological changes — fishes in river — case of sheep in Spain. Architectural instincts — a manufacturer's employee in making single articles extraordinary skill — often said seem to make it almost <<illegible>>, child born with such a notion of playing — we can fancy tailoring acquired in same perfection — mixture of reason — water-ouzel — taylorbird — gradation of simple nest to most complicated.

Bees again, distinction of faculty — how they make a hexagon — Waterhouse's theory — the impulse to use whatever faculty they possess — the taylor-bird has the faculty of sewing with beak, instinct impels him to do it.

Last case of parent feeding young with different food (take case of Galapagos birds, gradation from Hawfinch to Sylvia) selection and habit might lead old birds to vary taste <<?>> and form, leaving their instinct of feeding their young with same food — or I see no difficulty in parents being forced or induced to vary the food brought, and selection adapting the young ones to it, and thus by degree any amount of diversity might be arrived at. Although we can never hope to see the course revealed by which different instincts have been acquired, for we have only present animals (not well known) to judge of the course of gradation, yet once grant the principle of habits, whether congenital or acquired by experience, being inherited and I can see no limit to the <amount of variation> extraordinariness <<?>> of the habits thus acquired.

SUMMING UP THIS DIVISION

If variation be admitted to occur occasionally in some wild animals, and how can we doubt it, when we see <all> thousands <<of>> organisms, for whatever use taken by man, do vary. If we admit such variations tend to be hereditary, and how can we doubt it when we <<remember>> resemblances of features and character — disease and monstrosities inherited and endless races produced (1,200 cabbages). If we admit selection is steadily at work, and who will doubt it, when he considers amount of food on an average fixed and reproductive powers act in geometrical ratio. If we admit that external conditions vary, as all geology proclaims, they have done and are now doing — then, if no law of nature be opposed, there must occasionally be formed races, <slightly> differing from the parent races. So then any such law, none is known, but in all works it is assumed, in <<?>> flat contradiction to all known facts, that the amount of possible variation is soon acquired. Are not all the most varied species, the oldest domesticated: who <<would>> think that horses or corn could be produced? Take dahlia and potato, who will pretend in 5,000 years <<that great changes might not be effected>>: perfectly adapted to conditions and then again brought into varying conditions. Think what has been done in few last years, look at pigeons, and cattle. With the amount of food man can produce he may have arrived at limit of fatness or size, or thickness of wool <<?>>, but these are the most trivial points, but even in these I conclude it is impossible to say we know the limit of variation. And therefore with the <adapting> selecting power of nature, infinitely wise compared to those of man, <<I conclude>> that it is impossible to say we know the limit of races, which would be true <<to their>> kind; if of different constitutions would probably be infertile one with another, and which might be adapted in the most singular and admirable manner, according to their wants, to external nature and to other surrounding organisms — such races would be species. But is there any evidence <<that>> species <<have>> been thus produced, this is a question wholly independent of all previous points, and which on examination of the kingdom of nature <<we>> ought to answer one way or another.

1844 ESSAY

VARIATION OF ORGANIC BEINGS IN THE WILD STATE

Having treated of variation under domestication, we now come to it in a *state of nature*.

Most organic beings in a state of nature vary exceedingly little: I put out of the case variations (as stunted plants, etc., and sea-shells in brackish water) which are directly the effect of external agencies and which we do not *know are in the breed*, or are *hereditary*. The amount of hereditary variation is very difficult to ascertain, because naturalists (partly from the want of knowledge, and partly from the inherent difficulty of the subject) do not all agree whether certain forms are species or races. Some strongly marked races of plants, comparable with the decided sports of horticulturalists, undoubtedly exist in a state of nature, as is actually known by experiment, for instance in the primrose and cowslip, in two so-called species of dandelion, in two of foxglove, and I believe in some pines. Lamarck has observed that, as long as we confine our attention to one limited country, there is seldom much difficulty in deciding what forms to call species and what varieties: and that it is when collections flow in from all parts of the world that naturalists often feel at a loss to decide the limit of variation. Undoubtedly so it is, yet amongst British plants (and I may add land shells), which are probably better known than any in the world, the best naturalists differ very greatly in the relative proportions of what they call species and what varieties. In many genera of insects, and shells, and plants, it seems almost hopeless to establish which are which. In the higher classes there are less doubts: though we find considerable difficulty in ascertaining what deserve to be called species amongst foxes and wolves, and in some birds, for instance in the case of the white barn-owl. When specimens are brought from different parts of the world, how often do naturalists dispute this same question, as I found with respect to the birds brought from the Galapagos islands. Yarrell has remarked that the individuals of the same undoubted species of birds, from Europe and N. America, usually present slight, indefinable though preceptible differences. The recognition indeed of one animal by another of its kind seems to imply some difference. The disposition of wild animals undoubtedly differs. The variation, such as it is, chiefly affects the same parts in wild organisms as in domestic breeds; for instance, the size, colour, and the external and less important parts. In many species the variability of certain organs or qualities is even stated as one of the specific characters: thus, in plants, colour, size, hairiness, the number of the stamens and pistils, and even their presence, the form of the leaves; the size and form of the mandibles of the males of some insects; the length and curvature of the beak in some birds (as in Opetiorynchus) are variable characters in some species and quite fixed in others. I do not perceive that any just distinction can be drawn between this recognized variability of certain parts in many species and the more general variability of the whole frame in domestic races.

Although the amount of variation be exceedingly small in most organic beings in a state of nature, and probably quite wanting (as far as our senses serve) in the majority of cases: yet considering how many animals and plants, taken by mankind from different quarters of the world for the most diverse purposes, have varied under domestication in every country and in every age. I think we may safely conclude that all organic beings with few exceptions, if capable of being domesticated and bred for long periods, would vary. Domestication seems to resolve itself into a change from the natural conditions of the species <generally perhaps including an increase of food>: if this be so, organisms in a state of nature must *occasionally*, in the course of ages, be exposed to analogous influences; for geology clearly shows that many places must, in the course of time, become exposed to the widest range of climatic and other influences; and if such places be isolated, so that new and better adapted organic beings cannot freely emigrate, the old inhabitants will be exposed to new influences, probably far more varied, than man applies under the form of domestication. Although every species no doubt will soon breed up to the full number which the country will support, yet it is easy to conceive that, on an average, some species may receive an increase of food: for the times of dearth may be short, yet enough to kill, and recurrent only at long intervals. All such changes of conditions from geological causes would be exceedingly slow; what effect the slowness might have we are ignorant; under domestication it appears that the effects of change of conditions accumulate, and then break out. Whatever might be the result of these slow geological changes, we may feel sure, from the means of dissemination common in a lesser or greater degree to every organism taken conjointly with the changes of geology, which are steadily (and sometimes suddenly, as when an isthmus at last separates) in progress, that occasionally organisms must suddenly be introduced into new regions, where, if the conditions of existence are not so foreign as to cause its extermination, it will often be propagated under circumstances still more closely analogous to those of domestication; and therefore we expect will evince a tendency to vary. It appears to be quite *inexplicable* if this has never happened; but it can happen very rarely. Let us then suppose that an organism by some chance (which might be hardly repeated in 1,000 years) arrives at a modern volcanic island in process of formation and not fully stocked with the most appropriate organisms; the new organism might readily gain a footing, although the external conditions were considerably different from its native ones. The effect of this we might expect would influence in some small degree the size, colour, nature of covering, etc., and from inexplicable influences even special parts and organs of the body. But we might further (and <<this>> is far more important) expect that the reproductive system would be affected, as under domesticity, and the structure of the offspring rendered in some degree plastic. Hence almost every part of the body would tend to vary from the typical form in slight degrees, and in no determinate way, and therefore *without selection* the free crossing of these small variations (together with the tendency to reversion to the original form) would constantly be counteracting this unsettling effect of the extraneous conditions on the reproductive system. Such, I conceive, would be the unimportant result without selection. And here I must observe that the foregoing remarks are equally applicable to that small and admitted amount of variation which has been observed in some organisms in a state of nature; as well as to the above hypothetical variation consequent on changes of condition.

Let us now suppose a Being with penetration sufficient to perceive differences in the outer and innermost organization quite imperceptible to man, and with forethought extending over future centuries to watch with unerring care and select for any object the offspring of an organism produced under the foregoing circumstances: I can see no conceivable reason why he could not form a new race (or several were he to separate the stock of the original organism and work on several islands) adapted to new ends. As we assume his discrimination, and his forethought, and his steadiness of object, to be incomparably greater that those qualities in man, so we may suppose the beauty and complications of the adaptations of the new races and their differences from the original stock to be greater than in the domestic races produced by man's agency: the ground-work of his labours we may aid by supposing that the external conditions of the volcanic island, from its continued emergence and the occasional introduction of new immigrants, vary; and thus to act on the reproductive system of the organism, on which he is at work, and so keep its organization somewhat plastic. With time enough, such a Being might rationally (without some unknown law opposed him) aim at almost any result.

For instance, let this imaginary Being wish, from seeing a plant growing on the decaying matter in a forest and choked by other plants, to give it power of growing on the rotten stems of trees, he would commence selecting every seedling whose berries were in the smallest degree more attractive to tree-frequenting birds, so as to cause a proper dissemination of the seeds, and at the same time he would select those plants which had in the slightest degree more and more power of drawing nutriment from rotten wood; and he would destroy all other seedlings with less of this power. He might thus, in the course of century after century, hope to make the plant by degrees grow on rotten wood, even high up on trees, wherever birds dropped the non-digested seeds. He might then, if the organization of the plant was plastic, attempt by continued selection of chance seedlings to make it grow on less and less rotten wood, till it would grow on sound wood. Supposing again, during these changes the plant failed to seed quite freely from non-impregnation, he might begin selecting seedlings with a little sweeter (or) differently tasted honey or pollen, to tempt insects to visit the flowers regularly: having effected this, he might wish, if it profited the plant, to render abortive the stamens and pistils in different flowers, which he could do by continued selection. By such steps he might aim at making a plant as wonderfully related to other organic beings as is the mistletoe, whose existence absolutely depends on certain insects for impregnation, certain birds for transportal, and certain trees for growth. Furthermore, if the insect which had been induced regularly to visit this hypothetical plant profited much by it, our same Being might wish by selection to modify by gradual selection the insect's structure, so as to facilitate its obtaining the honey or pollen: in this manner he might adapt the insect (always presupposing its organization to be in some degree plastic) to the flower, and the impregnation of the flower to the insect; as is the case with many bees and many plants.

Seeing what blind capricious man has actually effected by selection during the few last years, and what in a ruder state he has probably effected
without any systematic plan during the last few thousand years, he will be
a bold person who will positively put limits to what the supposed Being could
effect during whole geological periods. In accordance with the plan by which
this universe seems governed by the Creator, let us consider whether there
exists any *secondary* means in the economy of nature by which the process
of selection could go on adapting, nicely and wonderfully, organisms, if in
ever so small a degree plastic, to diverse ends. I believe such secondary means
do exist.

NATURAL MEANS OF SELECTION

De Candolle, in an eloquent passage, has declared that all nature is at war,
one organism with another, or with external nature. Seeing the contented
face of nature, this may at first be well doubted; but reflection will inevitably
prove it is too true. The war, however, is not constant, but only recurrent in
a slight degree at short periods and more severely at occasional more distant
periods; and hence its effects are easily overlooked. It is the doctrine of
Malthus applied in most cases with ten-fold force. As in every climate there
are seasons for each of its inhabitants of greater and less abundance, so all
annually breed; and the moral restraint, which in some small degree checks
the increase of mankind, is entirely lost. Even slow-breeding mankind has
doubled in 25 years, and if he could increase his food with greater ease, he
would double in less time. But for animals, without artificial means, *on the
average* the amount of food for each species must be constant; whereas the
increase of all organisms tends to be geometrical and in a vast majority of
cases at an enormous ratio. Suppose in a certain spot there are eight pairs of
<robins> birds, and that *only* four pairs of them annually (including double
hatches) rear only four young; and that these go on rearing their young at
the same rate: then at the end of seven years (a short life, excluding violent
deaths, for any birds) there will be 2,048 robins, instead of the original sixteen; as this increase is quite impossible, so we must conclude either that
robins do not rear nearly half their young or that the average life of a robin
when reared is from accident not nearly seven years. Both checks probably
concur. The same kind of calculation applied to all vegetables and animals
produces results either more or less striking, but in scarcely a single instance
less striking than in man.

Many practical illustrations of this rapid tendency to increase are on record,
namely during peculiar seasons, in the extraordinary increase of certain animals, for instance during the years 1826 to 1828, in La Plata, when from
drought, some millions of cattle perished, the whole country *swarmed* with
innumerable mice: now I think it cannot be doubted that during the breed-

ing season all the mice (with the exception of a few males or females in excess) ordinarily pair; and therefore that this astounding increase during three years must be attributed to a greater than usual number surviving the first year, and then breeding, and so on, till the third year, when their numbers were brought down to their usual limits on the return of wet weather. Where man has introduced plants and animals into a new country favourable to them, there are many accounts in how surprisingly few years the whole country has become stocked with them. This increase would necessarily stop as soon as the country was fully stocked; and yet we have every reason to believe from what is known of wild animals that *all* would pair in the spring. In the majority of cases it is most difficult to imagine where the check falls, generally no doubt on the seeds, eggs, and young; but when we remember how impossible even in mankind (so much better known than any other animal) it is to infer from repeated casual observations what the average of life is, or to discover how different the percentage of deaths to the births in different countries, we ought to feel no legitimate surprise at not seeing where the check falls in animals and plants. It should always be remembered that in most cases the checks are yearly recurrent in a small regular degree, and in an extreme degree during occasionally unusually cold, hot, dry, or wet years, according to the constitution of the being in question. Lighten any check in the smallest degree, and the geometrical power of increase in every organism will instantly increase the average numbers of the favoured species. Nature may be compared to a surface, on which rest ten thousand sharp wedges touching each other and driven inwards by incessant blows. Fully to realize these views much reflection is requisite; Malthus on man should be studied; and all such cases as those of the mice in La Plata, of the cattle and horses when first turned out in S. America, of the robins by our calculation, etc., should be well considered; reflect on the enormous multiplying power *inherent and annually in action* in all animals; reflect on the countless seeds scattered by a hundred ingenious contrivances, year after year, over the whole face of the land; and yet we have every reason to suppose that the average percentage of every one of the inhabitants of a country will *ordinarily* remain constant. Finally, let it be borne in mind that this average number of individuals (the external conditions remaining the same) in each country is kept up by recurrent struggles against other species or against external nature (as on the borders of the arctic regions, where the cold checks life); and that ordinarily each individual of each species holds its place, either by its own struggle and capacity of acquiring nourishment in some period (from the egg upwards) of its life, or by the struggle of its parents (in short lived organisms, when the main check occurs at long intervals) against and compared with other individuals of the *same* or *different* species.

But let the external conditions of a country change; if in a small degree, the relative proportions of the inhabitants will in most cases simply be slightly changed; but let the number of inhabitants be small, as in an island, and free access to it from other countries be circumscribed; and let the change of condition continue progressing (forming new stations); in such case the original inhabitants must cease to be so perfectly adapted to the changed conditions as they originally were. It has been shown that probably such changes of external conditions would, from acting on the reproductive system, cause the organization of the beings most affected to become, as under domestication, plastic. Now can it be doubted from the struggle each individual (or its parents) has to obtain subsistence that any minute variation in structure, habits, or instincts, adapting that individual better to the new conditions, would tell upon its vigour and health? In the struggle it would have a better *chance* of surviving, and those of its offspring which inherited the variation, let it be ever so slight, would have a better *chance* to survive. Yearly more are bred than can survive; the smallest grain in the balance, in the long run, must tell on which death shall fall, and which shall survive. Let this work of selection, on the one hand, and death on the other, go on for a thousand generations; who would pretend to affirm that it would produce no effect, when we remember what in a few years Bakewell effected in cattle and Western in sheep, by this identical principle of selection.

To give an imaginary example, from changes in progress on an island, let the organization of a canine animal become slightly plastic, which animal preyed chiefly on rabbits, but sometimes on hares; let these same changes cause the number of rabbits very slowly to decrease and the number of hares to increase; the effect of this would be that the fox or dog would be driven to try to catch more hares, and his numbers would tend to decrease; his organization, however, being slightly plastic, those individuals with the lightest forms, longest limbs, and best eyesight (though perhaps with less cunning or scent) would be slightly favoured, let the difference be ever so small, and would tend to live longer and to survive during that time of the year when food was shortest; they would also rear more young, which young would tend to inherit these slight peculiarities. The less fleet ones would be rigidly destroyed. I can see no more reason to doubt but that these causes in a thousand generations would produce a marked effect, and adapt the form of the fox to catching hares instead of rabbits, than that greyhounds can be improved by selection and careful breeding. So would it be with plants under similar circumstances: if the number of individuals of a species with plumed seeds could be increased by greater powers of dissemination within its own area (that is if the check to increase fell chiefly on the seeds), those seeds which were provided with ever so little more down, or with a plume placed so as to be slightly more acted on by the winds, would in the long run tend to be most disseminated; and hence a greater number of seeds thus formed would germinate, and would tend to produce plants inheriting this slightly better adapted down.

Besides this natural means of selection, by which those individuals are preserved, whether in their egg or seed or in their mature state, which are best adapted to the place they fill in nature, there is a second agency at work in most bisexual animals tending to produce the same effect, namely the struggle of the males for the females. These struggles are generally decided by the law of battle; but in the case of birds, apparently, by the charms of their song, by their beauty or their power of courtship, as in the dancing rock-thrush of Guiana. Even in the animals which pair there seems to be an excess of males which would aid in causing a struggle: in the polygamous animals, however, as in deer, oxen, poultry, we might expect there would be severest struggle: is it not in the polygamous animals that the males are best formed for mutual war? The most vigorous males, implying perfect adaptation, must generally gain the victory in their several contests. This kind of selection, however, is less rigorous than the other; it does not require the death of the less successful, but gives to them fewer descendants. This struggle falls, moreover, at a time of year when food is generally abundant, and perhaps the effect chiefly produced would be the alteration of sexual characters, and the selection of individual forms, no way related to their power of obtaining food, or of defending themselves from their natural enemies, but of fighting one with another. This natural struggle amongst the males may be compared in effect, but in a less degree, to that produced by those agriculturalists who pay less attention to the careful selection of all the young animals which they breed and more to the occasional use of a choice male.

DIFFERENCES BETWEEN 'RACES' AND 'SPECIES': FIRST, IN THEIR TRUENESS OR VARIABILITY

Races produced by these natural means of selection we may expect would differ in some respects from those produced by man. Man selects chiefly by the eye, and is not able to perceive the course of every vessel and nerve, or the form of the bones, or whether the internal structure corresponds to the outside shape. He is unable to select shades of constitutional differences, and by the protection he affords and his endeavours to keep his property alive, in whatever country he lives, he checks, as much as lies in his power, the selecting action of nature, which will, however, go on to a lesser degree with all living things, even if their length of life is not determined by their own powers of endurance. He has bad judgement, is capricious, he does not, or his successors do not, wish to select for the same exact end for hundreds of generations. He cannot always suit the selected form to the properest conditions; nor does he keep those conditions uniform: he selects that which is useful to him, not that best adapted to those conditions in which each variety is placed by him: he selects a small dog, but feeds it highly; he selects a long-backed dog, but does not exercise it in any peculiar manner, at least not during every generation. He seldom allows the most vigorous males to struggle for themselves and propagate, but picks out such as he possesses, or such as he prefers, and not necessarily those best adapted to the existing conditions. Every agriculturalist and breeder knows how difficult it is to prevent an occasional cross with another beed. He often grudges to destroy an individual which departs considerably from the required type. He often begins his selection by a form or sport considerably departing from the parent form. Very differently does the natural law of selection act; the varieties selected differ only slightly from the parent forms; the conditions are constant for long periods and change slowly; rarely can there be a cross; the selection is rigid and unfailing, and continued through many generations; a selection can *never be made* without the form be *better* adapted to the conditions than the parent form: the selecting power goes on without caprice, and steadily for thousands of years adapting the form to these conditions. The selecting power is not deceived by external appearances, it tries the being during its whole life; and if less well <<?>> adapted than its *congeners*, without fail it is destroyed; every part of its structure is thus scrutinized and proved good towards the place in nature which it occupies.

We have every reason to believe that in proportion to the number of generations that a domestic race is kept free from crosses, and to the care employed in continued steady selection with one end in view, and to the care in not placing the variety in conditions unsuited to it; in such proportion does the new race become 'true' or subject to little variation. How incomparably 'truer' then would a race produced by the above rigid, steady, natural means of selection, excellently trained and perfectly adapted to its conditions, free from stains of blood or crosses, and continued during thousands of years, be compared with one produced by the feeble, capricious, misdirected and ill-adapted selection of man. Those races of domestic animals produced by savages, partly by the inevitable conditions of their life, and partly unintentionally by their greater care of the individuals most valuable to them, would probably approach closest to the character of a species; and I believe this is the case. Now the characteristic mark of a species, next, if not equal in importance to its sterility when crossed with another species, and indeed almost the only other character (without we beg the question and affirm the essence of a species, is its not having descended from a parent common to any other form), is the similarity of the individuals composing the species, or in the language of agriculturalists their 'trueness'.

DIFFERENCE BETWEEN 'RACES' AND 'SPECIES'
IN FERTILITY WHEN CROSSED

The sterility of species, or of their offspring, when crossed has, however, received more attention than the uniformity in character of the individuals composing the species. It is exceedingly natural that such sterility should have been long thought the certain characteristic of species. For it is obvious that if the allied different forms which we meet with in the same country could cross together, instead of finding a number of distinct species, we should have a confused and blending series. The fact however of a perfect gradation in the degree of sterility between species, and the circumstance of some species most closely allied (for instance many species of crocus and European heaths) refusing to breed together, whereas other species, widely different, and even belonging to distinct genera, as the fowl and the peacock, pheasant and grouse, Azalea and Rhododendron, Thuja and Juniperus, breeding together ought to have caused a doubt whether the sterility did not depend on other causes, distinct from a law, coincident with their creation. I may here remark that the fact whether one species will or will not breed with another is far less important than the sterility of the offspring when produced; for even some domestic races differ so greatly in size (as the great stag-greyhound and lap-dog, or cart-horse and Burmese ponies) that union is nearly impossible; and what is less generally known is, that in plants Kölreuter has shown by hundreds of experiments that the pollen of one species will fecundate the germen of another species, whereas the pollen of this latter will never act on the germen of the former; so that the simple fact of mutual impregnation certainly has no relation whatever to the distinctness in creation of the two forms. When two species are attempted to be crossed which are so distantly allied that offspring are never produced, it has been observed in some cases that the pollen commences its proper action by exserting its tube, and the germen commences swelling, though soon afterwards it decays. In the next stage in the series, hybrid offspring are produced though only rarely and few in number, and these are absolutely sterile: then we have hybrid offspring more numerous, and occasionally, though very rarely, breeding with either parent, as is the case with the common mule. Again, other hybrids, though infertile *inter se*, will breed *quite* freely with either parent, or with a third species, and will yield offspring generally infertile, but sometimes fertile; and these latter again will breed with either parent, or with a third or fourth species: thus Kölreuter blended together many forms. Lastly it is now admitted by those botanists who have longest contended against the admission, that in certain families the hybrid offspring of many of the species are sometimes perfectly fertile in the first generation when bred together: indeed in some few cases Mr. Herbert found that the hybrids were decidedly more fertile than either of their pure parents. There is no way to escape from the admission that the hybrids from some species of plants are fertile, except by declaring that no form shall be considered as a species, if it produces with another species fertile offspring; but this is begging the question. It has often been stated that different species of animals have a sexual repugnance towards each other; I can find no evidence of this; it appears as if they merely did not excite each others passions. I do not believe that in this respect there is any essential distinction between animals and plants; and in the latter there cannot be a feeling of repugnance.

CAUSES OF STERILITY IN HYBRIDS

The difference in nature between species which causes the greater or lesser degree of sterility in their offspring appears, according to Herbert and Kölreuter, to be connected much less with external form, size, or structure, than with constitutional peculiarities; by which is meant their adaptation to different climates, food and situation, etc.: these peculiarities of constitution probably affect the entire frame, and no one part in particular.

From the foregoing facts I think we must admit that there exists a perfect gradation in fertility between species which when crossed are quite fertile (as in Rhododendron, Calceolaria, etc.), and indeed in an extraordinary degree fertile (as in Crinum), and those species which never produce offspring, but which by certain effects (as the exsertion of the pollen-tube) evince their alliance. Hence, I conceive, we must give up sterility, although undoubtedly in a lesser or greater degree of very frequent occurrence, as an unfailing mark by which *species* can be distinguished from *races*, i.e. from those forms which have descended from a common stock.

INFERTILITY FROM CAUSES DISTINCT
FROM HYBRIDIZATION

Let us see whether there are any analogous facts which will throw any light on this subject, and will tend to explain why the offspring of certain species, when crossed, should be sterile, and not others, without requiring a distinct law connected with their creation to that effect. Great numbers, probably a large majority of animals when caught by man and removed from their natural conditions, although taken very young, rendered quite tame, living to a good old age, and apparently quite healthy, seem incapable under these circumstances of breeding. I do not refer to animals kept in menageries, such as at the Zoological Gardens, many of which, however, appear healthy and live long and unite but do not produce; but to animals caught and left partly at liberty in their native country. Rengger enumerates several caught young and rendered tame, which he kept in Paraguay, and which would not breed: the hunting leopard or cheetah and elephant offer other instances; as do bears in Europe, and the 25 species of hawks, belonging to different genera, thousands of which have been kept for hawking and have lived for long periods in perfect vigour. When the expense and trouble of procuring a succession of young animals in a wild state be borne in mind, one may feel sure that no trouble has been spared in endeavours to make them breed. So clearly marked is this difference in different kinds of animals, when captured by man, that St. Hilaire makes two great classes of animals useful to man: the *tame*, which will not breed, and the *domestic* which will breed in domestication. From certain singular facts we might have supposed that the non-breeding of animals was owing to some perversion of instinct. But we meet with exactly the same class of facts in plants: I do not refer to the large number of cases where the climate does not permit the seed or fruit to ripen, but where the flowers do not 'set', owing to some imperfection of the ovule or pollen. The latter, which alone can be distinctly examined, is often manifestly imperfect, as any one with a microscope can observe by comparing the pollen of the Persian and Chinese lilacs with the common lilac: the two former species (I may add) are equally sterile in Italy as in this country. Many of the American bog plants here produce little or no pollen, whilst the Indian species of the same genera freely produce it. Lindley observes that sterility is the bane of the horticulturist: Linnaeus has remarked on the sterility of nearly all alpine flowers when cultivated in a lowland district. Perhaps the immense class of double flowers chiefly owe their structure to an excess of food acting on parts rendered slightly sterile and less capable of performing their true function, and therefore liable to be rendered monstrous, which monstrosity, like any other disease, is inherited and rendered common. So far from domestication being in itself unfavourable to fertility, it is well known that when an organism is once capable of submission to such conditions <<its>> fertility is increased beyond the natural limit. According to agriculturalists, slight changes of conditions, that is of food or habitation, and likewise crosses with races slightly different, increase the vigour and probably the fertility of their offspring. It would appear also that even a great change of condition, for instance, transportal from temperate countries to India, in many cases does not in the least affect fertility, although it does health and length of life and the period of maturity. When sterility is induced by domestication it is of the same kind, and varies in degree, exactly as with hybrids: for be it remembered that the most sterile hybrid is no way monstrous; its organs are perfect, but they do not act, and minute microscopical investigations show that they are in the same state as those of pure species in the intervals of the breeding season. The defective pollen in the cases above alluded to precisely resembles that of hybrids. The occasional breeding of hybrids, as of the common mule, may be aptly compared to the most rare but occasional reproduction of elephants in captivity. The cause of many exotic Geraniums producing (although in vigorous health) imperfect pollen seems to be connected with the period when water is given them; but in the far greater majority of cases we cannot form any conjecture on what exact cause the sterility of organisms taken from their natural conditions depends. Why, for instance, the cheetah will not breed whilst the common cat and ferret (the latter generally kept shut up in a small box) do — why the elephant will not whilst the pig will abundantly — why the partridge and grouse in their own country will not, whilst several species of pheasants, the guinea-fowl from the deserts of Africa and the peacock from the jungles of India, will. We must, however, feel convinced that it depends on some constitutional peculiarities in these beings not suited to their new condition; though not necessarily causing an ill state of health. Ought we then to wonder much that those hybrids which have been produced by the crossing of species with different constitutional tendencies (which tendencies we know to be eminently inheritable) should be sterile: it does not seem improbable that the cross from an alpine and lowland plant should have its constitutional powers deranged, in nearly the same manner as when the parent alpine plant is brought into a lowland district. Analogy, however, is a deceitful guide and it would be rash to affirm, although it may appear probable, that the sterility of hybrids is due to the constitutional peculiarities of one parent being disturbed by being blended with those of the other parent in exactly the same manner as it is caused in some organic beings when placed by man out of their natural conditions. Although this would be rash, it would, I think, be still rasher, seeing that sterility is no more incidental to *all* cross-bred productions than it is to all organic beings when captured by man, to assert that the sterility of certain hybrids proved a distinct creation of their parents.

But it may be objected (however little the sterility of certain hybrids is connected with the distinct creations of species), how comes it, if species are only races produced by natural selection, that when crossed they so frequently produce sterile offspring, whereas in the offspring of those races confessedly produced by the arts of man there is no one instance of sterility. There is not much difficulty in this, for the races produced by the natural means above explained will be slowly but steadily selected; will be adapted to various and diverse conditions, and to these conditions they will be rigidly confined for immense periods of time; hence we may suppose that they would acquire different constitutional peculiarities adapted to the stations they occupy; and on the constitutional differences between species their sterility, according to the best authorities, depends. On the other hand man selects by external appearance; from his ignorance, and from not having any test at least comparable in delicacy to the natural struggle for food, continued at intervals through the life of each individual, he cannot eliminate fine shades of constitution, dependent on invisible differences in the fluids or solids of the body; again, from the value which he attaches to each individual, he asserts his utmost power in contravening the natural tendency of the most vigorous to survive. Man, moreover, especially in the earlier ages, cannot have kept his conditions of life constant, and in later ages his stock pure. Until man selects two varieties from the same stock, adapted to two climates or to other different external conditions, and confines each rigidly for one or several thousand years to such conditions, always selecting the individuals best adapted to them, he cannot be said to have even commenced the experiment. Moreover, the organic beings which man has longest had under domestication have been those which were of the greatest use to him, and one chief element of their usefulness, especially in the earlier ages, must have been their capacity to undergo sudden transportals into various climates, and at the same time to retain their fertility, which in itself implies that in such respects their constitutional peculiarities were not closely limited. If the opinion already mentioned be correct, that most of the domestic animals in their present state have descended from the fertile commixture of wild races or species, we have indeed little reason now to expect infertility between any cross of stock thus descended.

It is worthy of remark, that as many organic beings, when taken by man out of their natural conditions, have their reproductive system <<so>> affected as to be incapable of propagation, so, we saw in the first chapter, that although organic beings when taken by man do propagate freely, their offspring after some generations vary or sport to a degree which can only be explained by their reproductive system being <<in>> some way affected. Again, when species cross, their offspring are generally sterile: but it was found by Kölreuter that when hybrids are capable of breeding with either parent, or with other species, that their offspring are subject after some generations to excessive variation. Agriculturists, also, affirm that the offspring from mongrels, after the first generation, vary much. Hence we see that both sterility and variation in the succeeding generations are consequent both on the removal of individual species from their natural states and on species crossing. The connection between these facts may be accidental, but they certainly appear to elucidate and support each other — on the principle of the reproductive system of all organic beings being eminently sensitive to any disturbance, whether from removal or commixture, in their constitutional relations to the conditions to which they are exposed.

POINTS OF RESEMBLANCE BETWEEN 'RACES' AND 'SPECIES'

Races and reputed species agree in some respects, although differing from causes which, we have seen, we can in some degree understand, in the fertility and 'trueness' of their offspring. In the first place, there is no clear sign by which to distinguish races from species, as is evident from the great difficulty experienced by naturalists in attempting to discriminate them. As far as external characters are concerned, many of the races which are descended from the same stock differ far more than true species of the same genus; look at the willow-wrens, some of which skilful ornithologists can hardly distinguish from each other except by their nests; look at the wild swans, and compare the distinct species of these genera with the races of domestic ducks, poultry, and pigeons; and so again with plants, compare the cabbages, almonds, peaches and nectarines, etc. with the species of many genera. St. Hilaire has even remarked that there is a greater difference in size between races, as in dogs (for he believes all have descended from one stock), than between the species of any one genus; nor is this surprising, considering that amount of food and consequently of growth is the element of change over which man has most power. I may refer to a former statement, that breeders believe the growth of one part or strong action of one function causes a decrease in other parts; for this seems in some degree analogous to the law of 'organic compensation', which many naturalists believe holds good. To give an instance of this law of compensation — those species of Carnivora which have the canine teeth greatly developed have certain molar teeth deficient; or again, in that division of the Crustaceans in which the tail is much developed, the thorax is little so, and the converse. The points of difference between different races is often strikingly analogous to that between species of the same genus: trifling spots or marks of colour (as the bars on pigeons' wings) are often preserved in races of plants and animals, precisely in the same manner as similar trifling characters often pervade all the species of a genus, and even of a family. Flowers in varying their colours often become veined and spotted and the leaves become divided like true species: it is known that the varieties of the same plant never have red, blue and yellow flowers, though the hyacinth makes a very near approach to an exception; and different species of the same genus seldom, though sometimes they have flowers of these three colours. Dun-coloured horses having a dark stripe down their backs, and certain domestic asses having transverse bars on their legs, afford striking examples of a variation analogous in character to the distinctive marks of other species of the same genus.

EXTERNAL CHARACTERS OF HYBRIDS
AND MONGRELS

There is, however, as it appears to me, a more important method of comparison between species and races, namely the character of the offspring when species are crossed and when races are crossed: I believe, in no one respect, except in sterility, is there any difference. It would, I think, be a marvellous fact, if species have been formed by distinct acts of creation, that they should act upon each other in uniting, like races descended from a common stock.

In the first place, by repeated crossing one species can absorb and wholly obliterate the characters of another, or of several other species, in the same manner as one race will absorb by crossing another race. Marvellous, that one act of creation should absorb another or even several acts of creation! The offspring of species, that is hybrids, and the offspring of races, that is mongrels, resemble each other in being either intermediate in character (as is most frequent in hybrids) or in resembling sometimes closely one and sometimes the other parent: in both the offspring produced by the same act of conception sometimes differ in their degree of resemblance; both hybrids and mongrels sometimes retain a certain part or organ very like that of either parent, both, as we have seen, become in succeeding generations variable; and this tendency to vary can be transmitted by both; in both for many generations there is a strong tendency to reversion to their ancestral form. In the case of a hybrid laburnum and of a supposed mongrel vine different parts of the same plants took after each of their two parents. In the hybrids from some species, and in the mongrel of some races, the offspring differ according as which of the two species, or of the two races, is the father (as in the common mule and hinny) and which the mother. Some races will breed together, which differ so greatly in size, that the dam often perishes in labour; so it is with some species when crossed; when the dam of one species has borne offspring to the male of another species, her succeeding offspring are sometimes stained (as in Lord Morton's mare by the quagga, wonderful as the fact is) by this first cross; so agriculturists positively affirm is the case when a pig or sheep of one breed has produced offspring by the sire of another breed.

SUMMARY OF SECOND CHAPTER

Let us sum up this second chapter. If slight variations do occur in organic beings in a state of nature; if changes of condition from geological causes do produce in the course of ages effects analogous to those of domestication on any, however few, organisms; and how can we doubt it — from what is actually known, and from what may be presumed, since thousands of organisms taken by man for sundry uses, and placed in new conditions, have varied. If such variations tend to be hereditary; and how can we doubt it — when we see shades of expression, peculiar manners, monstrosities of the strangest kinds, diseases, and a multitude of other peculiarities, which characterize and form, being inherited, the endless races (there are 1,200 kinds of cabbages) of our domestic plants and animals. If we admit that every organism maintains its place by an almost periodically recurrent struggle; and how can we doubt it — when we know that all beings tend to increase in a geometrical ratio (as is instantly seen when the conditions become for a time more favourable); whereas on an average the amount of food must remain constant, if so, there will be a natural means of selection, tending to preserve those individuals with any slight deviations of structure more favourable to the then existing conditions, and tending to destroy any with deviations of an opposite nature. If the above propositions be correct, and there be no law of nature limiting the possible amount of variation, new races of beings will — perhaps only rarely, and only in some few districts — be formed.

LIMITS OF VARIATION

That a limit to variation does exist in nature is assumed by most authors, though I am unable to discover a single fact on which this belief is grounded. One of the commonest statements is that plants do not become acclimatized; and I have even observed that kinds not raised by seed, but propagated by cuttings, etc., are instanced. A good instance has, however, been advanced in the case of kidney beans, which it is believed are now as tender as when first introduced. Even if we overlook the frequent introduction of seed from warmer countries, let me observe that as long as the seeds are gathered promiscuously from the bed, without continual observation and *careful* selection of those plants which have stood the climate best during their whole growth, the experiment of acclimatization has hardly been begun. Are not all those plants and animals, of which we have the greatest number of races, the oldest domesticated? Considering the quite recent progress of systematic agriculture and horticulture, is it not opposed to every fact, that we have exhausted the capacity of variation in our cattle and in our corn — even if we have done so in some trivial points, as their fatness or kind of wool? Will any one say, that if horticulture continues to flourish during the next few centuries, that we shall not have numerous new kinds of the potato and Dahlia? But take two varieties of each of these plants, and adapt them to certain fixed conditions and prevent any cross for 5,000 years, and then again vary their conditions; try many climates and situations; and who will predict the number and degrees of difference which might arise from these stocks? I repeat that we know nothing of any limit to the possible amount of variation, and therefore to the number and differences of the races, which might be produced by the natural means of selection, so infinitely more efficient than the agency of man. Races thus produced would probably be very 'true'; and if from having been adapted to different conditions of existence, they possessed different constitutions, if suddenly removed to some new station, they would perhaps be sterile and their offspring would perhaps be infertile. Such races would be undistinguishable from species. But is there any evidence that the species, which surround us on all sides, have been thus produced? This is a question which an examination of the economy of nature we might expect would answer either in the affirmative or negative.

DARWIN'S VIEW OF THE
IMPORTANCE OF THE 1844 ESSAY

Darwin finished the Essay in July. He was, by then, so convinced of the importance of his theory that he took steps to ensure that his work would not be lost in the event of his sudden death. He expressed his wishes in a letter to his wife Emma, which is reproduced here. The letter is interesting not only for its portrayal of Darwin's opinion of the Essay, but also because it reveals the research method he used when writing the book—a kind of primitive information retrieval of "scraps" of either notes or pages torn from journals, together with key books, with his own annotations indexed.[1]

My Dear Emma.

   I have just finished my sketch of my species theory. If, as I believe that my theory is true & if it be accepted even by one competent judge, it will be a considerable step in science.

   I therefore write this, in case of my sudden death, as my most solemn & last request, which I am sure you will consider the same as if legally entered in my will, that you will devote 400£ to its publication & further will yourself, or through Hensleigh <Wedgwood>, take trouble in promoting it. I wish that my sketch be given to some competent person, with this sum to induce him to take trouble in its improvement & enlargement. I give to him all my Books on Natural History, which are either scored or have references at end to the pages, begging him carefully to look over & consider such passages, as actually bearing or by possibility bearing on this subject. I wish you to make a list of all such books, as some temptation to an Editor. I also request that you hand over to him all those scraps roughly divided in eight or ten brown paper Portfolios: The scraps with copied quotations from various works are those which may aid my Editor. I also request that you (or some amanuensis) will aid in deciphering any of the scraps which the Editor may think possibly of use. I leave to the Editor's judgement whether to interpolate these facts in the text, or as notes, or under appendices. As the looking over the references & scraps will be a long labour, & as the correcting & enlarging & altering my sketch will also take considerable time, I leave this sum of 400£ as some remuneration & any profits from the work. I consider that for this the Editor is bound to get the sketch published either at a Publishers or at his own risk. Many of the scraps in the Portfolios contain mere rude suggestions & early views now useless, & many of the facts will probably turn out as having no bearing on my theory.

   With respect to Editors. Mr. Lyell would be the best if he would undertake it: I believe he would find the work pleasant and he would learn some facts new to him. As the Editor must be a geologist, as well as Naturalist. The next best editor would be Professor <Edward> Forbes of London. The next best (& quite the best

---

1. On Darwin's research methods, see Barry G. Gale, *Evolution Without Evidence: Charles Darwin and the Origin of Species* (Albuquerque, University of New Mexico Press, 1982), 79–80.

in many respects) would be Professor *Henslow*?? Dr. Hooker would perhaps correct the Botanical Part probably he would do as Editor. Dr. Hooker would be very good. The next, Mr. [Hugh] Strickland. If no one of these would undertake it, I would request you to consult with Mr. Lyell, or some other capable man, for some Editor, a geologist & naturalist.

Should one other hundred pounds, make the difference of procuring a good Editor, I request earnestly that you will raise 500£.

My remaining collection in Natural History, may be given to anyone or any Museum, where it would be accepted.

My dear Wife/ Yours affect/C. R. Darwin[2]

2. *Correspondence*, 3: 43–44. Postscripts omitted.

Chapter 5

A Monograph of the Sub-Class Cirripedia (1851)

INTRODUCTION

While exploring in the Chonos Archipelago off the coast of Chile in January 1835, Darwin discovered an unusual barnacle—"an illformed little monster," he said—which he and Hooker later named *Arthrobalanus*, a strange creature with two penises but which otherwise appeared to be a degenerated form in comparison to the female. This was the beginning of Darwin's interest in barnacles, which led to a full-blown project in 1847, when Richard Owen convinced him of the scientific value of studying the group's comparative anatomy and John Edward Gray agreed to lend him the British Museum's entire barnacle collection, a bit of largesse that enabled Darwin to work in the quiet of his own laboratory at Down.[1]

Darwin's research soon came to focus on the unusual sexual relations of different subgroups of barnacles. The best-known barnacles were hermaphrodites, individuals that possessed both sexes. But early the following year, he discovered in two separate genera tiny, rudimentary males that in effect lived as parasites within the shell of the female. Soon afterward, he discovered another seemingly anomalous arrangement: a typical hermaphrodite barnacle which was also supplied with what Darwin called a "complemental" male, a tiny, separate individual which lived as a parasite on the female.

1. For a particularly clear description of Darwin's research on barnacles, see Janet Browne, *Charles Darwin. Voyaging* (New York, Knopf, 1995), 475–486. See also Adrian Desmond and James Moore, *Darwin: The Life of a Tormented Evolutionist* (New York, Time Warner, 1991), 339–343.

Soon, Darwin was impressed by the evolutionary significance of these discoveries, which seemed to suggest the evolution of sexuality in barnacles, beginning with a hermaphrodite mode, then progressing to partial hermaphroditism in the form of hermaphrodite individuals supplied with complemental males, and finally progressing to a species in which the sexes were represented by a female and a parasitic male.

The divergence of sexes in the barnacle suggested a broader evolutionary principle whereby, in different animal groups, male and female sexes evolved from originally hermaphroditic prototypes. A similar principle could be found in plants as well, as we will later observe (see chapter 11). The barnacle studies were published in four volumes between 1851 and 1854, comprising two volumes on living cirripedia[2] and two on fossil species.[3] These studies established Darwin's credentials as a meticulous anatomist, giving him a professional scientific credibility that would stand him in good stead some years later when he sought support for his controversial theory.

Here Darwin describes the striking polymorphism of male and female forms in closely related species of barnacles—*Ibla* and *Scalpellum*—whose sharply varying forms can be related to the different sexual functions mainly of the males.

COMPLEMENTAL MALES

Had the question been, whether the parasites which I have now described, were simply the males of the cirripedes to which they are attached, the present summary and discussion would perhaps have been superfluous; but it is so novel a fact, that there should exist in the animal kingdom hermaphrodites, aided in their sexual functions by independent and, as I have called them, complemental males, that a brief consideration of the evidence already advanced, and of some fresh points, will not be useless. These parasites are confined to the allied genera Ibla and Scalpellum; but they do not occur in Pollicipes — a genus still more closely allied to Scalpellum; and it deserves notice, that their presence is only occasional in those species of Scalpellum which come nearest to Pollicipes. In the general Ibla and Scalpellum, the facts present a singular parallelism; in both we have the simpler case of a female, with one or more males of an abnormal structure attached to her; and in both the far more extraordinary case of an hermaphrodite, with similarly attached complemental males. In the two species of Ibla, the comple-

2. *A Monograph of the Sub-Class Cirripedia*. Vol. 1, *The Lepadidae* (London, 1851). Vol. 2, *The Balanidae* (London, 1854).

3. *A Monograph of the Fossil Lepadidae* (London, 1851). *A Monograph of the Fossil Balanidae and Verrucidae* (London, 1854).

mental and ordinary males resemble each other, as closely as do the corresponding hemaphrodite and female forms; so it is with two sets of the species of Scalpellum. But the males of Ibla and the males of Scalpellum certainly present no special relations to each other, as might have been expected, had they been distinct parasites independent of the animals to which they are attached, and considering that they are all cirripedes having the same most unusual habits. On the contrary, it is certain that the animals which I consider to be the males and complemental males of the two species of Ibla, if classed by their own characters, would, from the reasons formerly assigned, form a new genus, nearer to Ibla than to the parasites of Scalpellum: so, again, the assumed males of the three latter species of Scalpellum would form two new genera, both of which would be more closely allied to Scalpellum, than to the parasites of Ibla. With respect to the parasites of the first three species of Scalpellum, they are in such an extraordinarily modified and embryonic condition, that they can hardly be compared with other cirripedes; but certainly they do not approach the parasites of Ibla, more closely than the parasites of Scalpellum; and in the one important character of the antennae, they are identical both with the parasitic and ordinary forms of Scalpellum. That two sets of parasites having closely similar habits, and belonging to the same subclass, should be more closely related in their whole organization to the animals to which they are respectively attached, than to each other, would, if the parasites were really distinct and independent creatures, be a most singular phenomenon; but on the view that they differ only sexually from the cirripedes on which they are parasitic, this relationship is obviously what might have been expected.

The two species of Ibla differ extremely little from each other, and so, as above remarked, do the two males. In Scalpellum the species differ more from each other, and so do the males. In this latter genus the species may be divided into two groups, the first containing *S. vulgare, S. ornatum* and *S. rutilum*, characterized by not having a sub-carina, by the rostrum being small, by the constant presence of four pair of latera, and by the peculiar shape of the carinal latera; the second group is characterized by having a sub-carina and a large rostrum, and may be subdivided into two little groups; viz., *S. rostratum* having four pairs of latera, and *S. Peronii* and *villosum* having only three pairs of latera: now the males, if classed by themselves, would inevitably be divided in exactly the same manner, namely, into two main groups — the one including the closely similar, sack-formed males of *S. vulgare, ornatum,* and *rutilum*, the other the pedunculated males of *S. rostratum, Peronii,* and *villosum*; but this latter group would have to be subdivided into two little sub-groups, the one containing the three-valved male of *S. rostratum*, and the other the six-valved males of *S. Peronii* and *S. villosum*. It should not, however, be overlooked, that the two main groups of parasites differ from each other, far more than do the two corresponding groups of species to which they are attached; and, on the other hand, that the parasitic males of *S. Peronii* and *S. villosum* resemble each other more closely, than do the two hermaphrodite forms; but it is very difficult to weigh the value of the differences in the different parts of species.

Besides these general, there are some closer relations between the parasites and the animals to which they are attached; thus the most conspicuous internal character by which *Ibla quadrivalvis* is distinguished from *I. Cumingii*, is the length of the caudal appendages and the greater size of the parts of the mouth; in the parasites, we have exactly corresponding differences. Out of the six species of Scalpellum in their ordinary state, *S. ornatum* is alone quite destitute of spines on the membrane connecting the valves; and had it not been for this circumstance, I should even have used the presence of spines as a generic character; on the other hand, *S. villosum*, in accordance with its specific name, has larger and more conspicuous spines than any other species. In the parasites we have an exactly parallel case; the parasite of *S. ornatum* being the only one without spines, and the spines on the parasite of *S. villosum* being much the largest! This latter species is highly singular in having no caudal appendages, and the parasite is destitute of these same organs, though present in the parasites of *S. rostratum* and *S. Peronii*. Again, *S. villosum* approaches, in all its characters, very closely to the genus Pollicipes, and the parasite in having prehensile antennae, with the disc but little pointed, and with spines at the further end, departs from Scalpellum and approaches Pollicipes! Will any one believe that these several parallel differences, between the cirripedial parasites and the cirripedes to which they are attached, are accidental, and without signification? yet, this must be admitted, if my view of their male sex and nature be rejected.

One more, and the most important, special relation between the parasites and the cirripedes to which they are attached, remains to be noticed, namely that of their prehensile larval antennae. I observed the antennae more or less perfectly in the males of all, and except in *S. villosum*, in all the species, though so utterly different in general appearance and structure, I found the peculiar, pointed, hoof-like discs, which are confined, I believe, to the genera Ibla and Scalpellum. In the hermaphrodite forms of Scalpellum, I was enabled to examine the antennae only in two species, *S. vulgare* and *S. Peronii* (belonging, fortunately, to the two most distinct sections of the genus), and after the most careful measurements of every part, I can affirm that, in *S. vulgare*, the antennae of the male and of the hermaphrodite are identical; but that they differ slightly in the proportional lengths of their segments, and in no other respect, from these same organs in *S. Peronii* — in which again the antennae of the male and of the hermaphrodite are identical. The importance of this agreement will be more fully appreciated, if the reader will consider the following table, in which the generic and specific differences of the antennae in the Lepadidae, as far as known to me, are given. These organs are of high functional importance; they serve the larva for crawling, and being furnished with long, sometimes plumose spines, they serve apparently as organs of touch; and lastly, they are indispensable as a means of permanent attachment, being adapted to the different objects, to which the larva adheres. Hence the antennae might, *à priori*, have been deemed of high importance for classification. They are, moreover, embryonic in their nature; and embryonic parts, as is well known, possess the highest classificatory value. From these considerations, and looking to the actual facts as exhibited in the following table, the improbability that the parasites of *S. vulgare* and *S. Peronii*, so utterly different in external structure and habits one from the other, and from the cirripedes to which they are attached, should yet have absolutely similar prehensile antennae with these cirripedes, appears to me, on the supposition of the parasites being really independent creatures, and not, as I fully believe, merely in a different state of sexual development, insurmountably great.

The parasites of *S. vulgare* take advantage of a pre-existing fold on the edge of the scutum, where the chitine border is thicker; and in this respect there is nothing different from what would naturally happen with an independent parasite; but in *S. ornatum* the case is very different, for here the two scuta are specially modified, *before the attachment of the parasites*, in a manner which it is impossible to believe can be of any service to the species itself, irrespectively of the lodgment thus afforded for the males. So again in *S. rutilum*, the shape of the scutum seems adapted for the reception of the male, in a manner which must be attributed to its own growth, and not to the pressure or attachment of a foreign body. Now there is a strong and manifest improbability in an animal being specially modified to favour the parasitism of another, though there are innumerable instances in which parasites take advantage of pre-existing structures in the animals to which they are attached. On the other hand, there is no great improbability in the female being modified for the attachment of the male, in a class in which all the individuals are attached to some object, than in the mutual organs of copulation being adapted to each other throughout the animal kingdom.

It should be observed that the evidence in this summary is of a cumulative nature. If we think it highly, or in some degree probable — from the ordinary form of *Ibla Cumingii* having been shown on good evidence to be exclusively female — from the absence of ova and ovaria in the assumed males of both species of Ibla, at the period when their vesiculae seminales were gorged with spermatozoa — from the close general resemblance between the parts of the mouth in the parasites and in the Iblas to which they are attached — from the differences between the two parasites being strictly analogous to the differences between the two species of Ibla — from the generic character of their prehensile antennae — and from other such points — if from these several considerations, we admit that these parasites really are the males of the two species to which they adhere, then in some degree the occurrence of parasitic males in the allied genus Scalpellum is rendered more probable. So the absolute similarity in the antennae of the males and hermaphrodites both in *S. vulgare* and *S. Peronii*; and such relations as that of the relative villosity of the several species in this same genus, all in return strengthen the case in Ibla. Again, the six-valved parasites of *S. Peronii* and *S. villosum* are so closely similar, that their nature, whatever it may be, must be the same; hence we may add up the evidence derived from the identity of the antennae in the parasite and hermaphrodite *S. Peronii*, with that from the antennae in the male *S. villosum*, approaching in character to Pollicipes, to which genus the hermaphrodite is so closely allied; and to this evidence, again, may be added the singular coincident absence of caudal appendages in the male and hermaphrodite *S. villosum*. If these two six-valved parasites be received as the complemental males of their respective species, no one, probably, will doubt regarding the nature of the parasite of *S. rostratum*, in which the direct evidence is the weakest; but even in this case, the particular point of attachment, and the state of development of the valves, form a link connecting in some degree, the parasites of the first three species with the last two species of Scalpellum, in accordance with the affinities of the hermaphrodites.

When first examining the parasites of *S. rostratum, S. Peronii*, and *S. villosum*, before the weight of the cumulative evidence had struck me, and noting their apparent state of immaturity, it occurred to me that possibly they were the young of their respective species, in their normal state of development, attached to old individuals, as may often be seen in Lepas; this, however, would be a surprising fact, considering that *S. rostratum* and *S. Peronii* are ordinarily attached, in a certain definite position, to horny corallines, and considering that the exact points of attachment in these three parasites (of which I have seen no other instance among common cirripedes), namely, between the scuta, would inevitably cause their early destruction, either directly or indirectly, by their living supports being destroyed. Nevertheless, I carefully examined a young specimen of *S. rostratum* only thrice as large as the parasite; and not having very young specimens of *S. Peronii* and *S. villosum*, I procured the young of close allied forms, namely, of *S. vulgare* (with a capitulum only $\frac{1}{100}$th of an inch in length), and of *Pollicipes polymerus* (with a capitulum of less size than that of one of the parasites), and there was not the least sign of anything abnormal in the development of the valves. In *S. vulgare*, at a period when the calcified scuta could have been only $\frac{1}{100}$th of an inch in length and therefore considerably less than the scuta in the parasites), the upper latera must have been as much as $\frac{4}{100}$ths of an inch in length, and the valves of the lower whorl certainly distinguishable.

To sum up the evidence on the sex of the parasites, I was not able to discover a vestige of ova or ovaria in the two male Iblas; and I can venture to affirm positively, that the parasites of *S. Peronii* and *S. vilossum* are not female. On the other hand, in the two male Iblas, I was enabled to demonstrate all the male organs, and I most distinctly saw spermatozoa. In the parasitic complemental male of *S. vulgare*, I also most plainly saw spermatozoa. In the parasites of *S. rostratum*, *S. Peronii*, and *S. villosum*, the external male organs were present. I may here just allude to the facts given in detail under Ibla, showing that it was hardly possible that I could be mistaken regarding the exclusively female sex of the ordinary form of *I. Cumingii*, seeing how immediately I perceived all the male organs in the hermaphrodite *I. quadrivalvis*; and as the parasite contained spermatozoa and no ova, the only possible way to escape from the conclusion that it was the male and *I. Cumingii* the female of the same species, was to invent two hypothetical creatures, of opposite sexes to the Ibla and its parasite, and which, though cirripedes, would have to be locomotive! I insisted upon this alternative, because if the parasite of *I. Cumingii* be the male of that species, then unquestionably we have in *I. quadrivalvis* a male, complemental to an hermaphrodite — a conclusion, as we have seen, hardly to be avoided in the genus Scalpellum, even if we trust exclusively to the facts therein exhibited.

With respect to the positions of the parasitic males, in relation to the impregnation of the ova in the females and hermaphrodites, it may be observed that in the two male Iblas, the elongated movable body seems perfectly adapted for this end; in the males of the first three species of Scalpellum, the spermatoza, owing to the manner in which the thorax is bent when protruded, would be easily discharged into the sack of the female or hermaphrodite; this would likewise probably happen with the complemental male of *S. rostratum*, considering its position within the orifice of the capitulum, between the mouth and the adductor scutorum muscle. The males of *S. Peronii* and *S. villosum* being fixed a little way beneath the orifice of the sack, below the adductor muscle, are less favourably situated, but the spermatozoa would probably be drawn into the sack by the ordinary action of the cirri of the hermaphrodite, and therefore would at least have as good a chance of fertilizing some of the ova, as the pollen of many dioecious plants, trusted to the wind, has of reaching the stigmas of the female plants. Regarding the final cause, both of the simpler case of the separation of the sexes, notwithstanding that the two individuals, after the metamorphosis of the male, become indissolubly united together, and of the much more singular fact of the existence of complemental males, I can throw no light; I will only repeat the observation made more than once, that in some of the hermaphrodites, the vesiculae seminales were small, and that in others the prosciformed penis was unusually short and thin.

Viewing the parasitic males, in relation to the structure and appearance of the species to which they belong, they present a singular series. In *S. Peronii* and *S. villosum*, the internal organs have the appearance of immaturity; the shape of the capitulum is specially modified for its reception between the scuta of the hermaphrodite, and several of the valves have not been developed. This atrophy of the valves, is carried much further in *S. rostratum*. In Ibla, many of the parts are embryonic in character, but others mature and perfect; some parts, as the capitulum, thorax, and cirri, are in a quite extraordinary state of atrophy; in fact, the parasitic males of Ibla consist almost exclusively of a mouth, mounted on the summit of the three anterior segments of the twenty-one normal segments of the archetype crustacean. In the males of the first three species of Scalpellum, some of the characters are embryonic — as the absence of the mouth, the presence of the abdominal lobe, and the position of the few existing internal organs; other characters, such as the general external form, the four bead-like valves, the narrow orifice, the peculiar thorax and limbs, are special developments. These three latter parasites, certainly, are wonderfully unlike the hermaphrodites or females to which they belong; if classed as independent animals, they would assuredly be placed not in another family, but in another order. When mature they may be said essentially to be mere bags of spermatozoa.

In looking for analogies to the facts here described, I have already referred to the minute male Lerneidae which cling to their females — to the worm-like males of certain Cephalopoda, parasitic on the females — and to certain Entozoons, in which the sexes cohere, or even are organically blended by one extremity of their bodies. The females in certain insects depart in structure, nearly or quite as widely from the order to which they belong, as do these male parasitic cirripedes; some of these females, like the males of the first three species of Scalpellum, do not feed, and some, I believe, have their mouths in a rudimentary condition; but in this latter respect, we have, among the Rotifera, a closely analogous case in the male of the Asplanchna of Gosse, which was discovered by Mr. Brightwell to be entirely destitute of mouth and stomach, exactly as I find to be the case with the parasitic male of *S. vulgare*, and doubtless with its two close allies. For any analogy to the existence of males, complemental to hermaphrodites, we must look to the vegetable kingdom.

Finally, the simple fact of the diversity in the sexual relations, displayed within the limits of the genera Ibla and Scalpellum, appears to me eminently curious; we have (1st) a female, with a male (or rarely two) permanently attached to her, protected by her, and nourished by any minute animals which may enter her sack; (2nd) a female, with successive pairs of short-lived males, destitute of mouth and stomach, inhabiting two pouches formed on the undersides of her valves; (3rd) an hermaphrodite, with from one or two, up to five or six similar short-lived males without mouth or stomach, attached to one particular spot on each side of the orifice of the capitulum; and (4th) hermaphrodites, with occasionally one, two, or three males, capable of seizing and devouring their prey in the ordinary cirripedial method, attached to two different parts of the capitulum, in both cases being protected by the closing of the scuta. As I am summing up the singularity of the phenomena here presented, I will allude to the marvellous assemblage of beings seen by me within the sack of an *Ibla quadrivalvis* — namely, an old and young male, both minute, worm-like, destitute of a capitulum, with a great mouth, and rudimentary thorax and limbs, attached to each other and to the hermaphrodite, which latter is utterly different in appearance and structure; secondly, the four or five, free, boat-shaped larvae, with their curious prehensile antennae, two great compound eyes, no mouth, and six natatory legs; and lastly, several hundreds of the larvae in their first stage of development, globular, with horn-shaped projections on their carapaces, minute single eyes, filiformed antennae, prosciformed mouths, and only three pair of natatory legs; what diverse beings, with scarcely anything in common, and yet all belonging to the same species!

DARWIN REFLECTS ON HIS CIRRIPEDE RESEARCH

I have been getting on quite well with my beloved cirripedia, & got more skilful in dissection: I have worked out the nervous system pretty well in several genera, & made out their ears and nostrils, which were quite unknown. I have lately got a bisexual cirripede, the male being microscopically small & parasitic within the sack of the female; I tell you this to boast of my species theory, for the nearest & closely allied genus to it is, as usual, hermaphrodite, but I had observed some minute parasites adhering to it, & these parasites, I now can show, are supplemental males, the male organs in the hermaphrodite being unusually small, though perfect & containing zoosperms: so we have almost a polygamous animal, simple females alone being wanting. I never should have made this out had not my species theory convinced me, that an hermaphrodite species must pass into a bisexual species by insensibly small stages, & here we have it, for the male organs in the hermaphrodite are beginning to fail, & independent males ready formed. But I can hardly explain what I mean, & you will perhaps wish my Barnacles & Species theory al Diabolo together. But I don't care what you say, my species theory is all gospel.[1]

1. Darwin to Hooker, 10 May 1848, *Correspondence,* 4: 140.

Chapter 6

NATURAL SELECTION: THE BIG SPECIES BOOK (1856–1858)

INTRODUCTION

Darwin vividly remembered the discovery of the principle of divergence.

> But at that time [1844] I overlooked one problem of great importance; and it is astonishing to me, except on the principle of Columbus and his egg, how I could have overlooked it and its solution. . . . I can remember the very spot in the road, whilst in my carriage, when to my joy the solution occurred to me . . .[1]

The long-overlooked problem was "the tendency of organic beings descended from the same stock to diverge in character as they become modified."[2] The joyful solution was "that the modified offspring of all dominant and increasing forms tend to become adapted to many and highly diversified places in the economy of nature."[3] The essence of this principle is that natural selection not only produces adaptation of organisms to their environment, but also selects for divergence. In other words, divergence can be adaptive under certain circumstances; hence there is selection in nature for ecological complexity, and that ultimately leads to taxonomic diversity—or so Darwin thought. The principle of divergence is thus a special and enhanced form of natural selection.

It is an intriguing question whether such an apparently vivid memory of discovery as Darwin's carriage ride is to be taken literally. Historians tend to

---

1. *The Autobiography of Charles Darwin*, Nora Barlow, ed. (New York: Norton, 1959), pp. 120–121.

2. *Ibid.*, p. 120.

3. *Ibid.*, p. 121.

be skeptical, and notable parallels—in Pasteur's career, for example—have been debunked.[4] Yet from the perspective of Darwin's intellectual development, it seems clear that Darwin did have a problem with natural selection in its original (1838) form. That formulation only addressed the origin of adaptations. In a sense, Darwin's critical advance in 1838 had been to reduce evolution to its smallest spatial and temporal scales: local adaptation, one generation at a time. Natural selection made transmutation an ecological problem—or, in eighteenth- and nineteenth-century parlance, a problem in the detailed working of the economy of nature. Yet for all that this was an advance, it left a major gap in Darwin's theory, for it left the intermediate and higher scales of evolution without explanation. As he put it:

> That they have diverged greatly is obvious from the manner in which species of all kinds can be classed under genera, genera under families, families under suborders, and so forth . . . (*Autobiography*, 120)

What was the explanation for this divergence—which included everything from the tendency to form varieties to the irregularly branched natural system of classification—whose very tree of life was emblematic of evolutionary descent with modification? If this issue had remained unresolved into the 1850s, a sudden "discovery" of a new principle would have been very welcome.

Yet scholars have never been able to date that famous carriage ride or to identify a moment of inspiration tied to a specific note or scrap of paper in Darwin's hand. Rather, the construction of the principle of divergence, through a gestation of some six years, from, say, 1852 to 1858, forms part of the writing of a unique, massive, and long-underestimated document: *Charles Darwin's Natural Selection, Being the Second Part of his Big Species Book Written from 1856 to 1858*. This work, published only in 1975 in R. C. Stauffer's edition, is aptly referred to as the long version of the *Origin*. The document is striking not only because it is 25 percent longer than the *Origin*, covering nearly the same material,[5] but because, unlike the *Origin*, it contains proper

4. Gerald L. Geison, *The Private Science of Louis Pasteur* (Princeton: Princeton University Press, 1995).

5. *Natural Selection* did not include the chapters on paleontology, morphology, and embryology that appear in the *Origin*, but it did include the long chapter "On the possibility of all organic beings occasionally crossing . . .," a topic ultimately treated in Darwin's later botanical writings (see *Orchids, Cross and Self Fertilization*, and *Forms of Flowers*, chapter 11, below) rather than in the *Origin*. Finally, two chapters on variation under domestication, carefully outlined for *Natural Selection*, were not written for, but formed the core of, *Variation of Plants and Animals* (see chapter 9).

scholarly apparatus, including rich footnotes and the citations for an impressive bibliography.

The story of the "long version" begins after Darwin completed and stored away the 1844 Essay. Throughout the late 1840s and 1850s, Darwin continued to make notes and collect material relevant to the species question. He no longer used notebooks. Instead, he jotted his references and observations on whatever piece of paper came to hand. At some unknown point, he began the system of keeping these notes in topical portfolios. Perhaps this was done after the completion of the *Cirripedes*, when:

> From September 1854 onwards I devoted all my time to arranging my huge pile of notes, to observing, and experimenting, in relation to the transmutation of species.[6]

The portfolios were eventually kept in pigeonholes in Darwin's study, just over his right shoulder as he sat at his writing chair. Clearly, the "huge pile" of notes had lain fallow for years. But once Darwin got organized, his system was impressive and up to the grand dimensions of his undertaking. Many notes were marked with numbers indicating the appropriate portfolio. The portfolios included, besides the accumulated reading notes, pieces cut out of his burgeoning scientific correspondence—a good letter from Hugh Falconer in India, for example, was often cut up into several pieces and marked for three or four portfolios. Also dropped into the portfolios were pages excised—frequently not for their more expansive ideas, but for their apt bibliographic reference—from his original transmutation notebooks. The portfolio system would form the basis first for the "long version" chapters, then for the *Origin*, and then for all the books Darwin wrote after the *Origin*. But the portfolios were just the center of a wider system that included annotations on books, journals, and offprints—most of which were methodically abstracted and indexed for easy retrieval.

No matter when in the 1850s Darwin conceived of the principle of divergence, by the time he wrote it up in chapter 7 of *Natural Selection*, two shifts had occurred from his thinking in 1844. Most important, Darwin had come to believe that there is considerable variation in natural species. This shift to the direct opposite of the position he had taken in 1844, presented most strikingly in chapter 4 of *Natural Selection*, is at least partly attributable to Darwin's hard-earned familiarity with the scope of hereditary variation gained by studying cirripede morphology for seven years:

> I may state . . . that in Lepas anatifera & Balanus tintinnabulum. I at first wrote out full descriptions of several supposed species; then after getting more speci-

6. *Ibid.*, p. 118.

mens from various parts of the world, I thought that I ought to run them all into one, & tore up my separate descriptions: after an interval of some months I looked over my specimens & could not persuade myself to call such different forms one species & rewrote separate descriptions; but lastly having got still more specimens, I had again to tear up those & finally concluded that it is impossible to separate them![7]

Darwin now believed that there was ample variation in nature. The significance of this shift is that selection for divergence implies that natural selection operates at an intense rate. But intense selection can be sustained only if there is a large pool of variability from which to select. Finally, Darwin, who in the 1830s had laid great stress on the power of mountain chains and oceans to produce new species, came to believe that new species could also be formed without geographic isolation. He came to believe that even minute differences in habitat within a geographically undivided area, when exploited by selection for divergence, could provide sufficient breeding barriers to permit speciation.

ON THE PRINCIPLE OF DIVERGENCE

Principle of Divergence. — This principle, which for want of a better name, I have called that of Divergence, has, I believe played a most important part in Natural Selection. To seek light, as in all other cases, by looking to our domestic productions, we may see in those which have varied most from long domestication or cultivation, something closely analogous to our principle. Each new peculiarity either strikes man's eye as curious or may be useful to him; & he goes on slowly & often unconsciously selecting the most extreme forms. He has made the race-horse as fleet & slim as possible & goes on trying to make it fleeter; the cart-horse he makes as powerful as he can: he selects his Dorking-fowls for weight & disregards plumage; the Bantam he tries to get as small as possible, with elegant plumage & erect carriage: a pigeon has been born with slightly smaller beak, another with slightly longer beak & wattle, another with a crop a little more inflated than usual, another with a somewhat larger & expanded tail &c; his eye is struck & he goes on selecting each of these peculiarities, & he makes his several breeds of improved tumblers, carriers, pouters, fantails &c, all as different or divergent as possible from their original parent-stock the rock-pigeon; the intermediate, & in his eyes inferior birds, having been neglected in each generation & now become extinct. It is the same with his dress, each new fashion ever fluctuating is carried to an extreme & displaces the last; but living productions will not so readily bend to his inordinate caprice. [Moreover, far

7. *Natural Selection*, R. C. Stauffer, ed. (Cambridge, 1993), p. 101.

more *fancy*-pigeons will be kept, (I do not mean those kept as food) after they have become broken up into very distinct breeds, than when fewer & more similar birds existed; for each fancier likes to keep several kinds, or one fancier keeps one kind & another becomes famous for another breed.]

Now in nature, I cannot doubt, that an analogous principle, not liable to caprice, is steadily at work, through a widely different agency; & that varieties of the same species, & species of the same genus, family or order are all, more or less, subjected to this influence. For in any country, a far greater number of individuals descended from the same parents can be supported, when greatly modified in different ways, in habits constitution & structure, so as to fill as many places, as possible, in the polity of nature, than when not at all or only slightly modified.

We may go further than this, &, independently of the case of forms supposed to have descended from common parents, assert that a greater absolute amount of life can be supported in any country or on the globe; when life is developed under many & widely different forms, than when under a few & allied forms; — the fairest measure of the amount of life, being probably the amount of chemical composition & decomposition within a given period. Imagine the case of an island, peopled with only three or four plants of the same order all well adapted to their conditions of life, & by three or four insects of the same order; the surface of the island would no doubt be pretty well clothed with plants & there would be many individuals of these species & of the few well adapted insects; but assuredly there would be seasons of the year, peculiar & intermediate stations & depths of the soil, decaying organic matter &c, which would not be well searched for food, & the amount of life would be consequently less, than if our island had been stocked with hundreds of forms, belonging to the most diversified orders.

Practice shows the same result; farmers all over the world find that they can raise within the period of their leases most vegetable matter by a rotation of crops; & they choose the most different plants for their rotation: the nurseryman often practices a sort of simultaneous rotation in his alternate rows of different vegetables. I presume that it will not be disputed that on a large farm, a greater weight of flesh, bones, and blood could be raised within a given time by keeping cattle, sheep, goats, horses, asses, pigs, rabbits & poultry, than if only cattle had been kept. In regard to plants this has been experimentally proved by Sinclair who found that land sown with only two species of grass, or one kind of grass with clover, bore on an average 470 plants to the square foot; but that when sown, with from 8 to 20 different species, it bore at the rate of about 1000 plants, "& the weight of produce in herbage & in hay was increased in proportion." It is important to observe that the same rule holds for different & not very distinct varieties of the same species when sown together; for M. L. Rousseau, a distinguished practical farmer, on sowing fifteen varieties of wheat separately, & the same kinds mixed together found on actual measurement that the latter "yielded a much heavier crop than that obtained on far better land on which the unmixed wheats were grown for the purpose of the comparative trial."

We see on a great scale, the same general law in the natural distribution of organic beings; if we look to an extremely small area supposing the conditions to be absolutely uniform & not very peculiar. Where the conditions are peculiar & the station small as compared with the whole area of the country, as Alpine summits; Heaths salt-marshes, or even common marshes, lakes & rivers, &c. — a great number of individual plants are often supported, belonging to very few species: so it is with Fresh-water shells; so it is with the marine inhabitants of the arctic seas. But even in these cases, though the individuals appear to be very numerous compared with the species, yet even in these cases, the coinhabitants belong to very different types; for instance Dr. Hooker has marked for us all the plants in Britain, which he thinks may be called truly aquatic: they are, [   ] in number, & they belong to [   ] genera and to [   ] orders. — With respect to the number of individuals to the species, we shall have to return to this subject in our chapter on geographical distribution, & I will here only say that I believe it mainly, but not wholly, depends, on the manufacturing, if I may so express myself, being small in size (& sometimes in duration); that is that the number of individuals is small in comparison with the numbers of individuals of the commoner species which inhabit ordinary stations: for we have seen in our 4th Ch. that it is species which most abound in individuals which oftenest present varieties, or incipient species. Supposing the conditions to be absolutely uniform & not very peculiar or unfavourable for life, we seldom find it occupied by any two or three closely allied & best adapted forms, but by a considerable number of extremely diversified forms. To give an example, I allowed the plants on a plot of my lawn three feet by four square which was quite uniform & had been treated for years uniformly, to run up to flower; I found the species 20 in number, & as these belonged [to] 18 genera & these to 8 orders & they were clearly much diversified. The most remarkable exception to this rule, under conditions not apparently very peculiar, is one given by Mr. C. A. Johns who says that he covered with his hat, (I presume broad-brimmed) near to Lands End six species of Trifolium, a Lotus & Anthyllis; & had the brim been a little wider it would have covered another Lotus & Genista; which would have made ten species of Leguminosae, belonging to only four genera! The wretched soil of Heaths, though covered thickly with one or two species of Erica, supports very little life, as judged by their extremely slow growth, & yet, selecting the very worst spots, I have very rarely been able to find a space two yards square, without one or two other plants, belonging to quite different orders, not to mention a good crop of Cryptogams.

To show the degree of diversity in our British plants on a small plot, I may mention, that I selected a field, in Kent, of $13\frac{1}{2}$ acres, which had been thrown out of cultivation for 15 years, & had been thinly planted with small trees most of which had failed: the field all consisted of heavy very bad clay, but one side sloped & was drier: there was no water or marsh: 142 phanerogamic plants were here collected by a friend during the course of a year; these belonged to 108 genera, & to 32 orders out of the 86 orders into which the plants of Britain have been classed. Another friend collected for me all the plants on about 40 uncultivated, very poor, acres of Ashdown Common in Sussex; these were 106 in number, & belonged to 82 genera & 34 orders; the greater proportional number of orders in this case being chiefly owing to the presence of water & marsh plants on the Common: the vegetation was, however, considerably different in other respects, no less than nine of the 34 orders, not being found on the field of thirteen acres in Kent. — To give another example of a small area having singularly uniform conditions of life; namely one of the low & quite flat, coral-islets having a wretched soil, composed exclusively of coral-debris, but with a fine climate; for instance Keeling Atoll, on which I collected nearly every phanerogamic plant, & these consisted of 20 species belonging to 19 genera & to no less than 16 different orders!

The extreme poverty of the floras of all such islets may be partly due to their isolation & the seeds arriving from lands having different floras, but chiefly to the poverty & peculiarity of the soil; for coral-islets, when lying close to large volcanic groups, have an almost equally poor & closely similar flora: the extreme diversity of the plants, the twenty in the case of Keeling islands, belonging to sixteen orders, can, I think, only be accounted for by the fact that of all the plants of which the seeds have been borne across the sea in the later periods of the natural colonisation of the island, those alone, which differed greatly from the earlier occupants, were able to come into competition with them & so lay hold of the ground & survive.

As with plants so with insects. I may premise that entomologists divide the Coleoptera into 13 grand sections, & then into families, sub-families &c. Mr. Wollaston carefully collected during several visits all the Beetles on the Dezerta Grande, a desert volcanic islet about four miles long, & in widest part only three-quarters broad, lying close to Madeira; & he found 57 species, belonging to 47 genera; & these to all 13 grand sections, except two, which being aquatic forms, could not exist on this waterless islet. Again on the Salvages, an extremely small volcanic is^{d}. between Madeira, & the Canaries, six beetles were collected, & these belonged to six genera, to six Families, & to three of the grand Sections! As a general rule, I think we may conclude, that the smaller the area, even though the conditions be remarkably uniform, the more widely diversified will its inhabitants be: for to this very diversity, the power of supporting the greatest possible number of living beings, all of which are struggling to live, will be due.

There is another way of looking at this subject; namely to consider the productions naturalised through man's agency in several countries; & see what relation they bear to each other & to the aboriginal productions of the country, i.e. Are they closely allied to, that is do they generally belong to the same genera with, the aboriginal inhabitants of the country? Do many species of the same genus become naturalised? If we looked only to the inorganic conditions of a country, we might have expected that species, belonging to genera already inhabiting it, & supposed on the common view to have [been] adapted by creation for such country, would have formed the main body of the colonists: or the many species of certain favoured genera would have been the successful intruders. On the other hand, the principle of diversity being favourable to the support of the greatest number of living beings would lead to the expectation, that land already well stocked by the hand of nature would support such new forms alone, as differed much from each other & from the aborigines. Alph. De Candolle has fully discussed the subject of naturalisaton: He shows that 64 plants have become naturalised in Europe (excluding species from neighbouring regions) during the last three centuries and a half; & these 64 species belong to 46 genera & 24 orders; of the genera, 21/46 are new to Europe. Again in N. America, 184 species have become naturalised & these belong to 120 genera & to 38 orders; of the genera, 56/120 are new to N. America. A list of the naturalised plants in Australia & on many islands would give similar, but much more striking results. The number of new genera naturalised in Europe & N. America, reciprocally from each other, is the more remarkable when we consider how much allied the two floras are; & that a very large proportion of the naturalised plants inhabit land, cultivated nearly in the same manner, which would favour the introduction of allied forms & many forms of the same groups. Hence, I think, we may conclude that naturalised productions are generally of a diversified nature; & as Alph. De. Candolle has remarked native floras gain by naturalisation, proportionally to their own numbers, far more in genera than in species.

If we turn to animals, we find, though our data are very scanty, the same general fact: no where in the world have more mammals become well naturalised than in S. America (cattle, horses, pigs, dogs, cats, rats & mice); & yet how extremely unlike is the native mammalian Fauna of S. America to that of the Old World!

The whole subject of naturalisation seems to me extremely interesting under this point of view, & would deserve to be treated at much greater length. It confirms the view that in natural colonisation, for instance in that of a coral-islet, diverse forms very different from the few previous occupants, would have the best chance of succeeding. It shows us, & by no other means can we form a conjecture on this head, what are the gaps or still open places in the polity of nature in any country: we see that these gaps are wide apart, & that they can be best filled up by organic beings, of which a large proportion are very unlike the aboriginal inhabitants of the country. Consequently we might perhaps from this alone infer, that natural selection by the preservation of the most diversified varieties & species, would in the long run tend, if immigration were prevented, to make the inhabitants, more & more diversified; though such modified forms would for immense periods plainly retain from heritage the stamp of their common parentage.

The view that the greatest number of organic beings (or more strictly the greatest amount of life) can be supported on any area, by the greatest amount of their diversification is, perhaps, most plainly seen by taking an imaginary case. This doctrine is in fact that of "the division of labour", so admirably propounded by Milne Edwards, who argues that a stomach will digest better, if it does not, as in many of the lowest animals, serve at the same time as a respiratory organ; that a stomach will get more nutriment out of vegetable or animal matter, if adapted to digest either separately instead of both. It is obvious that more descendants from a carnivorous animal could be supported in any country: if some were adapted, by long continued modification through natural selection, to hunt small prey, & others large prey living either on plains or in forests, in burrows, or on trees or in the water. So with the descendants of a vegetable feeder more could be supported, if some were adapted to feed on tender grass & others on leaves of trees or on aquatic plants & others on bark, roots, hard seeds or fruit.—

Perhaps I have already argued this point superfluously; but I consider it as of the utmost importance fully to recognise that the amount of life in any country, & still more that the number of modified descendants from a common parent, will in chief part depend on the amount of diversification which they have undergone, so as best to fill as many & as widely different places as possible in the great scheme of nature. Now let it be borne in mind that all the individuals of the same variety, and all the individuals of all the species of the same genus, family &c, are perpetually struggling to become more numerous by their high geometrical powers of increase. Under ordinary circumstances each species will in the briefest period have arrived at its fluctuating numerical maximum. Nor can it pass this point, without some other inhabitants of the same country suffer diminutions; or without all the descendants of one species becoming similarly modified in some respect so that they better fill the place of their parent-species; or without (& this would be the most effectual) several varieties & then several species are thus formed by modification, so as to occupy various new places, the more different the better, in the natural economy of one country. Although all the inhabitants of the country will be tending to increase in numbers by the preservation through natural selection of diverse modifications; but few will succeed; for variation must arise in the right direction & there must be an unfilled or less well-filled place in the polity of nature: the process, moreover, in all cases, as we shall presently see, must be slow in an extreme degree.

Let us take an imaginary case of the Ornithorhynchus; & suppose this strange animal to have an advantage over some of the other inhabitants of Australia, so as to increase in numbers & to vary: it could, we may feel pretty sure, increase to any *very great* extent, only by its descendants becoming modified, so that some could live on dry land, some could feed exclusively on vegetable matter in various stations, & some could prey on various animals, insects, fish or quadrupeds. In fact its descendants would have to become diversified, somewhat like the other Australian marsupials, which, as Mr. Waterhouse has remarked, typify in their several sub-families, our true carnivores, insectivores, ruminants & rodents. Moreover it can, I think, hardly be doubted, that these very marsupials would, profit by a still further division of physiological labour; that is by their structure becoming as perfectly carnivorous, ruminant & rodent as are our old-world forms; for it may well be doubted (not here considering the probable intellectual infirmity of the marsupialia in comparison with the other or placentate mammals) whether many marsupial vegetable feeders could long exist in free competition with true ruminants, & perhaps still less the carnivorous marsupials with true feline animals. And who can pretend to say that the mammals of the old world are diversified & have their organs adapted to different physiological labours to the extreme, which would be best for them under the conditions to which they are exposed? Had we known the existing mammals of S. America alone, we should no doubt have thought them perfect & diversified in structure & habit to the exact right degree; but the vast herds of feral cattle, horses, pigs & dogs, at least show that other animals, & some of them as the horse & solid-horned ruminants, very different from the endemic S. American mammals, could beat & take the place of the native occupants.

In Chapter IV we have seen on evidence, which seems to me in a fair degree satisfactory, that on an average the species in the larger genera in any country oftenest present varieties in some degree permanent, and likewise a greater average number of such varieties, than do the species of the smaller genera. It is not that all the species of the larger genera vary, but only some, & chiefly those which are wide-rangers, much diffused & numerous in in-

Diagram I

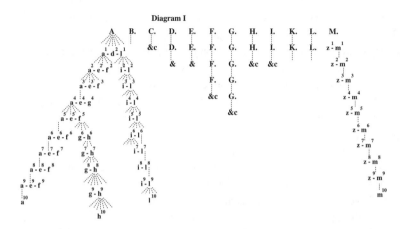

dividuals. In the same chapter we also saw that the species in the larger genera are thought by highly competent judges to be more closely related together, being clustered in little sub-groups round other species, than are the species in the smaller genera; & this closer affinity & grouping of the species in the larger genera, & the fact that there is no unfailing test by which to distinguish species & varieties, all to a certain extent confirm the view that varieties, when in some degree permanent, do not essentially differ from species, more especially from such species, as are closely allied together. Hence I look at varieties as incipient species.

I have lately remarked that the formation of new varieties & species through natural selection almost necessarily implies (as with our domestic productions) much extinction of the less altered, & therefore less favoured, de-

Diagram II

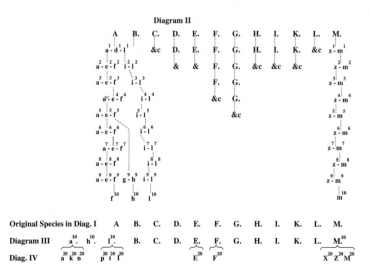

scendants from the same original parent-stock, whose places they occupy in the struggle for life. Hence, though the larger genera may be now varying most, & must, according to our theory, have varied largely, so as to have become modified into many specific forms, yet such large genera must have suffered a large amount of extinction, & very many intermediate & less modified forms have been wholly swept away. Nevertheless, I think we may infer that in any given country, on the whole, there will have been rather less extinction, proportionally to the whole amount of extinction within any given period, amongst the larger than amongst the smaller genera. For the species which vary most & thus give rise to new species, are chiefly the very common & much diffused species, & therefore the most favoured forms, which would naturally be the least liable to extinction; & such common & much diffused species tend to belong to the larger genera. Indeed it seems to me that the simple fact of a number of allied species, beyond the average number of allied species, inhabiting any country; shows that there is something in common in such groups of species, or genera, which is favourable to them, & consequently that they would suffer proportionally less from extinction than the smaller genera. Therefore, from the species of larger genera tending to vary most & so to give rise to more species, & from their being somewhat less liable to extinction, I believe that the genera now large in any area, are now generally tending to become still larger. But what will be the end of this? for we do not find in nature genera of indefinite size, with innumerable species. Here in one way comes in the importance of our so-called principle of divergence: as in the long run, more descendants from a common parent will survive, the more widely they become diversified in habits, constitution & structure so as to fill as many places as possible in the polity of nature, the extreme varieties & the extreme species will have a better chance of surviving or escaping extinction, than the intermediate & less modified varieties or species. But if in a large genus we destroy all the intermediate species, the remaining forms will constitute sub-genera or distinct genera, according to the almost arbitrary value put on these terms, — according to the number of intermediate forms which have been destroyed, — and according to the degree of difference between the extreme species of the original genus. Nevertheless the modified descendants from the common parent-stock, though no longer forming what is called the same genus, may still go on becoming more & more numerous, & more & more diversified.

The complex action of these several principles, namely, natural selection, divergence & extinction, may be best, yet very imperfectly, illustrated by the following Diagram, printed on a folded sheet for convenience of reference.

Darwin uses a similar branching diagram in the *Origin of Species*, in the chapter on Natural Selection (p. 199, below), together with a shorter description of the process than the one provided here.

This diagram will show the manner, in which I believe species descend from each other & therefore shall be explained in detail: it will, also, clearly show several points of doubt & difficulty; Let A to M represent the species of a genus, numerically large compared with the other genera of the same class in the same country, & arranged as naturally as can be done, so that A & M are the two most distinct forms in all respects. The unequal distances of the letters may represent the ordinary way in which the species, even when as in this imaginary case all are closely related together, yet stand unequally related in little sub-groups. This genus may have one, two or even more varying species. Any of the species may vary; but it will generally be those species which are most numerous in individuals & most diffused; & this shows that such species have already some advantages over the other inhabitants of the country. From our principle of divergence, the extreme varieties of any of the species, & more especially of those species which are now extreme in some characters, will have the best chance, after a vast lapse of time, of surviving; for they will tend to occupy new places in the economy of our imaginary country. I do not mean that any of these points are of invariable occurrence, but that in the long run such cases will prevail. The extreme species A and M will differ in very many respects; but for convenience sake we may look to any one character, & suppose A the most moisture-loving & M the least moisture-loving species.

We will first take the simplest case. Let M inhabit a continuous area, not separated by barriers, & let it be a very common & widely diffused & varying plant. From the fact of M. being very common & widely diffused, it clearly has some advantages in comparison with most of the other inhabitants of the same country; but, we will suppose, that it might become still more common, if retaining the advantages which it already has, it could endure still more drought. It is a varying species; & let $z^1 - m^1$ represent numerous, extremely slight variations of many kinds, produced at intervals, of which m^1 alone is a more drought-enduring variety. As m^1 tends to inherit all the advantages of its parent M, with the additional advantage of enduring somewhat more drought, it will have an advantage over it, & will probably first be a thriving local variety, which will spread & become extremely common & ultimately, supplant its own parent. We may now repeat the process, & let the variety m^1 vary in a similar manner; perhaps many thousands of generations may pass before m^1 will produce another variety m^2, still more drought-enduring & yet inheriting the common advantages of m^1 & M; but if this should ever occur, the same results, as before, will follow; & ultimately, by repeating the process, there may be produced m^{10}, which may either be, according to the amount of difference thus acquired, a very strongly marked variety, or a sub-species, or good species, enduring far more drought than M & probably with correlated differences of structure. In each stage of descent, there will be a tendency in the new forms to supplant its parent, though probably, as we shall see, very slowly, & so ultimately cause, its extinction. But if M had originally inhabited a country separated by barriers into distinct districts, in one or more of which the varieties M^{1-10} had never originated or had never been able to enter, M and m^1 & ultimately m^{10} might be living contemporaneously, but separated: or, again, if m^{1-10} had been produced, capable of enduring more drought, but not at the same time enduring an equal amount of moisture with the parent M, both parent & modified offspring might coexist: the parent (with perhaps a more restricted range) in the dryer stations, & m^{1-10} in the very dryest stations.

It should always, be borne in mind that there is a wide distinction between mere variations & the formation of permanent varieties. Variation is due to the action of external or internal causes on the generative systems, causing the child to be in some respects unlike its parent; & the differences thus produced may be advantageous or disadvantageous to the child. The formation of a permanent variety, implies not only that the modifications are inherited, but that they are not disadvantageous, generally that they are in some degree advantageous to the variety, otherwise it could not compete with its parent when inhabiting the same area. The formation of a permanent variety must <can> be effected by natural selection; or it may be the result, generally in unimportant respects, of the direct action of peculiar external conditions on all the individuals & their off-spring exposed to such conditions. We shall best perceive the importance of the difference by glancing at our domestic breeds: in our truest breeds, innumerable slight differences are continually occurring & can be detected by measurement, but only those differences which improve the breed in the often fanciful eyes of the Fancier are rendered permanent by the animals so characterised being carefully preserved, matched & largely bred from; all other slight differences being lost, by the animals not being largely bred from, & from indiscriminate crossing. If, however, the process of selection were continued for a long time by two Fanciers, under very different conditions of climate or food, some subordinate differences would probably arise between the two lots, owing to the direct action of such conditions. Now in our diagram, the letters $z^1 - m^1$, $z^2 - m^2$ &c represent all sorts of successive slight variations, of which m^{1-10}, the most drought-enduring varieties alone have been naturally selected & been rendered permanent.

This natural selection has been possible, owing to there having been a place in the economy of our imaginary country, which the descendants of M, from inheriting all or some of the advantages over the other inhabitants which made M a very common species, could seize on, when rendered more drought-enduring.

With respect to the process by which each new & improved variety supplants its parent, this must often have gone on in two slightly different manners, differing, however, only in degree. In those animals which are highly locomotive & of which two individuals unite for each birth, there can only seldom have arisen as we shall hereafter see, within the same continuous area, especially if of not very large size, distinct varieties, for they would become blended by such free crossing. In such cases, modifications must be effected quite insensibly by the natural selection of mere individual differences; nearly in the same way as many of our domestic breeds throughout whole districts have been insensibly changed from their ancient state. So that in our diagram the letters m^{1-10} may represent in the case of the higher animals, not recognizable varieties, but mere ideal steps in a real, yet insensibly gradual, change of structure. In organic beings which do not cross freely & which are more stationary, & which are capable of propagating at a great rate, a variety might easily be formed in one spot (more especially if in some slight degree isolated) & might not spread & supplant its parent-stock, until it had become developed by the continud natural selection of similar extremely slight or individual differences into a distinct & plainly recognizable variety. I am inclined to think from the frequency of local varieties, though the subject must remain very doubtful, that this latter process has been a very common one, for a variety would often be unable to supplant its parent, until it had become considerably modified so as to have a decided advantage over it. For instance in the imaginary case of the varieties m^{1-10} which are supposed to inherit all the characters of M, with the addition of enduring more drought; these varieties would inhabit stations, where M could not exist, but in the less dry stations m^{1-10} would have very little power of supplanting their parent M; nevertheless during unusually dry seasons m^{1-10} would have a great advantage over M & would spread; but in damper seasons M, would not have a corresponding advantage over m^{1-10} for these latter varieties are supposed to inherit all the characters of their parent. So there would be a tendency in m^{1-10} to supplant M, but at an excessively slow rate. It would be easy to show that the same thing might occur in the case of many other new characters thus acquired; but the subject is far too doubtful & speculative to be worth pursuing. I will only add that with the most freely crossing & locomotive animals, when inhabiting an area, separated by barriers only to be passed after geographical changes or through some most rare accident, a similar process must often have occurred; for in such cases, distinct & plainly marked varieties might have been insensibly formed in the different districts by the selection of mere individual differences; & when these districts became united, so that the varieties could mingle & come into competition, the best variety would supplant the other varieties or the parent-stock.

To return to our diagram. I do not suppose the process generally to have been so simple as represented under M, where a simple variety m^{1-10} in each stage of descent has been naturally selected. We have seen that not only more species, especially the very common species, in the larger genera in any country present varieties in some degree permanent, but that each such species on an average tends to present a greater number of varieties, than do the species, especially the rarer species, in the smaller & less flourishing genera. As varieties from a species tend to inherit the advantages which made the parent common, these varieties will ultimately tend to be common & to vary; moreover they descend from a variable stock, & are still exposed to the conditions which made their parents vary, hence for this cause they will be liable to vary. Consequently there will be a tendency in the original varying species, after a vast number of generations to produce an almost infinite number of varieties; but our principle of divergence explains how the most diversified varieties will generally have decided advantages over the less diversified & intermediate varieties, causing their extinction & thus reducing the number of varieties living at any one time. These remarks are illustrated in our diagram under A., which species, after many generations represented by dots, is supposed to have varied largely, & to have produced these varieties a^1, d^1 l^1 in some degree permanent; of these, again after many generations & much variation; the two extreme varieties a^1 and l^1, are supposed to have produced other varieties in some degree permanent; of which the extreme varieties have again reproduced others, represented finally by a^{10} & l^{10}. In the diagram I have been able to represent only one other branch proceeding from f^5, & giving rise to a third variety h^{10} which being the extreme form in its own branch has the best chance of surviving & seizing on some place in the natural economy of the country inhabited by the genus.

By continuing the process represented in the diagram, the forms marked a^{10}, h^{10}, l^{10}, may be made different in any degree, till they would be universally be [sic] ranked as good species; & the number of such new forms would continually tend to increase. These new species will generally have supplanted, perhaps by a very slow process their several parents in each stage of descent & their original common parent A, — that is if formed in one continuous area, or as soon as they came into competition with each other if formed in different areas. The original species A. was supposed to be the most moisture loving plant; & l^1 less moisture loving, but inheriting some of the advantages which made A in the great & complex battle for life a very common species; & the offspring of these varieties to be continually selected on the same principle, a^{10} will have been rendered so moisture loving as to have become semi-aquatic, & 1^{10} far less moisture loving than A; & in the third branch, h^{10}, about as moisture loving as A, for it has descended from f^5 which was more moisture-loving than A, and subsequently has become less so. Not that I at all suppose the diversity is ever thus confined to one point; for as a^{1-10} becomes moisture-loving & as 1^{10} becomes less moisture-loving both would under the extremely complex conditions to which all organic beings are exposed, come to be exposed to new dangers & would have to gain some other advantages over other organic beings with which they would have to compete. So that in love of moisture & in many other respects, a^{1-10}, h^{1-10}, 1^{1-10} would come to differ or diverge more & more from each other & their original parent-stock.

A little reflexion will show the extreme importance of this principle of divergence for our theory. I believe all the species of the same genus have descended from a common parent; & we may call the average amount of difference between the species, x; but if we look at the contemporaneous varieties of any one species, the amount of difference between them is comparatively extremely slight & may be called a. How thus can the slight difference a be augmented into the greater difference x; which must be on our theory be continually occurring in nature, if varieties are converted into good species? The process feebly illustrated in our diagram, I believe, explains this; namely the continued natural selection or preservation of those varieties, which diverge most in all sorts of respects from their parent-type, (but still largely inheriting those advantages which made their parents generally dominant & common species) so as to fill as many, as new, & as widely different places in the economy of nature, as possible.

A glance at Diagram 2. will perhaps render this plainer. The varieties a^{1-10}, 1^{1-10} may be here again for simplicity be [*sic*] looked at as more & less moisture loving plants; & everything is the same as in diagram I (the third branch h^{6-10} cannot be introduced) except that it is left to mere chance in each stage of descent, whether the more or less moisture loving varieties are perserved; & the result is, as graphically shown, that a^{10} & 1^{10} differ in this respect; & so in other respects, hardly more than did the first varieties (a^1 1^1) which were produced.

In regard to the difference between varieties & species, I may add that varieties differ from each other & their parents, chiefly in what naturalists call unimportant respects, as size, colour proportions &c; but species differ from each other in these same respects, only generally in a greater degree, & in addition in what naturalists consider more important respects. But we have seen in Ch. IV, that varieties do occasionally, though rarely, vary slightly in such important respects; and in so far as differences in important physiological characters generally stand in direct relation to different habits of life, modifications however slight in such characters would be very apt to be picked out by natural selection & so augmented, thus to fit the modified descendants from the same parent to fill as many & as widely different places in nature as possible. We shall, also, see in a future chapter that a large part of the differences in structure between species may be accounted for by the mysterious laws of correlation; by which, I mean, that when one part is modified, (or the whole animal at one age, as with the larvae of insects) other parts necessarily become altered through the correlated laws of growth. That there is no obvious & unmistakeable difference between the differential characters of species & varieties, is plainly shown by the number of debateable forms in the best known countries, which are ranked by one good naturalist as true species, & by another as mere varieties.

Our principle of divergence has another very important bearing. In the diagram, A. has given rise to three new species, & M to one. The other species of the genus, B to L, are supposed to have transmitted unaltered descendents. Hence, even supposing that A & M have been supplanted as I believe will usually have been the case, by their modified & improved descendants, the genus will have become not only more divergent in character (a^{10} more aquatic than A; and m^{10} more drought-enduring than M.) but numerically larger. The original species A to M were supposed to be closely allied, but yet to exhibit traces, as is so general, of being divided into sub-groups. The sub-groups, after the formation of the new species, will be slightly altered & increased in number; for a^{10} & h^{10} will be closely related together from common descent from f^{5}, & closely yet less closely with l^{10} from descent from their common ancestor A.; and they will all differ as much, generally more from B, than did A. So again m^{10} having constantly diverged from the characters of M will now stand more distant from L, than M originally stood. This is represented in the Diagram III. And from the reasons already given, I believe there will be a constant tendency in the modified descendants of A & of M to go on thus producing more & more new specific forms & thus more & more modified or divergent.

What will be the limit to this process in nature? Though many genera are large, they do not include an indefinite number of species. I believe that there is no limit to the number of species tending to be formed from the most favoured forms in any country (or those which have any [sic] the greatest advantages over the coinhabitants), except the number of species which the country is capable of supporting; but such modified descendants, or new species, after a long period will have to be ranked not in the same genera, but in distinct genera, families or orders. For if we suppose the process illustrated in diagram I. to have long continued & the modified descendants of A to have become extremely much multiplied and diversified in many ways, they will tend to take the places of & thus exterminate the species B.C.D. &c, which originally were nearest related to A. & were not then such common & flourishing species. So if M had left several modified & divergent descendants, it would have been with L, K. &c. It may be here worth observing, that although the new species in taking the place of the old (their great uncle) may have acquired through natural selection, some of their characters; this kind of resemblence would be called by naturalists that of analogy, & the real affinity of the new species would be with their real parents: thus 1^{10} might come to simulate some of the character of B, from occupying its place in nature yet the real affinity of 1^{10} would be with A. — Continue this process, & all, or nearly all the original species (A to M) will become extinct. In Diagram IV. this is represented, E & F alone now having descendents, whether or not modified. And the final result will be, that we shall have two large groups of modified descendants, coming from the two species, generally the extreme species, (A & M) of the original genus, and differing as much as natural selection could make them from each other & from their two parents, which at the first start differed much: assuredly these two new groups of new species would be ranked in different genera, which would be very distinct, if all the original intermediate species from B to L. had been exterminated, but somewhat less distinct if some of these species (as represented in Diagram IV.) had left descendants, whether or not modified.

Now for a moment let us go back many stages in descent: on our theory the original twelve species A to M are supposed to have descended & diverged from some one species, which may be called Z, of a former genus. But now, according to the result given in the last paragraph, Z will have become the ancestor of two or three very distinct groups of new species; & such groups, naturalists call genera. By continuing the same process, namely the natural selection of generally the most divergent forms, with the extinction of those which have been less modified & are intermediate, Z may become the ancestor of two very distinct groups of genera; & such groups of genera, naturalists call Families or even Orders. But to this subject, we shall have to return in our Chapter on Classification.

I have previously remarked that there seems to be no limit to the number of modified descendants, likely to proceed from the most favoured form in any country,—the most favoured always tending to diverge in structure & take the place of & exterminate the less favoured & intermediate forms,— except the total number of species, which the country is capable of supporting. But it may be objected that as natural selection, extinction & divergence must have been going on since the dawn of Life, why have we not an infinite number of species, almost as many species as individuals? We shall presently see that natural selection can act only with extreme slowness. Nor do we by any means know that the maximum number of species, which any country would be best fitted to support, has anywhere been as yet produced: the fact that there is no country which does not support several, often many, organic beings naturalised by man, without, as far as we know, having caused the extinction of an equal number of the indigenous productions, renders it probable that such countries were capable of supporting a greater number of specific forms than nature has supplied them with. Even the Cape of Good Hope, which is apparently the richest district in the world in different kinds of plants has received, as I am informed by Prof. Haney from [   ] to [   ] naturalised species. Many geologists, indeed believe that the number of species in the world has gone on increasing from the earliest geological days; but I am sorry to own that the evidence on this head seems to me quite insufficient.

It might indeed be argued from the enormous list of shells, found in the eocene Paris basin, & even in the ancient Silurian system of Bohemia, as so admirably worked out by Barrande, that at these periods & in these places, a greater number of species existed than anywhere at the present day. But it may be doubted how far such comparisons are in any instance trustworthy; for we have reason to suppose that the duration of each sub-division of each geological formation is so enormous, that it is not fair to compare all the species found in one such sub-division with all existing within an area at the present day. Barrande's "colonies" show, according to Sir C. Lyell's explanation of them, what changes of climate or currents must have taken place within certain definite periods: the Glacial epoch within what may be called the present period, should teach us caution, for far lesser changes than the glacial epoch, not easily to be detected in ancient geological formations, might alternately bring in & displace, & apparently mingle many organic beings, which never really coinhabited the same area.

But if the time has not yet arrived, may it not at some epoch come, when there will be almost as many specific forms as individuals? I think we can clearly see that this would never be the case. Firstly, there would be no apparent benefit in a greater amount of modification than would adapt organic beings to different places in the polity of nature; for although the structure of each organism stands in the most direct & important relation to many other organic beings, and as these latter increase in number & diversity of organisation, the conditions of the one will tend to become more & more complex, & its descendants might well profit by a further division of labour; yet all organisms are fundamentally related to the inorganic conditions of the world, which do not tend to become infinitely more varied. Secondly as the amount of life & number of individual beings, whether or not much diversified, also primarily depends on such inorganic conditions; if there exist in any country, a vast number of species (although a greater amount of life could be supported) the average number of individuals of each species must be somewhat less than if there were not so many species; & any species, represented by but few individuals, during the fluctuation in number to which all species must be subject from fluctuations in seasons, number of enemies &c, would be extremely liable to total extinction. Moreover, whenever the number of individuals of any species becomes very small, the ill-effects, as I believe, of close inter breeding would come into play. Lastly we have seen in our Chap. IV & shall presently again see, that the amount of variations, & consequently of variation in a right or beneficial direction for natural selection to seize on & preserve, will bear some relation within any given period, to the number of individuals living & liable to variation during such period: consequently when the descendants from any one species have become modified into very many species, without all become numerous in individuals, which [we] see hardly ever to be the case with all the species of the same genus or family, there will be a check amongst the less common species to their further modification: the lesser number of the individuals serving as a regulator or fly-wheel to the increasing rate of further modification, or the production of new specific forms.

Subject to these restraining influences, I can see no limit to the number of modified descendants, which might proceed from the most favoured forms, whatever they may be, now living in the world. If we return to look to the future, as far into the remotest futurity as the Silurian system lies in the remote past, our theory would lead to the conclusion that all organic beings which will live at that far distant period, will be descendants from a very few of our contemporaries; perhaps from not so few, owing to the increasing complexity of the organic world, as our existing organisms have descended from; for our animals have descended, according to our theory, from four or five ancestral types & our plants from apparently still fewer; & if we rashly dare trust to mere analogy, all our plants & animals from some one form, into which life was first breathed.

Taking a more modest glance into futurity, we may predict that the dominant genera, now abounding with common & widely difused species, will tend to be still more dominant for at least some considerable lapse of time, & will give rise to new groups of species, always diverging in character, & seizing on the places occupied by the less favoured forms, whether or not their near blood relations, supplanting them & causing their extermination. The great & flourishing genera both of plants & animals, which now play so important a part in nature, thus viewed become doubly interesting, for they include the ancestors of future conquering races. In the great scheme of nature, to that which has much, much will be given.

Finally, then, in regard to our principle of Divergence, which regulates the natural Selection of variations, & causes the Extinction of intermediate

& less favoured forms, I believe it to be all important as explaining why the average difference between two species of the same genus, the parents of which by our theory once existed as mere varieties, is greater than the average difference between two such varieties. It bears on, & I think explains, the classification or natural affinities during all times of all organic beings, which seeming to diverge from common stems are yet grouped like families within the same tribes, tribes within the same nations, & nations within the same sections of the human race. We shall, also, hereafter see that these views bear on palaeontology & explain why extinct forms either fall within existing groups, or, as is so frequently the case, are in some slight degree intermediate between them.

The relation of all past & present beings may be loosely compared with the growth of a few gigantic trees; that is if we suppose that from each of the innumerable twigs, innumerable buds are trying to sprout forth, & that the other buds, twigs & branches have the best chance of growing from getting more light. The buds & twigs may represent existing species, & all beneath their living extremities may represent extinct forms. We know that the twigs proceed from lesser branches, these from larger & these from main linbs, from the trunk, & that the several branches & limbs are of very unequal sizes; & this grouping of the branches may represent the natural classification of organic beings. In our living trees we can trace in the gnarled & leafless branches the connecting links; but so imperfect are our palaeontological records, that we can only here & there find a form which may be called a forked branch, with its two arms directed towards two now distinct groups of organisms. As we know that the gnarled branches were at successive periods tender twigs crowded with buds, so we may believe that every organic class, whether or not now having lineal descendants on the earth, swarmed at each stage of descent under diversified forms of life. Many a smaller & larger branch, & even some main limbs have utterly perished, from being over topped by the ever diverging budding twigs; so it has been with whole groups of organic beings. Here & there a branch is still alive, carrying only a few twigs & buds; & these will represent the organic groups having few species & fewer genera, which are now on the road to extinction. As buds give rise by growth to fresh buds, & these, if vigorous, branch out & give rise to many a diverging branch still branching out, & causing the death of many a feebler twig & branch on all sides & beneath, so by generations I believe it has been with the great Tree of Life, which fills the crust of the earth with fragments of its dead & broken branches, & covers with its ever living, ever diverging & marvellous ramifications, the face of the earth.

Long ere this, a crowd of difficulties will have arisen in the reader's mind, overwhelming my theory of natural selection, more especially when applied to organs or beings widely different in the same great classes. Some of these difficulties are indeed great enough almost to crush my belief; but many, I think, are only apparent. It is possible to believe that the eye with its admirable correction for spherical & chromatic aberration, & with its power of adapting the focus to the distance, could have been formed from the simplest conceivable eye, by natural selection? It is possible for the instinct of a bee, which produces a cell constructed on the highest geometrical principles, to be thus perfected? I confess that my mind recoils from such an admission; yet, reflecting on the known gradations in so wonderful an organ as the eye amongst existing animals,—a mere small fraction of those which have lived,—I can see no logical impossibility; & as far as probability is concerned, a safe conclusion can be drawn, as it seems to me, only from the general phenomena of organic beings, as indicative whether each being has been simply created or has been produced by the common laws of generation with superadded modification. But these questions, & likewise the general subject of instinct shall be discussed in separate chapters.

What shall we say of small & apparently trifling organs, yet most useful to the animal possessing them, as the eye-lash, or a tail serving as a fly-brush; could these have been produced by natural selection, which is in fact selection for life & death? But I have already shown how cautious we should be in deciding what trifle may turn the nicely-suspended balance of life in the great struggle for existence. Again how could a swimming animal be turned into a crawler, or a walking animal into a flyer: how could they live in an intermediate state? Undoubtedly nothing can be effected through natural selection except by the addition of infinitesimally small changes; & if it could be shown that in cases like the foregoing, transitional states were impossible, the theory would be overthrown. This being so, it may be further asked, do we not meet in certain members of a class organs, which, as far as we can see, are absolutely new creations, & which cannot be some other part or organ modified by natural selection in accordance with the laws of morphology? We shall see that such cases are surprisingly few & hard to find.

Again it has often been urged that if species were subject to change all nature would be in confusion & the limits of no species distinct; but this argument depends on the assumption that the change is rapid & that many species are simultaneously undergoing change. If species were as distinctly defined, as some authors pretend, systematic natural history would be a far less difficult subject, that those authors will find if they will take up for description almost any group, especially a varying group of species; but to this subject I shall immediately recur. So again it has been said, if species were subject to change, we should find plain evidence of such change in our collections of fossil remains; but the force of this objection, in main part, lies in the supposition that the records of geology are as ancient as the first commencement of life, & that they are far more perfect than some of our most experienced geologists have shown good reason for believing that they are in truth. I will here only ask those who make this objection, can they believe that at some future geological epoch, fossil remains will tell that which we do not now know, namely what were the exact steps by which the various British breeds of sheep & oxen have descended from some one or two parent stocks. It should be remembered we do not mean forms intermediate between horse and tapir, but between both of them & some unknown common parent.

Lastly why do two species when crossed, either yield few or no offspring, & these more or less sterile, & why do those varieties which we may safely conclude are descended from a single species yield abundantly fertile offspring: To this important subject I will devote a chapter. And all the foregoing great difficulties, & some curious special cases shall be stated in detail, as fairly as I can, & be discussed. That some difficulties remain to be mastered will not be thought surprising by those who will make allowance for our ignorance on what is daily passing round us in the living world; & our incomparably greater ignorance of the many past worlds which have teemed with life.

DARWIN'S PLANS FOR THE "BIG SPECIES BOOK"

The "Big Species Book" occupied Darwin from 1856 until circumstances forced him to abandon it and proceed with the abridged version—that is, *On the Origin of Species*. In 1858, he reported on his progress to a number of correspondents. To his former servant during and after the *Beagle* voyage, Syms Covington, he wrote:

> I have for some years been preparing a work for publication which I commenced twenty years ago, and for which I sometimes find extracts in your handwriting! This work will be my biggest; it treats on the origin of varieties of our domestic animal and plants, and on the origin of species in a state of nature. I have to discuss every branch of natural history, and the work is beyond my strength and tries me sorely.[1]

Naturalist Leonard Jenyns got a somewhat more detailed description:

> I have myself been for the last two years and suppose I shall be two more, very hard, (too hard) at work on my species Book, getting it ready for press. . . . You ask what my Book is about, I fear it is almost de rebus omnibus:[2] my attempt is to look at all facts in Nat. Hist & Geology under the two points of view,—has each species

1. Darwin to Covington, 18 May 1858, *Correspondence,* 7: 95.
2. Latin: "about everything."

been created independently or have species, like varieties, descended from other species? And the upshot is, that I have become dreadfully heterodox about the immutability of species.[3]

3. Darwin to Jenyns, 1 April 1858, *Correspondence*, 7: 60.

Chapter 7

ABSTRACT OF DARWIN'S THEORY SENT TO ASA GRAY (1857)

INTRODUCTION

On 5 September 1857, Darwin sent a brief abstract of his theory to the Harvard botanist Professor Asa Gray. It was a shrewd move to include Gray in the small but growing circle of influential scientists, including Joseph Hooker and Charles Lyell, who knew of Darwin's long-nurtured work on evolution. Gray became an early recruit and, after the *Origin* was published in 1859, an energetic supporter of Darwinism in America. Gray was primed and ready to defend evolution, which he did in public debates with Louis Agassiz, his Harvard colleague and an ardent opponent of evolution.

Beyond strengthening his own hand in the public battle over the *Origin*, by sending the abstract to Asa Gray—a man of unimpeachable reputation living three thousand miles from England—Darwin guaranteed the certification of his priority in the discovery of evolution by natural selection. Events would soon prove that such certification was barely established in time. Darwin had been working on his theory since 1837. He had his notebooks and his 1844 Essay as well as five long chapters of *Natural Selection* to vouch for his twenty-year intellectual investment in evolution. But since none of these documents had been published, there was no certifiable guarantee of their date. Sending the abstract in a letter to Gray established September 1857 as a minimum date for his priority. Nine months later, in June 1858, Darwin's writing of *Natural Selection* was "interrupted" by a letter from Alfred Russel Wallace that included a manuscript for publication that explained evolution by natural selection. Wallace's manuscript expressed the theory almost in Darwin's own terms.

Darwin moved quickly to have Wallace's manuscript published. With the help of his colleagues Lyell and Hooker, he also moved to protect his own claims. The result was that Wallace's paper and Darwin's abstract (as well as a small section of the 1844 Essay) were read jointly at the Linnean Society of London on 1 July 1858 and published jointly in the Society's proceedings in 1859. Thus, Darwin established himself and Wallace as independent co-discoverers of natural selection and, thanks to the Gray Abstract, established his own priority in that discovery.

ABSTRACT

(1) It is wonderful what the principle of Selection by Man, that is the picking out of individuals with any desired quality, & breeding from them, & again picking out, can do. Even Breeders have been astonished at their own results. They can act on differences inappreciable to an uneducated eye. Selection has been *methodically* followed in *Europe* for only the last half century. But it has occasionally & even in some degree methodically been followed in the most ancient times. There must have been, also, a kind of unconscious selection from the most ancient times namely in the preservation of the individual animals (without any thought of their offspring) most useful to each race of man in his particular circumstances. The "roguing", as nurserymen call the destroying of varieties, which depart from their type is a kind of selection. I am convinced that intentional & occasional selection has been the main agent in making our domestic races. But, however, this may be, its great power of modification has been indisputably shown in late times. Selection acts only by the accumulation of very slight or greater variations, caused by external conditions, or by the mere fact that in generation the child is not absolutely similar to its parent. Man by this power of accumulating variations adapts living beings to his wants, — *may be said* to make the wool of one sheep good for carpet & another for cloth &c.

(2) Now suppose there was a being, who did not judge by mere external appearance, but could study the whole internal organization — who never was capricious, — who should go on selecting for one end during millions of generations, who will say what he might not effect! In nature we have some *slight* variation, occasionally in all parts: & I think it can be shown that changed conditions of existence is the main cause of the child not exactly resembling its parents; & in nature geology shows us what changes have taken place & are taking place. We have almost unlimited time: no one but a practical geologist can fully appreciate this: think of the Glacial period, during the whole of which the same species of shells at least have existed: there must have been during this period, millions on millions of generations.

(3) I think it can be shown that there is such an unerring power at work on *natural selection* (the title of my Book), which selects exclusively for the good of each organic being. The elder Decandolle, W. Herbert, & Lyell have written strongly on the struggle for life; but even they have not written strongly enough. Reflect that every being (even the Elephant) breeds at such a rate, that in a few years, at most a few centuries or thousands of years the surface of the earth would hold the progeny of any one species. I have found it hard constantly to bear in mind that the increase of every single species is checked during some part of its life, or during some shortly recurrent generation. Only a few of those annually born can live to propagate their kind. What a trifling difference must often determine which shall survive & which perish. —

(4) Now take the case of a country undergoing some change: this will tend to cause some of its inhabitants to vary slightly; not but what I believe most beings vary at all times enough for selection to act on. Some of its inhabitants will be exterminated, & the remainder will be exposed to the mutual action of a different set of inhabitants, — which I believe to be more important to the life of each being than mere climate. Considering the infinitely various ways, beings have to obtain food by struggling with other beings, to escape danger at various times of life, to have their eggs or seeds disseminated &c &c, I cannot doubt that during millions of generations individuals of a species will be born with some slight variation profitable to some part of its economy: such will have a better chance of surviving, propagating this variation, which will be slowly increased by the accumulative action of natural selection; and the variety thus formed will either coexist with or more commonly will exterminate its parent form. An organic being like the woodpecker or missletoe may thus come to be adapted to a score of contingencies: natural selection, accumulating those slight variations in all parts of its structure, which are in any way useful to it, during any part of its life.

(5) Multiform difficulties will occur to everyone on this theory. Most can, I think, be satisfactorily answered. — "Natura non facit saltum" answers some of the most obvious. — The slowness of the change & only a very few undergoing change at any one time answers others. The extreme imperfection of our geological records answers others. —

(6) One other principle, which may be called the principle of divergence plays, I believe, an important part in the origin of species. The same spot will support more life if occupied by very diverse forms: we see this in the many generic forms in a square yard of turf — or in the plants & insects on any little uniform islet belonging almost to as many genera & families as to species. — We can understand this with the higher animals, whose habits we best understand. We know that it has been experimentally shown that a plot of land will yield a greater weight if cropped with several species of grasses

than with 2 or 3 species. Now every single organic being, by propagating so rapidly, may be said to be striving its utmost to increase in numbers. So it will be with the offspring of any species after it has broken into varieties or sub-species or true species. And it follows, I think, from the foregoing facts, that the varying offspring of each species will try (only few will succeed to seize on as many & as diverse places in the economy of nature, as possible. Each new variety or species, when formed will generally take the place of & so exterminate its less well-fitted parent. This, I believe, to be the origin of the classification or arrangement of all organic beings at all times. These always seem to branch & sub-branch like a tree from a common trunk; the flourishing twigs destroying the less vigorous, — the dead & lost branches rudely representing extinct genera & families. —

This sketch is *most* imperfect; but in so short a space I cannot make it better. Your imagination must fill up very wide blanks. — Without some reflexion it will appear all rubbish; perhaps it will appear so after reflexion. — C. D.

Chapter 8

ON THE ORIGIN OF SPECIES
(1859)

INTRODUCTION

We present here six selections from *On the Origin of Species*, all from the first edition of 1859. The first is Darwin's introduction, which sets forth the context of the book's publication and lists the main points to be discussed. Next are the two key theoretical chapters, "The Struggle for Existence" and "Natural Selection," chapters 3 and 4, respectively. These are followed by excerpts from Chapter 6, "Difficulties on Theory" and Chapter 7, "Instinct." Finally, we present the last section of his concluding chapter (chapter 14), or recapitulation, in which he reflects on the entire volume and predicts what its effect on natural history might be.

DARWIN'S INTRODUCTION

Darwin begins by recalling the origin of his doubts regarding the fixity of species, first entertained during the voyage of the *Beagle*. He publishes the *Origin* (an "abstract," he calls it) now, rather than the fuller form he had originally planned, because of the independent discovery of natural selection by Alfred Russel Wallace. He then disabuses the reader of commonly held presuppositions concerning external causes of variation and the notion that organic structures might be caused by the volition of the organism itself. This is a critique of Lamarck and his followers, although Darwin does not name names. He then alludes to Robert Chambers's notorious book *Vestiges of the Natural History of Creation* (1844), which he attacks for not accounting for adaptation. In 1859, Chambers's identity as the volume's author had not yet been clearly established. In 1876, however, in a historical sketch of the idea of evolution that Darwin included in the sixth edition of the *Origin*, he is more generous, praising the still anonymous author for the great service he had rendered in preparing the climate of public opinion for Darwin's own book of 1859: "The work, from its powerful and brilliant style, though displaying in the earlier editions little accurate knowledge and a great want of scientific caution, immediately had a very wide circulation. In my opinion it has done excellent service in this country in calling attention to the subject, in removing prejudice, and in thus preparing the ground for the reception of analogous views."[1]

THE STRUGGLE FOR EXISTENCE

The stimulus for Darwin's theory, as is well known, came from his reading "for amusement" (on 28 September 1838) Thomas Malthus's famous *Essay on Population*. Darwin describes the event in his *Autobiography*, and from his notebook entries for that day (see pp. 73–74), we can conjecture that his intuition was focused on the "warring of species" (Notebook D:134) resulting from the superfecundity that Malthus stresses in the very first chapter of his book.

Chapter 3 of the *Origin* is Darwin's tribute to Malthus, a very direct Malthusian interpretation of competition in nature, fleshed out with familiar observations designed to appeal to the general reader as well as evidence gathered from Darwin's own observations and experiments.

Early on, Darwin states the basic axiom: In the struggle for existence, success is measured in terms of the number of progeny left behind. In this sense, he makes clear, the image he had borrowed from Herbert Spencer was a metaphor embracing a variety of processes, including the struggle for survival under harsh environmental circumstances, intra- and interspecies competition for resources, and the everyday course of life in which the survival of a predator is predicated on the death of its prey.

Here the narrative is pure Malthus. Darwin describes the consequences of the geometrical growth of populations being limited by the availability of food. The observed stability of animal and plant populations suggested a struggle for existence either among individuals or groups or between them and the physical environment. Darwin gives a selection of evidence bearing on the overproduction of seeds and plants alike, or of the young of animals, all with the same result: that overpopulation would result, even among the slow-breeding elephant, unless mechanisms that limited population—*checks*, in Malthus's terminology—came into play. He supplies not only his own field observations and those of others, but also the results of his experiments on a famous weed patch—three feet by two feet—that he planted at Down. He observed the rate at which slugs and insects destroyed seedlings, and

---

1. *Origin of Species*, 6th ed., 6.

drew an analogy between this kind of devastation and that which might have been caused, in the same patch, by mowing or grazing.

Darwin compared the great natural power represented in population dynamics to the hammering of sharp wedges into the face of nature. This dramatic metaphor conveys how populations are forced, by the struggle for survival, to exploit specific ecological niches. It is a metaphor that struck Darwin on his first fateful reading of Malthus (Notebook D:135). The result of this "wedging" is divergence of character, which is the mechanism whereby natural selection leads to the creation of new species.

NATURAL SELECTION

Darwin begins his discussion of natural selection, defined as "the preservation of favourable variations and the rejection of injurious" ones, by alluding to the artificial selection that animal breeders and horticulturalists practice, and asking whether the same principles can apply under natural conditions. (Indeed, there are constant references to breeders in this chapter, as if to keep the analogy constantly before the reader.) Natural selection is a more subtle process than the methodical selection practiced by breeders, because the most minute variations may prove useful in the struggle for life. It also works for the "improvement" or "good" of organisms. Favorable variations, of course, are "good" by definition, since deleterious ones lead to extinction.

The nature of these variations is the key to the process of speciation. Natural selection works most effectively, Darwin argues, in large populations with large ranges whose individual members differ only slightly from one another. The larger the range, the more habitats there are to be exploited by variant members of the species. Darwin views the increasing specialization that competition among members of a large population generates as *good* because it leads to the production of subpopulations with a higher capacity for survival, and also as *progressive*, in the sense of better adapted.[2] Darwin calls the process whereby slightly differing varieties of the same species begin to exploit different habitats divergence of character. This tendency of large populations to diversify explains the success that many different species have in exploiting the same habitats simultaneously—as Darwin showed experimentally in his famous weed patch. Slight differences separating varieties become more significant as intervening varieties—which are not as successful in exploiting their habitats—die out. For this reason, populations

2. On the relationship between natural selection and progress, and on Darwin's conception of "competitive highness," see Dov Ospovat, *The Development of Darwin's Theory: Natural History, Natural Theology, and Natural Selection, 1838–1859* (Cambridge, Cambridge University Press, 1981), 210–25.

tend to become more diverse as successful variations are preserved and accumulated through natural selection. The result is a branching tree of life. (See chapter 6 for Darwin's extended discussion of divergence in the long version of the *Origin*.)

DIFFICULTIES ON THEORY

In the section *What Natural Selection Can Do* from the *Difficulties* chapter, Darwin undermines Paley's teleological argument from design by showing that natural selection produces adapted organisms to a far less exalted standard than "perfect adaptation" would require. Adaptations are tools for survival, and organisms need be only slightly better adapted than their competitors to survive.

INSTINCT

Darwin's discussion of Neuter Insects in the *Instinct* chapter deals with the apparent challenge to natural selection posed by castes of sterile worker insects. Such castes might be "good" for the species, but natural selection is supposed to work only through individual advantage. How could natural selection produce the ultimate self-sacrifice of sterility for the sake of the group? Here Darwin's thinking on the evolution of behavior came close to presaging the concept of kin selection, which explains how certain forms of altruistic behavior can be selected for within a closely related group. The importance of kin selection, which was not recognized until 1963, has since proved influential in the emergence of a strongly Darwinian school of behavioral biology and evolutionary psychology.[3]

CONCLUSION

The *Origin*'s conclusion is notable for Darwin's insight into what the volume accomplishes in natural science and his prescience regarding its impact on its future. The excerpt reproduced here begins with Darwin's declaration that he has, in essence, freed the species question from metaphysics. Naturalists need no longer waste their time on conjectures about what the "essence" of a species might be. Metaphorical terms will now have "plain significance," making natural history a more interesting pursuit because it is more realistic (in that real relationships among species, living and dead, can now be established, not merely, for example, some relationship with a pu-

3. William D. Hamilton, "The Evolution of Altruistic Behavior," *American Naturalist* 97 (1963): 354–356, and "The Genetical Evolution of Social Behaviour," parts 1 and 2, *Journal of Theoretical Biology* 7 (1964): 1–52. See Robert J. Richards, *Darwin and the Emergence of Evolutionary Theories of Mind and Behavior* (Chicago, 1987), 142–52.

tative, but imaginary, archetype). Even more striking are his predictions: His theory would stimulate a search for the causes of variation, presaging the future connection of genetics and evolutionary theory. Classification will become genealogical rather than follow some artificial scheme, and the result will be the revelation of the *true* plan of creation. To underline the new realism he predicts for biology, he invokes Lyell's uniformitarian view that geological and life history can now be addressed through the study of causes currently operating, and fictitious catastrophes can be dispensed with. He notes that his view of the origin of species is ennobling to those few ancient progenitors of the rich variety of living species, a much richer view of life than that afforded by the theory of special creation. He invokes a kind of teleology when he concludes, "as natural selection works solely by and for the good of each being, all corporeal and mental endowments will tend to progress towards perfection." This phrase seems to invoke perfect adaptation as well as a faith in progressivism that many contemporary evolutionists deny, although they continue to use similar language.

There follows the famous passage on the "entangled bank," Darwin's metaphor for the ecological interrelationship of living organisms, which recalls the lush jungle environment of his early field work in Brazil and the dense forests that line the banks of the Beagle Channel in Tierra del Fuego. His closing, the famous invocation of the "grandeur" of the evolutionary perspective, is a restatement of his belief in the operation of secondary laws of nature and his awe at the complexity and beauty of the life-forms created in accordance with them.

INTRODUCTION

When on board H.M.S. 'Beagle,' as naturalist, I was much struck with certain facts in the distribution of the inhabitants of South America, and in the geological relations of the present to the past inhabitants of that continent. These facts seemed to me to throw some light on the origin of species—that mystery of mysteries, as it has been called by one of our greatest philosophers. On my return home, it occurred to me, in 1837, that something might perhaps be made out on this question by patiently accumulating and reflecting on all sorts of facts which could possibly have any bearing on it. After five years' work I allowed myself to speculate on the subject, and drew up some short notes; these I enlarged in 1844 into a sketch of the conclusions, which then seemed to me probable: from that period to the present day I have steadily pursued the same object. I hope that I may be excused for entering on these personal details, as I give them to show that I have not been hasty in coming to a decision.

My work is now nearly finished; but as it will take me two or three more years to complete it, and as my health is far from strong, I have been urged to publish this Abstract. I have more especially been induced to do this, as Mr. Wallace, who is now studying the natural history of the Malay archipelago, has arrived at almost exactly the same general conclusions that I have on the origin of species. Last year he sent to me a memoir on this subject, with a request that I would forward it to Sir Charles Lyell, who sent it to the Linnean Society, and it is published in the third volume of the Journal of that Society. Sir C. Lyell and Dr. Hooker, who both knew of my work—the latter having read my sketch of 1844 — honoured me by thinking it advisable to publish, with Mr. Wallace's excellent memoir, some brief extracts from my manuscripts.

This Abstract, which I now publish, must necessarily be imperfect. I cannot here give references and authorities for my several statements; and I must trust to the reader reposing some confidence in my accuracy. No doubt errors will have crept in, though I hope I have always been cautious in trusting to good authorities alone. I can here give only the general conclusions at which I have arrived, with a few facts in illustration, but which, I hope, in most cases will suffice. No one can feel more sensible than I do of the necessity of hereafter publishing in detail all the facts, with references, on which my conclusions have been grounded; and I hope in a future work to do this. For I am well aware that scarcely a single point is discussed in this volume on which facts cannot be adduced, often apparently leading to conclusions directly opposite to those at which I have arrived. A fair result can be obtained only by fully stating and balancing the facts and arguments on both sides of each question; and this cannot possibly be here done.

I much regret that want of space prevents my having the satisfaction of acknowledging the generous assistance which I have received from very many naturalists, some of them personally unknown to me. I cannot, however, let this opportunity pass without expressing my deep obligations to Dr. Hooker, who for the last fifteen years has aided me in every possible way by his large stores of knowledge and his excellent judgement.

In considering the Origin of Species, it is quite conceivable that a naturalist, reflecting on the mutual affinities of organic beings, on their embryological relations, their geographical distribution, geological succession, and other such facts, might come to the conclusion that each species had not been independently created, but had descended, like varieties, from other species. Nevertheless, such a conclusion, even if well founded, would be unsatisfactory, until it could be shown how the innumerable species inhabiting this world have been modified so as to acquire that perfection of structure and coadaptation which most justly excites our admiration. Naturalists continually refer to external conditions, such as climate, food, &c., as the only possible cause of variation. In one very limited sense, as we shall hereafter see, this may be true; but it is preposterous to attribute to mere external conditions, the structure, for instance, of the woodpecker, with its feet, tail, beak, and tongue, so admirably adapted to catch insects under the bark of trees. In the case of the misseltoe, which draws its nourishment from certain trees, which has seeds that must be transported by certain birds, and which has flowers with separate sexes absolutely requiring the agency of certain insects to bring pollen from one flower to the other, it is equally preposterous to account for the structure of this parasite, with its relations to several distinct organic beings, by the effects of external conditions, or of habit, or of the volition of the plant itself.

The author of the 'Vestiges of Creation' would, I presume, say that, after a certain unknown number of generations, some bird had given birth to a woodpecker, and some plant to the misseltoe, and that these had been produced perfect as we now see them; but this assumption seems to me to be no explanation, for it leaves the case of the coadaptations of organic beings to each other and to their physical conditions of life, untouched and unexplained.

It is, therefore, of the highest importance to gain a clear insight into the means of modification and coadaptation. At the commencement of my observations it seemed to me probable that a careful study of domesticated animals and of cultivated plants would offer the best chance of making out this obscure problem. Nor have I been disappointed; in this and in all other perplexing cases I have invariably found that our knowledge, imperfect though it be, of variation under domestication, afforded the best and safest clue. I may venture to express my conviction of the large value of such studies, although they have been very commonly neglected by naturalists.

From these considerations, I shall devote the first chapter of this Abstract to Variation under Domestication. We shall thus see that a large amount of hereditary modification is at least possible; and, what is equally or more important, we shall see how great is the power of man in accumulating by his Selection successive slight variations. I will then pass on to the variability of species in a state of nature; but I shall, unfortunately, be compelled to treat this subject far too briefly, as it can be treated properly only by giving long catalogues of facts. We shall, however, be enabled to discuss what circumstances are most favourable to variation. In the next chapter, the Struggle for Existence amongst all organic beings throughout the world, which inevitably follows from their high geometrical powers of increase, will be treated of. This is the doctrine of Malthus, applied to the whole animal and vegetable kingdoms. As many more individuals of each species are born than can possibly survive; and as, consequently, there is a frequently recurring struggle for existence, it follows that any being, if it vary however slightly in any manner, profitable to itself, under the complex and sometimes varying conditions of life, will have a better chance of surviving, and thus be *naturally selected*. From the strong principle of inheritance, any selected variety will tend to propagate its new and modified form.

This fundamental subject of Natural Selection will be treated at some length in the fourth chapter; and we shall then see how Natural Selection almost inevitably causes much Extinction of the less improved forms of life, and induces what I have called Divergence of Character. In the next chapter I shall discuss the complex and little known laws of variation and of correlation of growth. In the four succeeding chapters, the most apparent and gravest difficulties on the theory will be given: namely, first, the difficulties of transitions, or understanding how a simple being or a simple organ can be changed and perfected into a highly developed being or elaborately constructed organ; secondly the subject of Instinct, or the mental powers of animals; thirdly, Hybridism, or the infertility of species and the fertility of varieties when intercrossed; and fourthly, the imperfection of the Geological Record. In the next chapter I shall consider the geological succession of organic beings throughout time; in the eleventh and twelfth, their geographical distribution throughout space; in the thirteenth, their classification or mutual affinities, both when mature and in an embryonic condition. In the last chapter I shall give a brief recapitulation of the whole work, and a few concluding remarks.

No one ought to feel surprise at much remaining as yet unexplained in regard to the origin of species and varieties, if he makes due allowance for our profound ignorance in regard to the mutual relations of all the beings which live around us. Who can explain why one species ranges widely and is very numerous, and why another allied species has a narrow range and is rare? Yet these relations are of the highest importance, for they determine the present welfare, and, as I believe, the future success and modification of every inhabitant of this world. Still less do we know of the mutual relations of the innumerable inhabitants of the world during the many past geological epochs in its history. Although much remains obscure, and will long remain obscure, I can entertain no doubt, after the most deliberate study and dispassionate judgement of which I am capable, that the view which most naturalists entertain, and which I formerly entertained—namely, that each species has been independently created—is erroneous. I am fully convinced that species are not immutable; but that those belonging to what are called the same genera are lineal descendants of some other and generally extinct species, in the same manner as the acknowledged varieties of any one species are the descendants of that species. Furthermore, I am convinced that Natural Selection has been the main but not exclusive means of modification.

THE STRUGGLE FOR EXISTENCE

Before entering on the subject of this chapter, I must make a few preliminary remarks, to show how the struggle for existence bears on Natural Selection. It has been seen in the last chapter that amongst organic beings in a state of nature there is some individual variability; indeed I am not aware that this has ever been disputed. It is immaterial for us whether a multitude of doubtful forms be called species or sub-species or varieties; what rank, for instance, the two or three hundred doubtful forms of British plants are entitled to hold, if the existence of any well-marked varieties be admitted. But the mere existence of individual variability and of some few well-marked varieties, though necessary as the foundation for the work, helps us but little in understanding how species arise in nature. How have all those exquisite adaptations of one part of the organisation to another part, and to the conditions of life, and of one distinct organic being to another being, been perfected? We see these beautiful co-adaptations most plainly in the woodpecker and misseltoe; and only a little less plainly in the humblest parasite which clings to the hairs of a quadruped or feathers of a bird; in the structure of the beetle which dives through the water; in the plumed seed which is wafted by the gentlest breeze; in short, we see beautiful adaptations everywhere and in every part of the organic world.

Again, it may be asked, how is it that varieties, which I have called incipient species, become ultimately converted into good and distinct species, which in most cases obviously differ from each other far more than do the varieties of the same species? How do those groups of species, which constitute what are called distinct genera, and which differ from each other more than do the species of the same genus, arise? All these results, as we shall more fully see in the next chapter, follow inevitably from the struggle for life. Owing to this struggle for life, any variation, however slight and from whatever cause proceeding, if it be in any degree profitable to an individual of any species, in its infinitely complex relations to other organic beings and to external nature, will tend to the preservation of that individual, and will generally be inherited by its offspring. The offspring, also, will thus have a better chance of surviving, for, of the many individuals of any species which are periodically born, but a small number can survive. I have called this principle, by which each slight variation, if useful, is preserved, by the term of Natural Selection, in order to mark its relation to man's power of selection. We have seen that man by selection can certainly produce great results, and can adapt organic beings to his own uses, through the accumulation of slight but useful variations, given to him by the hand of Nature. But Natural Selection, as we shall hereafter see, is a power incessantly ready for action, and is as immeasurably superior to man's feeble efforts, as the works of Nature are to those of Art.

We will now discuss in a little more detail the struggle for existence. In my future work this subject shall be treated, as it well deserves, at much greater length. The elder De Candolle and Lyell have largely and philosophically shown that all organic beings are exposed to severe competition. In regard to plants, no one has treated this subject with more spirit and ability than W. Herbert, Dean of Manchester, evidently the result of his great horticultural knowledge. Nothing is easier than to admit in words the truth of the universal struggle for life, or more difficult—at least I have found it so—than constantly to bear this conclusion in mind. Yet unless it be thoroughly engrained in the mind, I am convinced that the whole economy of nature, with every fact on distribution, rarity, abundance, extinction, and variation, will be dimly seen or quite misunderstood. We behold the face of nature bright with gladness, we often see superabundance of food; we do not see, or we forget, that the birds which are idly singing round us mostly live on insects or seeds, and are thus constantly destroying life; or we forget how largely these songsters, or their eggs, or their nestlings are destroyed by birds and beasts of prey; we do not always bear in mind, that though food may be now superabundant, it is not so at all seasons of each recurring year.

I should premise that I use the term Struggle for Existence in a large and metaphorical sense, including dependence of one being on another, and including (which is more important) not only the life of the individual, but success in leaving progeny. Two canine animals in a time of dearth, may be truly said to struggle with each other which shall get food and live. But a plant on the edge of a desert is said to struggle for life against the drought, though more properly it should be said to be dependent on the moisture. A plant which annually produces a thousand seeds, of which on an average only one comes to maturity, may be more truly said to struggle with the plants of the same and other kinds which already clothe the ground. The missletoe is dependent on the apple and a few other trees, but can only in a far-fetched sense be said to struggle with these trees, for if too many of these parasites grow on the same tree, it will languish and die. But several seedling missletoes, growing close together on the same branch, may more truly be said to struggle with each other. As the missletoe is disseminated by birds, its existence depends on birds; and it may metaphorically be said to struggle with other fruit-bearing plants, in order to tempt birds to devour and thus disseminate its seeds rather than those of other plants. In these several senses, which pass into each other, I use for convenience sake the general term of struggle for existence.

A struggle for existence inevitably follows from the high rate at which all organic beings tend to increase. Every being, which during its natural lifetime produces several eggs or seeds, must suffer destruction during some period of its life, and during some season or occasional year, otherwise, on the principle of geometrical increase, its numbers would quickly become so inordinately great that no country could support the product. Hence, as more individuals are produced than can possibly survive, there must in every case be a struggle for existence, either one individual with another of the same species, or with the individuals of distinct species, or with the physical conditions of life. It is the doctrine of Malthus applied with manifold force to the whole animal and vegetable kingdoms; for in this case there can be no artificial increase of food, and no prudential restraint from marriage. Although some species may be now increasing, more or less rapidly, in numbers, all cannot do so, for the world would not hold them.

There is no exception to the rule that every organic being naturally increases at so high a rate, that if not destroyed, the earth would soon be covered by the progeny of a single pair. Even slow-breeding man has doubled in twenty-five years, and at this rate, in a few thousand years, there would literally not be standing room for his progeny. Linnaeus has calculated that if an annual plant produced only two seeds—and there is no plant so unproductive as this—and their seedlings next year produced two, and so on, then in twenty years there would be a million plants. The elephant is reckoned to be the slowest breeder of all known animals, and I have taken some pains to estimate its probable minimum rate of natural increase: it will be under the mark to assume that it breeds when thirty years old, and goes on breeding till ninety years old, bringing forth three pairs of young in this interval; if this be so, at the end of the fifth century there would be alive fifteen million elephants, descended from the first pair.

Darwin's reading of Malthus's *Essay on Population* caused him to focus on the "superfecundity" of animals and plants. Even slowly breeding species like human beings and elephants display geometrical rates of increase over scant generations. See Darwin's initial reaction to Malthus in his notebook entry for September 28, 1838 (pp. 73–74) and the lines of the *Essay* that may most have impressed him on that occasion (p. 327).

But we have better evidence on this subject than mere theoretical calculations, namely, the numerous recorded cases of the astonishingly rapid increase of various animals in a state of nature, when circumstances have been favourable to them during two or three following seasons. Still more striking is the evidence from our domestic animals of many kinds which have run wild in several parts of the world: if the statements of the rate of increase of slow-breeding cattle and horses in South-America, and latterly in Australia, had not been well authenticated, they would have been quite incredible. So it is with plants: cases could be given of introduced plants which have become common throughout whole islands in a period of less than ten years. Several of the plants now most numerous over the wide plains of La Plata, clothing square leagues of surface almost to the exclusion of all other plants, have been introduced from Europe; and there are plants which now range in India, as I hear from Dr. Falconer, from Cape Comorin to the Himalaya, which have been imported from America since its discovery. In such cases, and endless instances could be given, no one supposes that the fertility of these animals or plants has been suddenly and temporarily increased in any sensible degree. The obvious explanation is that the conditions of life have been very favourable, and that there has consequently been less destruction of the old and young, and that nearly all the young have been enabled to breed. In such cases the geometrical ratio of increase, the result of which never fails to be surprising, simply explains the extraordinarily rapid increase and wide diffusion of naturalised productions in their new homes.

In a state of nature almost every plant produces seed, and amongst animals there are very few which do not annually pair. Hence we may confidently assert, that all plants and animals are tending to increase at a geometrical ratio, that all would most rapidly stock every station in which they could any how exist, and that the geometrical tendency to increase must be checked by destruction at some period of life. Our familiarity with the larger domestic animals tends, I think, to mislead us: we see no great destruction falling on them, and we forget that thousands are annually slaughtered for food, and that in a state of nature an equal number would have somehow to be disposed of.

The only difference between organisms which annually produce eggs or seeds by the thousand, and those which produce extremely few, is, that the slow-breeders would require a few more years to people, under favourable conditions, a whole district, let it be ever so large. The condor lays a couple of eggs and the ostrich a score, and yet in the same country the condor may be the more numerous of the two: the Fulmar petrel lays but one egg, yet it is believed to be the most numerous bird in the world. One fly deposits hundreds of eggs, and another, like the hippobosca, a single one; but this difference does not determine how many individuals of the two species can be supported in a district. A large number of eggs is of some importance to those species, which depend on a rapidly fluctuating amount of food, for it allows them rapidly to increase in number. But the real importance of a large number of eggs or seeds is to make up for much destruction at some period of life; and this period in the great majority of cases is an early one. If an animal can in any way protect its own eggs or young, a small number may be produced, and yet the average stock be fully kept up; but if many eggs or young are destroyed, many must be produced, or the species will become extinct. It would suffice to keep up the full number of a tree, which lived on an average for a thousand years, if a single seed were produced once in a thousand years, supposing that this seed were never destroyed, and could be ensured to germinate in a fitting place. So that in all cases, the average number of any animal or plant depends only indirectly on the number of its eggs or seeds.

In looking at Nature, it is most necessary to keep the foregoing considerations always in mind—never to forget that every single organic being around us may be said to be striving to the utmost to increase in numbers; that each lives by a struggle at some period of its life; that heavy destruction inevitably falls either on the young or old, during each generation or at recurrent intervals. Lighten any check, mitigate the destruction ever so little, and the number of the species will almost instantaneously increase to any amount. The face of Nature may be compared to a yielding surface, with ten thousand sharp wedges packed close together and driven inwards by incessant blows, sometimes one wedge being struck, and then another with greater force.

What checks the natural tendency of each species to increase in number is most obscure. Look at the most vigorous species; by as much as it swarms in numbers, by so much will its tendency to increase be still further increased. We know not exactly what the checks are in even one single instance. Nor will this surprise any one who reflects how ignorant we are on this head, even in regard to mankind, so incomparably better known than any other animal. This subject has been ably treated by several authors, and I shall, in my future work, discuss some of the checks at considerable length, more especially in regard to the feral animals of South America. Here I will make only a few remarks, just to recall to the reader's mind some of the chief points. Eggs or very young animals seem generally to suffer most, but this is not invariably the case. With plants there is a vast destruction of seeds, but, from some observations which I have made, I believe that it is the seedlings which suffer most from germinating in ground already thickly stocked with other plants. Seedlings, also, are destroyed in vast numbers by various enemies; for instance, on a piece of ground three feet long and two wide, dug and cleared, and where there could be no choking from other plants, I marked all the seedlings of our native weeds as they came up, and out of the 357 no less than 295 were destroyed, chiefly by slugs and insects. If turf which has long been mown, and the case would be the same with turf closely browsed by quadrupeds, be let to grow, the more vigorous plants gradually kill the less vigorous, though fully grown, plants: thus out of twenty species growing on a little plot of turf (three feet by four) nine species perished from the other species being allowed to grow up freely.

The amount of food for each species of course gives the extreme limit to which each can increase; but very frequently it is not the obtaining food, but the serving as prey to other animals, which determines the average numbers of a species. Thus, there seems to be little doubt that the stock of partridges, grouse, and hares on any large estate depends chiefly on the destruction of vermin. If not one head of game were shot during the next twenty years in England, and, at the same time, if no vermin were destroyed, there would, in all probability, be less game than at present, although hundreds of thousands of game animals are now annually killed. On the other hand, in some cases, as with the elephant and rhinoceros, none are destroyed by beasts of prey: even the tiger in India most rarely dares to attack a young elephant protected by its dam.

Climate plays an important part in determining the average numbers of a species, and periodical seasons of extreme cold or drought, I believe to be the most effective of all checks. I estimated that the winter of 1854–55 destroyed four-fifths of the birds in my own grounds; and this is a tremendous destruction, when we remember that ten per cent. is an extraordinarily severe mortality from epidemics with man. The action of climate seems at first sight to be quite independent of the struggle for existence; but in so far as climate chiefly acts in reducing food, it brings on the most severe struggle between the individuals, whether of the same or of distinct species, which subsist on the same kind of food. Even when climate, for instance extreme cold, acts directly, it will be the least vigorous, or those which have got least food through the advancing winter, which will suffer most. When we travel from south to north, or from a damp region to a dry, we invariably see some species gradually getting rarer and rarer, and finally disappearing; and the change of climate being conspicuous, we are tempted to attribute the whole effect to its direct action. But this is a very false view: we forget that each species, even where it most abounds, is constantly suffering enormous destruction at some period of its life, from enemies or from competitors for the same place and food; and if these enemies or competitors be in the least degree favoured by any slight change of climate, they will increase in numbers, and, as each area is already fully stocked with inhabitants, the other species will decrease. When we travel southward and see a species decreasing in numbers, we may feel sure that the cause lies quite as much in other species being favoured, as in this one being hurt. So it is when we travel northward, but in a somewhat lesser degree, for the number of species of all kinds, and therefore of competitors, decreases northwards; hence in going northward, or in ascending a mountain, we far oftener meet with stunted forms, due to the *directly* injurious action of climate, than we do in proceeding southwards or in descending a mountain. When we reach the Arctic regions, or snow-capped summits, or absolute deserts, the struggle for life is almost exclusively with the elements.

That climate acts in main part indirectly by favouring other species, we may clearly see in the prodigious number of plants in our gardens which can perfectly well endure our climate, but which never become naturalised, for they cannot compete with our native plants, nor resist destruction by our native animals.

When a species, owing to highly favourable circumstances, increases inordinately in numbers in a small tract, epidemics—at least, this seems generally to occur with our game animals—often ensue: and here we have a limiting check independent of the struggle for life. But even some of these so-called epidemics appear to be due to parasitic worms, which have from some cause, possibly in part through facility of diffusion amongst the crowded animals, been disproportionably favoured: and here comes in a sort of struggle between the parasite and its prey.

On the other hand, in many cases, a large stock of individuals of the same species, relatively to the numbers of its enemies, is absolutely necessary for its preservation. Thus we can easily raise plenty of corn and rape-seed, &c., in our fields, because the seeds are in great excess compared with the number of birds which feed on them; nor can the birds, though having a superabundance of food at this one season, increase in number proportionally to the supply of seed, as their numbers are checked during winter: but any one who has tried, knows how troublesome it is to get seed from a few wheat or other such plants in a garden; I have in this case lost every single seed. This view of the necessity of a large stock of the same species for its preservation, explains, I believe, some singular facts in nature, such as that of very rare plants being sometimes extremely abundant in the few spots where they do occur; and that of some social plants being social, that is, abounding in individuals, even on the extreme confines of their range. For in such cases, we may believe, that a plant could exist only where the conditions of its life were so favourable that many could exist together, and thus save each other from utter destruction. I should add that the good effects of frequent intercrossing, and the ill effects of close interbreeding, probably came into play in some of these cases; but on this intricate subject I will not here enlarge.

Many cases are on record showing how complex and unexpected are the checks and relations between organic beings, which have to struggle together in the same country. I will give only a single instance, which, though a simple one, has interested me. In Staffordshire, on the estate of a relation where I had ample means of investigation, there was a large and extremely barren heath, which had never been touched by the hand of man; but several hundred acres of exactly the same nature had been enclosed twenty-five years previously and planted with Scotch fir. The change in the native vegetation of the planted part of the heath was most remarkable, more than is generally seen in passing from one quite different soil to another: not only the proportional numbers of the heath-plants were wholly changed, but twelve species of plants (not counting grasses and carices) flourished in the plantations, which could not be found on the heath. The effect on the insects must have been still greater, for six insectivorous birds were very common in the plantations, which were not to be seen on the heath; and the heath was frequented by two or three distinct insectivorous birds. Here we see how potent has been the effect of the introduction of a single tree, nothing whatever else having been done, with the exception that the land had been enclosed, so that cattle could not enter. But how important an element enclosure is, I plainly saw near Farnham, in Surrey. Here there are extensive heaths, with a few clumps of old Scotch firs on the distant hill-tops: within the last ten years large spaces have been enclosed, and self-sown firs are now springing up in multitudes, so close together that all cannot live. When I ascertained that these young trees had not been sown or planted, I was so much surprised at their numbers that I went to several points of view, whence I could examine hundreds of acres of the unenclosed heath, and literally I could not see a single Scotch fir, except the old planted clumps. But on looking closely between the stems of the heath, I found a multitude of seedlings and little trees, which had been perpetually browsed down by the cattle. In one square yard, at a point some hundred yards distant from one of the old clumps, I counted thirty-two little trees; and one of them, judging from the rings of growth, had during twenty-six years tried to raise its head above the stems of the heath, and had failed. No wonder that, as soon as the land was enclosed, it became thickly clothed with vigorously growing young firs. Yet the heath was so extremely barren and so extensive that no one would ever have imagined that cattle would have so closely and effectually searched it for food.

Here we see that cattle absolutely determine the existence of the Scotch fir; but in several parts of the world insects determine the existence of cattle. Perhaps Paraguay offers the most curious instance of this; for here neither cattle nor horses nor dogs have ever run wild, though they swarm southward and northward in a feral state; and Azara and Rengger have shown that this is caused by the greater number in Paraguay of a certain fly, which lays its eggs in the navels of these animals when first born. The increase of these flies, numerous as they are, must be habitually checked by some means, probably by birds. Hence, if certain insectivorous birds (whose numbers are probably regulated by hawks or beasts of prey) were to increase in Paraguay, the flies would decrease — then cattle and horses would become feral, and this would certainly greatly alter (as indeed I have observed in parts of South

America) the vegetation: this again would largely affect the insects; and this, as we just have seen in Staffordshire, the insectivorous birds, and so onwards in ever-increasing circles of complexity. We began this series by insectivorous birds, and we have ended with them. Not that in nature the relations can ever be as simple as this. Battle within battle must ever be recurring with varying success; and yet in the long-run the forces are so nicely balanced, that the face of nature remains uniform for long periods of time, though assuredly the merest trifle would often give the victory to one organic being over another. Nevertheless so profound is our ignorance, and so high our presumption, that we marvel when we hear of the extinction of an organic being; and as we do not see the cause, we invoke cataclysms to desolate the world, or invent laws on the duration of the forms of life!

I am tempted to give one more instance showing how plants and animals, most remote in the scale of nature, are bound together by a web of complex relations. I shall hereafter have occasion to show that the exotic Lobelia fulgens, in this part of England, is never visited by insects, and consequently, from its peculiar structure, never can set a seed. Many of our orchidaceous plants absolutely require the visits of moths to remove their pollen-masses and thus to fertilise them. I have, also, reason to believe that humble-bees are indispensable to the fertilisation of the heartsease (Viola tricolor), for other bees do not visit this flower. From experiments which I have tried, I have found that the visits of bees, if not indispensable, are at least highly beneficial to the fertilisation of our clovers; but humble-bees alone visit the common red clover (Trifolium pratense), as other bees cannot reach the nectar. Hence I have very little doubt, that if the whole genus of humble-bees became extinct or very rare in England, the heartsease and red clover would become very rare, or wholly disappear. The number of humble-bees in any district depends in a great degree on the number of field-mice, which destroy their combs and nests; and Mr. H. Newman, who has long attended to the habits of humble-bees, believes that "more than two-thirds of them are thus destroyed all over England." Now the number of mice is largely dependent, as every one knows, on the number of cats; and Mr. Newman says, "Near villages and small towns I have found the nests of humble-bees more numerous than elsewhere, which I attribute to the number of cats that destroy the mice." Hence it is quite credible that the presence of a feline animal in large numbers in a district might determine, through the intervention first of mice and then of bees, the frequency of certain flowers in that district!

In the case of every species, many different checks, acting at different periods of life, and during different seasons or years, probably come into play; some one check or some few being generally the most potent, but all concurring in determining the average number or even the existence of the species. In some cases it can be shown that widely-different checks act on the same species in different districts. When we look at the plants and bushes clothing an entangled bank, we are tempted to attribute their proportional numbers and kinds to what we call chance. But how false a view is this! Every one has heard that when an American forest is cut down, a very different vegetation springs up; but it has been observed that the trees now growing on the ancient Indian mounds, in the Southern United States, display the same beautiful diversity and proportion of kinds as in the surrounding virgin forests. What a struggle between the several kinds of trees must here have gone on during long centuries, each annually scattering its seeds by the thousand; what war between insect and insect — between insects, snails, and other animals with birds and beasts of prey—all striving to increase, and all feeding on each other or on the trees or their seeds and seedlings, or on the other plants which first clothed the ground and thus checked the growth of the trees! Throw up a handful of feathers, and all must fall to the ground according to definite laws; but how simple is this problem compared to the action and reaction of the innumerable plants and animals which have determined, in the course of centuries, the proportional numbers and kinds of trees now growing on the old Indian ruins!

The dependency of one organic being on another, as of a parasite on its prey, lies generally between beings remote in the scale of nature. This is often the case with those which may strictly be said to struggle with each other for existence, as in the case of locusts and grass-feeding quadrupeds. But the struggle almost invariably will be most severe between the individuals of the same species, for they frequent the same districts, require the same food, and are exposed to the same dangers. In the case of varieties of the same species, the struggle will generally be almost equally severe, and we sometimes see the contest soon decided: for instance, if several varieties of wheat be sown together, and the mixed seed be resown, some of the varieties which best suit the soil or climate, or are naturally the most fertile, will beat the others and so yield more seed, and will consequently in a few years quite supplant the other varieties. To keep up a mixed stock of even such extremely close varieties as the variously coloured sweet-peas, they must be each year harvested separately, and the seed then mixed in due proportion, otherwise the weaker kinds will steadily decrease in numbers and disappear. So again with the varieties of sheep: it has been asserted that certain mountain-varieties will starve out other mountain-varieties, so that they cannot be kept together. The same result has followed from keeping together different varieties of the medicinal leech. It may even be doubted whether the varieties of any one of our domestic plants or animals have so exactly the same strength, habits, and constitution, that the original proportions of a mixed stock could be kept up for half a dozen generations, if they were allowed to struggle together, like beings in a state of nature, and if the seed or young were not annually sorted.

As species of the same genus have usually, though by no means invariably, some similarity in habits and constitution, and always in structure, the struggle will generally be more severe between species of the same genus, when they come into competition with each other, than between species of distinct genera. We see this in the recent extension over parts of the United States of one species of swallow having caused the decrease of another species. The recent increase of the missel-thrush in parts of Scotland has caused the decrease of the song-thrush. How frequently we hear of one species of rat taking the place of another species under the most different climates! In Russia the small Asiatic cockroach has everywhere driven before it its great congener. One species of charlock will supplant another, and so in other cases. We can dimly see why the competition should be most severe between allied forms, which fill nearly the same place in the economy of nature; but probably in no one case could we precisely say why one species has been victorious over another in the great battle of life.

A corollary of the highest importance may be deduced from the foregoing remarks, namely, that the structure of every organic being is related, in the most essential yet often hidden manner, to that of all other organic beings, with which it comes into competition for food or residence, or from which it has to escape, or on which it preys. This is obvious in the structure of the teeth and talons of the tiger; and in that of the legs and claws of the parasite which clings to the hair on the tiger's body. But in the beautifully plumed seed of the dandelion, and in the flattened and fringed legs of the water-beetle, the relation seems at first confined to the elements of air and water. Yet the advantage of plumed seeds no doubt stands in the closest relation to the land being already thickly clothed by other plants; so that the seeds may be widely distributed and fall on unoccupied ground. In the water-beetle, the structure of its legs, so well adapted for diving, allows it to compete with other aquatic insects, to hunt for its own prey, and to escape serving as prey to other animals.

The store of nutriment laid up within the seeds of many plants seems at first sight to have no sort of relation to other plants. But from the strong growth of young plants produced from such seeds (as peas and beans), when sown in the midst of long grass, I suspect that the chief use of the nutriment in the seed is to favour the growth of the young seedling, whilst struggling with other plants growing vigorously all around.

Look at a plant in the midst of its range, why does it not double or quadruple its numbers? We know that it can perfectly well withstand a little more heat or cold, dampness or dryness, for elsewhere it ranges into slightly hotter or colder, damper or drier districts. In this case we can clearly see that if we wished in imagination to give the plant the power of increasing in number, we should have to give it some advantage over its competitors, or over the animals which preyed on it. On the confines of its geographical range, a change of constitution with respect to climate would clearly be an advantage to our plant; but we have reason to believe that only a few plants or animals range so far, that they are destroyed by the rigour of the climate alone. Not until we reach the extreme confines of life, in the arctic regions or on the borders of an utter desert, will competition cease. The land may be extremely cold or dry, yet there will be competition between some few species, or between the individuals of the same species, for the warmest or dampest spots.

Hence, also, we can see that when a plant or animal is placed in a new country amongst new competitors, though the climate may be exactly the same as in its former home, yet the conditions of its life will generally be changed in an essential manner. If we wished to increase its average numbers in its new home, we should have to modify it in a different way to what we should have done in its native country; for we should have to give it some advantage over a different set of competitors or enemies.

It is good thus to try in our imagination to give any form some advantage over another. Probably in no single instance should we know what to do, so as to succeed. It will convince us of our ignorance on the mutual relations of all organic beings; a conviction as necessary, as it seems to be difficult to acquire. All that we can do, is to keep steadily in mind that each organic being is striving to increase at a geometrical ratio; that each at some period of its life, during some season of the year, during each generation or at intervals, has to struggle for life, and to suffer great destruction. When we reflect on this struggle, we may console ourselves with the full belief, that the war of nature is not incessant, that no fear is felt, that death is generally prompt, and that the vigorous, the healthy, and the happy survive and multiply.

NATURAL SELECTION

How will the struggle for existence, discussed too briefly in the last chapter, act in regard to variation? Can the principle of selection, which we have seen is so potent in the hands of man, apply in nature? I think we shall see that it can act most effectually. Let it be borne in mind in what an endless number of strange peculiarities our domestic productions, and, in a lesser degree, those under nature, vary; and how strong the hereditary tendency is. Under domestication, it may be truly said that the whole organisation becomes in some degree plastic. Let it be borne in mind how infinitely complex and close-fitting are the mutual relations of all organic beings to each other and to their physical conditions of life. Can it, then, be thought improbable, seeing that variations useful to man have undoubtedly occurred, that other variations useful in some way to each being in the great and complex battle of life, should sometimes occur in the course of thousands of generations? If such do occur, can we doubt (remembering that many more individuals are born than can possibly survive) that individuals having any advantage, however slight, over others, would have the best chance of surviving and of procreating their kind? On the other hand, we may feel sure that any variation in the least degree injurious would be rigidly destroyed. This preservation of favourable variations and the rejection of injurious variations, I call Natural Selection. Variations neither useful nor injurious would not be affected by natural selection, and would be left a fluctuating element, as perhaps we see in the species called polymorphic.

We shall best understand the probable course of natural selection by taking the case of a country undergoing some physical change, for instance, of climate. The proportional numbers of its inhabitants would almost immediately undergo a change, and some species might become extinct. We may conclude, from what we have seen of the intimate and complex manner in which the inhabitants of each country are bound together, that any change in the numerical proportions of some of the inhabitants, independently of the change of climate itself, would most seriously affect many of the others. If the country were open on its borders, new forms would certainly immigrate, and this also would seriously disturb the relations of some of the former inhabitants. Let it be remembered how powerful the influence of a single introduced tree or mammal has been shown to be. But in the case of an island, or of a country partly surrounded by barriers, into which new and better adapted forms could not freely enter, we should then have places in the economy of nature which would assuredly be better filled up, if some of the original inhabitants were in some manner modified; for, had the area been open to immigration, these same places would have been seized on by intruders. In such case, every slight modification, which in the course of ages chanced to arise, and which in any way favoured the individuals of any of the species, by better adapting them to their altered conditions, would tend to be preserved; and natural selection would thus have free scope for the work of improvement.

We have reason to believe, as stated in the first chapter, that a change in the conditions of life, by specially acting on the reproductive system, causes or increases variability; and in the foregoing case the conditions of life are supposed to have undergone a change, and this would manifestly be favourable to natural selection, by giving a better chance of profitable variations occurring; and unless profitable variations do occur, natural selection can do nothing. Not that, as I believe, any extreme amount of variability is necessary; as man can certainly produce great results by adding up in any given direction mere individual differences, so could Nature, but far more easily, from having incomparably longer time at her disposal. Nor do I believe that any great physical change, as of climate, or any unusual degree of isolation to check immigration, is actually necessary to produce new and unoccupied places for natural selection to fill up by modifying and improving some of the varying inhabitants. For as all the inhabitants of each country are struggling together with nicely balanced forces, extremely slight modifications in the structure or habits of one inhabitant would often give it an advantage over others; and still further modifications of the same kind would often still further increase the advantage. No country can be named in which all the native inhabitants are now so perfectly adapted to each other and to the physical conditions under which they live, that none of them could anyhow be improved; for in all countries, the natives have been so far conquered by naturalised productions, that they have allowed foreigners to take firm possession of the land. And as foreigners have thus everywhere beaten some of the natives, we may safely conclude that the natives might have been modified with advantage, so as to have better resisted such intruders.

As man can produce and certainly has produced a great result by his methodical and unconscious means of selection, what may not nature effect? Man can act only on external and visible characters: nature cares nothing for appearances, except in so far as they may be useful to any being. She can act on every internal organ, on every shade of constitutional difference, on the whole machinery of life. Man selects only for his own good; Nature only for that of the being which she tends. Every selected character is fully exercised by her; and the being is placed under well-suited conditions of life. Man keeps the natives of many climates in the same country; he seldom exercises each selected character in some peculiar and fitting manner; he feeds a long and a short beaked pigeon on the same food; he does not exercise a long-backed or long-legged quadruped in any peculiar manner; he exposes sheep with long and short wool to the same climate. He does not allow the most vigorous males to struggle for the females. He does not rigidly destroy all inferior animals, but protects during each varying season, as far as lies in his power, all his productions. He often begins his selection by some half-monstrous form; or at least by some modification prominent enough to catch his eye, or to be plainly useful to him. Under nature, the slightest difference of structure or constitution may well turn the nicely-balanced scale in the struggle for life, and so be preserved. How fleeting are the wishes and efforts of man! how short his time! and consequently how poor will his products be, compared with those accumulated by nature during whole geological periods. Can we wonder, then, that nature's productions should be far "truer" in character than man's productions; that they should be infinitely better adapted to the most complex conditions of life, and should plainly bear the stamp of far higher workmanship?

It may be said that natural selection is daily and hourly scrutinising, throughout the world, every variation, even the slightest; rejecting that which is bad, preserving and adding up all that is good; silently and insensibly working, whenever and wherever opportunity offers, at the improvement of each organic being in relation to its organic and inorganic conditions of life. We see nothing of these slow changes in progress, until the hand of time has marked the long lapse of ages, and then so imperfect is our view into long past geological ages, that we only see that the forms of life are now different from what they formerly were.

Although natural selection can act only through and for the good of each being, yet characters and structures, which we are apt to consider as of very trifling importance, may thus be acted on. When we see leaf-eating insects green, and bark-feeders mottled-grey; the alpine ptarmigan white in winter, the red-grouse the colour of heather, and the black-grouse that of peaty earth, we must believe that these tints are of service to these birds and insects in preserving them from danger. Grouse, if not destroyed at some period of their lives, would increase in countless numbers; they are known to suffer largely from birds of prey; and hawks are guided by eyesight to their prey,— so much so, that on parts of the Continent persons are warned not to keep white pigeons, as being the most liable to destruction. Hence I can see no reason to doubt that natural selection might be most effective in giving the proper colour to each kind of grouse, and in keeping that colour, when once acquired, true and constant. Nor ought we to think that the occasional destruction of an animal of any particular colour would produce little effect: we should remember how essential it is in a flock of white sheep to destroy every lamb with the faintest trace of black. In plants the down on the fruit and the colour of the flesh are considered by botanists as characters of the most trifling importance: yet we hear from an excellent horticulturist, Downing, that in the United States smooth-skinned fruits suffer far more from a beetle, a curculio, than those with down; that purple plums suffer far more from a certain disease than yellow plums; whereas another disease attacks yellow-fleshed peaches far more than those with other coloured flesh. If, with all the aids of art, these slight differences make a great difference in cultivating the several varieties, assuredly, in a state of nature, where the trees would have to struggle with other trees and with a host of enemies, such differences would effectually settle which variety, whether a smooth or downy, a yellow or purple fleshed fruit, should succeed.

In looking at many small points of difference between species, which, as far as our ignorance permits us to judge, seem to be quite unimportant, we must not forget that climate, food, &c., probably produce some slight and direct effect. It is, however, far more necessary to bear in mind that there are many unknown laws of correlation of growth, which, when one part of the organisation is modified through variation, and the modifications are accumulated by natural selection for the good of the being, will cause other modifications, often of the most unexpected nature.

As we see that those variations which under domestication appear at any particular period of life, tend to reappear in the offspring at the same period;— for instance, in the seeds of the many varieties of our culinary and agricultural plants; in the caterpillar and cocoon stages of the varieties of the silkworm; in the eggs of poultry, and in the colour of the down of their chickens; in the horns of our sheep and cattle when nearly adult; — so in a state of nature, natural selection will be enabled to act on and modify organic beings at any age, by the accumulation of profitable variations at that age, and by their inheritance at a corresponding age. If it profit a plant to have its seeds more and more widely disseminated by the wind, I can see no greater difficulty in this being effected through natural selection, than in the cotton-planter increasing and improving by selection the down in the pods on his cotton-trees. Natural selection may modify and adapt the larva of an insect to a score of contingencies, wholly different from those which concern the mature insect. These modifications will no doubt affect, through the laws of correlation, the structure of the adult; and probably in the case of those insects which live only for a few hours, and which never feed, a large part of their structure is merely the correlated result of successive changes in the structure of their larvae. So, conversely, modifications in the adult will probably often affect the structure of the larva; but in all cases natural selection will ensure that modifications consequent on other modifications at a different period of life, shall not be in the least degree injurious: for if they became so, they would cause the extinction of the species.

Natural selection will modify the structure of the young in relation to the parent, and of the parent in relation to the young. In social animals it will adapt the structure of each individual for the benefit of the community; if each in consequence profits by the selected change. What natural selection cannot do, is to modify the structure of one species, without giving it any advantage, for the good of another species; and though statements to this effect may be found in works of natural history, I cannot find one case which will bear investigation. A structure used only once in an animal's whole life, if of high importance to it, might be modified to any extent by natural selection; for instance, the great jaws possessed by certain insects, and used exclusively for opening the cocoon—or the hard tip to the beak of nestling birds, used for breaking the egg. It has been asserted, that of the best short-beaked tumbler-pigeons more perish in the egg than are able to get out of it; so that fanciers assist in the act of hatching. Now, if nature had to make the beak of a fullgrown pigeon very short for the bird's own advantage, the process of modification would be very slow, and there would be simultaneously the most rigorous selection of the young birds within the egg, which had the most powerful and hardest beaks, for all with weak beaks would inevitably perish: or, more delicate and more easily broken shells might be selected, the thickness of the shell being known to vary like every other structure.

Sexual Selection.—Inasmuch as peculiarities often appear under domestication in one sex and become hereditarily attached to that sex, the same fact probably occurs under nature, and if so, natural selection will be able to modify one sex in its functional relations to the other sex, or in relation to wholly different habits of life in the two sexes, as is sometimes the case with insects. And this leads me to say a few words on what I call Sexual Selection. This depends, not on a struggle for existence, but on a struggle between the males for possession of the females; the result is not death to the unsuccessful competitor, but few or no offspring. Sexual selection is, therefore, less rigorous than natural selection. Generally, the most vigorous males, those which are best fitted for their places in nature, will leave most progeny. But in many cases, victory will depend not on general vigour, but on having special weapons, confined to the male sex. A hornless stag or spurless cock would have a poor chance of leaving offspring. Sexual selection by always allowing the victor to breed might surely give indomitable courage, length to the spur, and strength to the wing to strike in the spurred leg, as well as the brutal cock-fighter, who knows well that he can improve his breed by careful selection of the best cocks. How low in the scale of nature this law of battle descends, I know not; male alligators have been described as fighting. bellowing, and whirling round, like Indians in a wardance, for the possession of the females; male salmons have been seen fighting all day long; male stag-beetles often bear wounds from the huge mandibles of other males. The war is, perhaps, severest between the males of polygamous animals, and these seem oftenest provided with special weapons. The males of carnivorous animals are already well armed; though to them and to others, special means of defence may be given through means of sexual selection, as the mane to the lion, the shoulder-pad to the boar, and the hooked jaw to the male salmon; for the shield may be as important for victory, as the sword or spear.

Amongst birds, the contest is often of a more peaceful character. All those who have attended to the subject, believe that there is the severest rivalry between the males of many species to attract by singing the females. The rock-thrush of Guiana, birds of Paradise, and some others, congregate; and successive males display their gorgeous plumage and perform strange antics before the females, which standing by as spectators, at last choose the most attractive partner. Those who have closely attended to birds in confinement well know that they often take individual preferences and dislikes: thus Sir R. Heron has described how one pied peacock was eminently attractive to all his hen birds. It may appear childish to attribute any effect to such apparently weak means: I cannot here enter on the details necessary to support this view; but if man can in a short time give elegant carriage and beauty to his bantams, according to his standard of beauty, I can see no good reason to doubt that female birds, by selecting, during thousands of generations, the most melodious or beautiful males, according to their standard of beauty, might produce a marked effect. I strongly suspect that some well-known laws with respect to the plumage of male and female birds, in comparison with the plumage of the young, can be explained on the view of plumage having been chiefly modified by sexual selection, acting when the birds have come to the breeding age or during the breeding season; the modifications thus produced being inherited at corresponding ages or seasons, either by the males alone, or by the males and females; but I have not space here to enter on this subject.

Thus it is, as I believe, that when the males and females of any animal have the same general habits of life, but differ in structure, colour, or ornament, such differences have been mainly caused by sexual selection; that is, individual males have had, in successive generations, some slight advantage over other males, in their weapons, means of defence, or charms; and have transmitted these advantages to their male offspring. Yet, I would not wish to attribute all such sexual differences to this agency: for we see peculiarities arising and becoming attached to the male sex in our domestic animals (as the wattle in male carriers, horn-like protuberances in the cocks of certain fowls, &c.), which we cannot believe to be either useful to the males in battle, or attractive to the females. We see analogous cases under nature, for instance, the tuft of hair on the breast of the turkey-cock, which can hardly be either useful or ornamental to this bird;—indeed, had the tuft appeared under domestication, it would have been called a monstrosity.

Illustrations of the action of Natural Selection.—In order to make it clear how, as I believe, natural selection acts, I must beg permission to give one or two imaginary illustrations. Let us take the case of a wolf, which preys on various animals, securing some by craft, some by strength, and some by fleetness; and let us suppose that the fleetest prey, a deer for instance, had from any change in the country increased in numbers, or that other prey had decreased in numbers, during that season of the year when the wolf is hardest pressed for food. I can under such circumstances see no reason to doubt that the swiftest and slimmest wolves would have the best chance of surviving, and so be preserved or selected,—provided always that they retained strength to master their prey at this or at some other period of the year, when they might be compelled to prey on other animals. I can see no more reason to doubt this, than that man can improve the fleetness of his greyhounds by careful and methodical selection, or by that unconscious selection which results from each man trying to keep the best dogs without any thought of modifying the breed.

Even without any change in the proportional numbers of the animals on which our wolf preyed, a cub might be born with an innate tendency to pursue certain kinds of prey. Nor can this be thought very improbable; for we often observe great differences in the natural tendencies of our domestic animals; one cat, for instance, taking to catch rats, another mice; one cat, according to Mr. St John, bringing home winged game, another hares or rabbits, and another hunting on marshy ground and almost nightly catching woodcocks or snipes. The tendency to catch rats rather than mice is known to be inherited. Now, if any slight innate change of habit or of structure benefited an individual wolf, it would have the best chance of surviving and of leaving offspring. Some of its young would probably inherit the same habits or structure, and by the repetition of this process, a new variety might be formed which would either supplant or coexist with the parent-form of wolf. Or, again, the wolves inhabiting a mountainous district, and those frequenting the lowlands, would naturally be forced to hunt different prey; and from the continued preservation of the individuals best fitted for the two sites, two varieties might slowly be formed. These varieties would cross and blend where they met; but to this subject of intercrossing we shall soon have to return. I may add, that, according to Mr. Pierce, there are two varieties of the wolf inhabiting the Catskill Mountains in the United States, one with a light greyhound-like form, which pursues deer, and the other more bulky, with shorter legs, which more frequently attacks the shepherd's flocks.

Let us now take a more complex case. Certain plants excrete a sweet juice, apparently for the sake of eliminating something injurious from their sap: this is effected by glands at the base of the stipules in some Leguminosae, and at the back of the leaf of the common laurel. This juice, though small in quantity, is greedily sought by insects. Let us now suppose a little sweet juice or nectar to be excreted by the inner bases of the petals of a flower. In this case insects in seeking the nectar would get dusted with pollen, and would certainly often transport the pollen from one flower to the stigma of another flower. The flowers of two distinct individuals of the same species would thus get crossed; and the act of crossing, we have good reason to believe (as will hereafter be more fully alluded to), would produce very vigorous seedlings, which consequently would have the best chance of flourishing and surviving. Some of these seedlings would probably inherit the nectar-excreting power. Those individual flowers which had the largest glands or nectaries, and which excreted most nectar, would be oftenest visited by insects, and would be oftenest crossed; and so in the long-run would gain the upper hand. Those flowers, also, which had their stamens and pistils placed, in relation to the size and habits of the particular insects which visited them, so as to favour in any degree the transportal of their pollen from flower to flower, would likewise be favoured or selected. We might have taken the case of insects visiting flowers for the sake of collecting pollen instead of nectar; and as pollen is formed for the sole object of fertilisation, its destruction appears a simple loss to the plant; yet if a little pollen were carried, at first occasionally and then habitually, by the pollen-devouring insects from flower to flower, and a cross thus effected, although nine-tenths of the pollen were destroyed, it might still be a great gain to the plant; and those individuals which produced more and more pollen, and had larger and larger anthers, would be selected.

When our plant, by this process of the continued preservation or natural selection of more and more attractive flowers, had been rendered highly attractive to insects, they would, unintentionally on their part, regularly carry pollen from flower to flower; and that they can most effectually do this, I could easily show by many striking instances. I will give only one—not as a very striking case, but as likewise illustrating one step in the separation of the sexes of plants, presently to be alluded to. Some holly-trees bear only male flowers, which have four stamens producing rather a small quantity of pollen, and a rudimentary pistil; other holly-trees bear only female flowers; these have a full-sized pistil, and four stamens with shrivelled anthers, in which not a grain of pollen can be detected. Having found a female tree exactly sixty yards from a male tree, I put the stigmas of twenty flowers, taken from different branches, under the microscope, and on all, without exception, there were pollen-grains, and on some a profusion of pollen. As the wind had set for several days from the female to the male tree, the pollen could not thus have been carried. The weather had been cold and boisterous, and therefore not favourable to bees, nevertheless every female flower which I examined had been effectually fertilised by the bees, accidentally dusted with pollen, having flown from tree to tree in search of nectar. But to return to our imaginary case: as soon as the plant had been rendered so highly attractive to insects that pollen was regularly carried from flower to flower, another process might commence. No naturalist doubts the advantage of what has been called the "physiological division of labour;" hence we may believe that it would be advantageous to a plant to produce stamens alone in one flower or on one whole plant, and pistils alone in another flower or on another plant. In plants under culture and placed under new conditions of life, sometimes the male organs and sometimes the female organs become more or less impotent; now if we suppose this to occur in ever so slight a degree under nature, then as pollen is already carried regularly from flower to flower, and as a more complete separation of the sexes of our plant would be advantageous on the principle of the division of labour, individuals with this tendency more and more increased, would be continually favoured or selected, until at last a complete separation of the sexes would be effected.

Let us now turn to the nectar-feeding insects in our imaginary case: we may suppose the plant of which we have been slowly increasing the nectar by continued selection, to be a common plant; and that certain insects depended in main part on its nectar for food. I could give many facts, showing how anxious bees are to save time; for instance, their habit of cutting holes and sucking the nectar at the bases of certain flowers, which they can, with a very little more trouble, enter by the mouth. Bearing such facts in mind, I can see no reason to doubt that an accidental deviation in the size and form of the body, or in the curvature and length of the proboscis, &c., far too slight to be appreciated by us, might profit a bee or other insect, so that an individual so characterised would be able to obtain its food more quicky, and so have a better chance of living and leaving descendants. Its descendants would probably inherit a tendency to a similar slight deviation of structure. The tubes of the corollas of the common red and incarnate clovers (Trifolium pratense and incarnatum) do not on a hasty glance appear to differ in length; yet the hive-bee can easily suck the nectar out of the incarnate clover, but not out of the common red clover, which is visited by humble-bees alone; so that whole fields of the red clover offer in vain an abundant supply of precious nectar to the hive-bee. Thus it might be a great advantage to the hive-bee to have a slightly longer or differently constructed proboscis. On the other hand, I have found by experiment that the fertility of clover greatly depends on bees visiting and moving parts of the corolla, so as to push the pollen on to the stigmatic surface. Hence, again, if humble-bees were to become rare in any country, it might be a great advantage to the red clover to have a shorter or more deeply divided tube to its corolla, so that the hive-bee could visit its flowers. Thus I can understand how a flower and a bee might slowly become, either simultaneously or one after the other, modified and adapted in the most perfect manner to each other, by the continued preservation of individuals presenting mutual and slightly favourable deviations of structure.

I am well aware that this doctrine of natural selection, exemplified in the above imaginary instances, is open to the same objections which were at first urged against Sir Charles Lyell's noble views on "the modern changes of the earth, as illustrative of geology;" but we now very seldom hear the action, for instance, of the coast-waves, called a trifling and insignificant cause, when applied to the excavation of gigantic valleys or to the formation of the longest lines of inland cliffs. Natural selection can act only by the preservation and accumulation of infinitesimally small inherited modifications, each profitable to the preserved being; and as modern geology has almost banished such views as the excavation of a great valley by a single diluvial wave, so will natural selection, if it be a true principle, banish the belief of the continued creation of new organic beings, or of any great and sudden modification in their structure.

On the Intercrossing of Individuals.—I must here introduce a short digression. In the case of animals and plants with separated sexes, it is of course obvious that two individuals must always unite for each birth; but in the case of hermaphrodites this is far from obvious. Nevertheless I am strongly inclined to believe that with all hermaphrodites two individuals, either occasionally or habitually, concur for the reproduction of their kind. This view, I may add, was first suggested by Andrew Knight. We shall presently see its importance; but I must here treat the subject with extreme brevity, though I have the materials prepared for an ample discussion. All vertebrate animals, all insects, and some other large groups of animals, pair for each birth. Modern research has much diminished the number of supposed hermaphrodites, and of real hermaphrodites a large number pair; that is, two individuals regularly unite for reproduction, which is all that concerns us. But still there are many hermaphrodite animals which certainiy do not habitually pair, and a vast majority of plants are hermaphrodites. What reason, it may be asked, is there for supposing in these cases that two individuals ever concur in reproduction? As it is impossible here to enter on details, I must trust to some general considerations alone.

In the first place, I have collected so large a body of facts, showing, in accordance with the almost universal belief of breeders, that with animals and plants a cross between different varieties, or between individuals of the same variety but of another strain, gives vigour and fertility to the offspring; and on the other hand, that *close* interbreeding diminishes vigour and fertility; that these facts alone incline me to believe that it is a general law of nature (utterly ignorant though we be of the meaning of the law) that no organic being self-fertilises itself for an eternity of generations; but that a cross with another individual is occasionally—perhaps at very long intervals— indispensable.

On the belief that this is a law of nature, we can, I think, understand several large classes of facts, such as the following, which on any other view are inexplicable. Every hybridizer knows how unfavourable exposure to wet is to the fertilisation of a flower, yet what a multitude of flowers have their anthers and stigmas fully exposed to the weather! but if an occasional cross be indispensable, the fullest freedom for the entrance of pollen from another individual will explain this state of exposure, more especially as the plant's own anthers and pistil generally stand so close together that self-fertilisation seems almost inevitable. Many flowers, on the other hand, have their organs of fructification closely enclosed, as in the great papilionaceous or pea-family; but in several, perhaps in all, such flowers, there is a very curious adaptation between the structure of the flower and the manner in which bees suck the nectar; for, in doing this, they either push the flower's own pollen on the stigma, or bring pollen from another flower. So necessary are the visits of bees to papilionaceous flowers, that I have found, by experiments published elsewhere, that their fertility is greatly diminished if these visits be prevented. Now, it is scarcely possible that bees should fly from flower to flower, and not carry pollen from one to the other, to the great good, as I believe, of the plant. Bees will act like a camel-hair pencil, and it is quite sufficient just to touch the anthers of one flower and then the stigma of another with the same brush to ensure fertilisation; but it must not be supposed that bees would thus produce a multitude of hybrids between distinct species; for if you bring on the same brush a plant's own pollen and pollen from another species, the former will have such a prepotent effect, that it will invariably and completely destroy, as has been shown by Gärtner, any influence from the foreign pollen.

When the stamens of a flower suddenly spring towards the pistil, or slowly move one after the other towards it, the contrivance seems adapted solely to ensure self-fertilisation; and no doubt it is useful for this end: but, the agency of insects is often required to cause the stamens to spring forward, as Köl-reuter has shown to be the case with the barberry; and curiously in this very genus, which seems to have a special contrivance for self-fertilisation, it is well known that if very closely-allied forms or varieties are planted near each other, it is hardly possible to raise pure seedlings, so largely do they natu-rally cross. In many other cases, far from there being any aids for self-fertilisation, there are special contrivances, as I could show from the writings of C. C. Sprengel and from my own observations, which effectually prevent the stigma receiving pollen from its own flower: for instance, in Lo-belia fulgens, there is a really beautiful and elaborate contrivance by which every one of the infinitely numerous pollen-granules are swept out of the conjoined anthers of each flower, before the stigma of that individual flower is ready to receive them; and as this flower is never visited, at least in my garden, by insects, it never sets a seed, though by placing pollen from one flower on the stigma of another, I raised plenty of seedlings; and whilst another species of Lobelia growing close by, which is visited by bees, seeds freely. In very many other cases, though there be no special mechanical contrivance to prevent the stigma of a flower receiving its own pollen, yet, as C. C. Sprengel has shown, and as I can confirm, either the anthers burst before the stigma is ready for fertilisation, or the stigma is ready before the pollen of that flower is ready, so that these plants have in fact separated sexes, and must habitually be crossed. How strange are these facts! How strange that the pollen and stigmatic surface of the same flower, though placed so close together, as if for the very purpose of self-fertilisation, should in so many cases be mutually useless to each other! How simply are these facts explained on the view of an occasional cross with a distinct individual being advantageous or indispensable!

If several varieties of the cabbage, radish, onion, and of some other plants, be allowed to seed near each other, a large majority, as I have found, of the seedlings thus raised will turn out mongrels: for instance, I raised 233 seedling cabbages from some plants of different varieties growing near each other, and of these only 78 were true to their kind, and some even of these were not perfectly true. Yet the pistil of each cabbage-flower is surrounded not only by its own six stamens, but by those of the many other flowers on the same plant. How, then, comes it that such a vast number of the seedlings are mongrelized? I suspect that it must arise from the pollen of a distinct *variety* having a prepotent effect over a flower's own pollen; and that this is part of the general law of good being derived from the intercrossing of distinct individuals of the same species. When distinct *species* are crossed the case is directly the reverse, for a plant's own pollen is always prepotent over foreign pollen; but to this subject we shall return in a future chapter.

In the case of a gigantic tree covered with innumerable flowers, it may be objected that pollen could seldom be carried from tree to tree, and at most only from flower to flower on the same tree, and that flowers on the same tree can be considered as distinct individuals only in a limited sense. I believe this objection to be valid, but that nature has largely provided against it by giving to trees a strong tendency to bear flowers with separated sexes. When the sexes are separated, although the male and female flowers may be produced on the same tree, we can see that pollen must be regularly carried from flower to flower; and this will give a better chance of pollen being occasionally carried from tree to tree. That trees belonging to all Orders have their sexes more often separated than other plants, I find to be the case in this country; and at my request Dr. Hooker tabulated the trees of New Zealand, and Dr. Asa Gray those of the United States, and the result was as I anticipated. On the other hand, Dr. Hooker has recently informed me that he finds that the rule does not hold in Australia; and I have made these few remarks on the sexes of trees simply to call attention to the subject.

Turning for a very brief space to animals: on the land there are some hermaphrodites, as land-mollusca and earth-worms; but these all pair. As yet I have not found a single case of a terrestrial animal which fertilises itself. We can understand this remarkable fact, which offers so strong a con-

trast with terrestrial plants, on the view of an occasional cross being indispensable, by considering the medium in which terrestrial animals live, and
the nature of the fertilising element; for we know of no means, analogous to
the action of insects and of the wind in the case of plants, by which an occasional cross could be effected with terrestrial animals without the concurrence of two individuals. Of aquatic animals, there are many self-fertilising
hermaphrodites; but here currents in the water offer an obvious means for
an occasional cross. And, as in the case of flowers, I have as yet failed, after
consultation with one of the highest authorities, namely, Professor Huxley,
to discover a single case of an hermaphrodite animal with the organs of reproduction so perfectly enclosed within the body, that access from without
and the occasional influence of a distinct individual can be shown to be physically impossible. Cirripedes long appeared to me to present a case of very
great difficulty under this point of view; but I have been enabled, by a fortunate chance, elsewhere to prove that two individuals, though both are self-
fertilising hermaphrodites, do sometimes cross.

Darwin's understanding of the importance of sexual crossing to the
general validity of evolution convinced him to study intercrossing in
hermaphroditic species. See E:50 (p. 75), his discussions of "complemental" males in cirripedes (pp. 119–26) and of heterostyly in *Lythrum salicaria*, (pp. 295–99).

It must have struck most naturalists as a strange anomaly that, in the case
of both animals and plants, species of the same family and even of the same
genus, though agreeing closely with each other in almost their whole organisation, yet are not rarely, some of them hermaphrodites, and some of
them unisexual. But if, in fact, all hermaphrodites do occasionally intercross
with other individuals, the difference between hermaphrodites and unisexual species, as far as function is concerned, becomes very small.

From these several considerations and from the many special facts which
I have collected, but which I am not here able to give, I am strongly inclined
to suspect that, both in the vegetable and animal kingdoms, an occasional
intercross with a distinct individual is a law of nature. I am well aware that
there are, on this view, many cases of difficulty, some of which I am trying
to investigate. Finally then, we may conclude that in many organic beings,
a cross between two individuals is an obvious necessity for each birth; in
many others it occurs perhaps only at long intervals; but in none, as I suspect, can self-fertilisation go on for perpetuity.

Circumstances favourable to Natural Selection.—This is an extremely intricate subject. A large amount of inheritable and diversified variability is favourable, but I believe mere individual differences suffice for the work. A large number of individuals, by giving a better chance for the appearance within any given period of profitable variations, will compensate for a lesser amount of variability in each individual, and is, I believe, an extremely important element of success. Though nature grants vast periods of time for the work of natural selection, she does not grant an indefinite period; for as all organic beings are striving, it may be said, to seize on each place in the economy of nature, if any one species does not become modified and improved in a corresponding degree with its competitors, it will soon be exterminated.

In man's methodical selection, a breeder selects for some definite object, and free intercrossing will wholly stop his work. But when many men, without intending to alter the breed, have a nearly common standard of perfection, and all try to get and breed from the best animals, much improvement and modification surely but slowly follow from this unconscious process of selection, notwithstanding a large amount of crossing with inferior animals. Thus it will be in nature; for within a confined area, with some place in its polity not so perfectly occupied as might be, natural selection will always tend to preserve all the individuals varying in the right direction, though in different degrees, so as better to fill up the unoccupied place. But if the area be large, its several districts will almost certainly present different conditions of life; and then if natural selection be modifying and improving a species in the several districts, there will be intercrossing with the other individuals of the same species on the confines of each. And in this case the effects of intercrossing can hardly be counterbalanced by natural selection always tending to modify all the individuals in each district in exactly the same manner to the conditions of each; for in a continuous area, the conditions will generally graduate away insensibly from one district to another. The intercrossing will most affect those animals which unite for each birth, which wander much, and which do not breed at a very quick rate. Hence in animals of this nature, for instance in birds, varieties will generally be confined to separated countries; and this I believe to be the case. In hermaphrodite organisms which cross only occasionally, and likewise in animals which unite for each birth, but which wander little and which can increase at a very rapid rate, a new and improved variety might be quickly formed on any one spot, and might there maintain itself in a body, so that whatever intercrossing took place would be chiefly between the individuals of the same new variety. A local variety when once thus formed might subsequently slowly spread to other districts. On the above principle, nurserymen always prefer getting seed from a large body of plants of the same variety, as the chance of intercrossing with other varieties is thus lessened.

Even in the case of slow-breeding animals, which unite for each birth, we must not overrate the effects of intercrosses in retarding natural selection; for I can bring a considerable catalogue of facts, showing that within the same area, varieties of the same animal can long remain distinct, from haunting different stations, from breeding at slightly different seasons, or from varieties of the same kind preferring to pair together.

Intercrossing plays a very important part in nature in keeping the individuals of the same species, or of the same variety, true and uniform in character. It will obviously thus act far more efficiently with those animals which unite for each birth; but I have already attempted to show that we have reason to believe that occasional intercrosses take place with all animals and with all plants. Even if these take place only at long intervals, I am convinced that the young thus produced will gain so much in vigour and fertility over the offspring from long-continued self-fertilisation, that they will have a better chance of surviving and propagating their kind; and thus, in the long run, the influence of intercrosses, even at rare intervals, will be great. If there exist organic beings which never intercross, uniformity of character can be retained amongst them, as long as their conditions of life remain the same, only through the principle of inheritance, and through natural selection destroying any which depart from the proper type; but if their conditions of life change and they undergo modification, uniformity of character can be given to their modified offspring, solely by natural selection preserving the same favourable variations.

Isolation, also, is an important element in the process of natural selection. In a confined or isolated area, if not very large, the organic and inorganic conditions of life will generally be in a great degree uniform; so that natural selection will tend to modify all the individuals of a varying species throughout the area in the same manner in relation to the same conditions. Intercrosses, also, with the individuals of the same species, which otherwise would have inhabited the surrounding and differently circumstanced districts, will be prevented. But isolation probably acts more efficiently in checking the immigration of better adapted organisms, after any physical change, such as of climate or elevation of the land, &c.; and thus new places in the natural economy of the country are left open for the old inhabitants to struggle for, and become adapted to, through modifications in their structure and constitution. Lastly, isolation, by checking immigration and consequently competition, will give time for any new variety to be slowly improved; and this may sometimes be of importance in the production of new species. If, however, an isolated area be very small, either from being surrounded by barriers, or from having very peculiar physical conditions, the total number of the individuals supported on it will necessarily be very small; and fewness of individuals will greatly retard the production of new species through natural selection, by decreasing the chance of the appearance of favourable variations.

If we turn to nature to test the truth of these remarks, and look at any small isolated area, such as an oceanic island, although the total number of the species inhabiting it, will be found to be small, as we shall see in our chapter on geographical distribution; yet of these species a very large proportion are endemic,—that is, have been produced there, and nowhere else. Hence an oceanic island at first sight seems to have been highly favourable for the production of new species. But we may thus greatly deceive ourselves, for to ascertain whether a small isolated area, or a large open area like a continent, has been most favourable for the production of new organic forms, we ought to make the comparison within equal times; and this we are incapable of doing.

Although I do not doubt that isolation is of considerable importance in the production of new species, on the whole I am inclined to believe that largeness of area is of more importance, more especially in the production of species, which will prove capable of enduring for a long period, and of spreading widely. Throughout a great and open area, not only will there be a better chance of favourable variations arising from the large number of individuals of the same species there supported, but the conditions of life are infinitely complex from the large number of already existing species; and if some of these many species become modified and improved, others will have to be improved in a corresponding degree or they will be exterminated. Each new form, also, as soon as it has been much improved, will be able to spread over the open and continuous area, and will thus come into competition with many others. Hence more new places will be formed, and the competition to fill them will be more severe, on a large than on a small and isolated area. Moreover, great areas, though now continuous, owing to oscillations of level, will often have recently existed in a broken condition, so that the good effects of isolation will generally, to a certain extent, have concurred. Finally, I conclude that, although small isolated areas probably have been in some respects highly favourable for the production of new species, yet that the course of modification will generally have been more rapid on large areas; and what is more important, that the new forms produced on large areas, which already have been victorious over many competitors, will be those that will spread most widely, will give rise to most new varieties and species, and will thus play an important part in the changing history of the organic world.

We can, perhaps, on these views, understand some facts which will be again alluded to in our chapter on geographical distribution; for instance, that the productions of the smaller continent of Australia have formerly yielded, and apparently are now yielding, before those of the larger Europæo-Asiatic area. Thus, also, it is that continental productions have everywhere become so largely naturalised on islands. On a small island, the race for life will have been less severe, and there will have been less modification and less extermination. Hence, perhaps, it comes that the flora of Madeira, according to Oswald Heer, resembles the extinct tertiary flora of Europe. All fresh-water basins, taken together, make a small area compared with that of the sea or of the land; and, consequently, the competition between fresh-water productions will have been less severe than elsewhere; new forms will have been more slowly formed, and old forms more slowly exterminated. And it is in fresh water that we find seven genera of Ganoid fishes, remnants of a once preponderant order: and in fresh water we find some of the most anomalous forms now known in the world, as the Ornithorhynchus and Lepidosiren, which, like fossils, connect to a certain extent orders now widely separated in the natural scale. These anomalous forms may almost be called living fossils; they have endured to the present day, from having inhabited a confined area, and from having thus been exposed to less severe competition.

To sum up the circumstances favourable and unfavourable to natural selection, as far as the extreme intricacy of the subject permits. I conclude, looking to the future, that for terrestrial productions a large continental area, which will probably undergo many oscillations of level, and which consequently will exist for long periods in a broken condition, will be the most favourable for the production of many new forms of life, likely to endure long and to spread widely. For the area will first have existed as a continent, and the inhabitants, at this period numerous in individuals and kinds, will have been subjected to very severe competition. When converted by subsidence into large separate islands, there will still exist many individuals of the same species on each island: intercrossing on the confines of the range of each species will thus be checked: after physical changes of any kind, immigration will be prevented, so that new places in the polity of each island will have to be filled up by modifications of the old inhabitants; and time will be allowed for the varieties in each to become well modified and perfected. When, by renewed elevation, the islands shall be re-converted into a continental area, there will again be severe competition: the most favoured or improved varieties will be enabled to spread: there will be much extinction of the less improved forms, and the relative proportional numbers of the various inhabitants of the renewed continent will again be changed; and again there will be a fair field for natural selection to improve still further the inhabitants, and thus produce new species.

That natural selection will always act with extreme slowness, I fully admit. Its action depends on there being places in the polity of nature, which can be better occupied by some of the inhabitants of the country undergoing modification of some kind. The existence of such places will often depend on physical changes, which are generally very slow, and on the immigration of better adapted forms having been checked. But the action of natural selection will probably still oftener depend on some of the inhabitants becoming slowly modified; the mutual relations of many of the other inhabitants being thus disturbed. Nothing can be effected, unless favourable variations occur, and variation itself is apparently always a very slow process. The process will often be greatly retarded by free intercrossing. Many will exclaim that these several causes are amply sufficient wholly to stop the action of natural selection. I do not believe so. On the other hand, I do believe that natural selection will always act very slowly, often only at long intervals of time, and generally on only a very few of the inhabitants of the same region at the same time. I further believe, that this very slow, intermittent action of natural selection accords perfectly well with what geology tells us of the rate and manner at which the inhabitants of this world have changed.

Slow though the process of selection may be, if feeble man can do much by his powers of artificial selection, I can see no limit to the amount of change, to the beauty and infinite complexity of the coadaptations between all organic beings, one with another and with their physical conditions of life, which may be effected in the long course of time by nature's power of selection.

Extinction.—This subject will be more fully discussed in our chapter on Geology; but it must be here alluded to from being intimately connected with natural selection. Natural selection acts solely through the preservation of variations in some way advantageous, which consequently endure. But as from the high geometrical powers of increase of all organic beings, each area is already fully stocked with inhabitants, it follows that as each selected and favoured form increases in number, so will the less favoured forms decrease and become rare. Rarity, as geology tells us, is the precursor to extinction. We can, also, see that any form represented by few individuals will, during fluctuations in the seasons or in the number of its enemies, run a good chance of utter extinction. But we may go further than this; for as new forms are continually and slowly being produced, unless we believe that the number of specific forms goes on perpetually and almost indefinitely increasing, numbers inevitably must become extinct. That the number of specific forms has not indefinitely increased, geology shows us plainly; and indeed we can see reason why they should not have thus increased, for the number of places in the polity of nature is not indefinitely great,—not that we have any means of knowing that any one region has as yet got its maximum of species. Probably no region is as yet fully stocked, for at the Cape of Good Hope, where more species of plants are crowded together than in any other quarter of the world, some foreign plants have become naturalised, without causing, as far as we know, the extinction of any natives.

Furthermore, the species which are most numerous in individuals will have the best chance of producing within any given period favourable variations. We have evidence of this, in the facts given in the second chapter, showing that it is the common species which afford the greatest number of recorded varieties, or incipient species. Hence, rare species will be less quickly modified or improved within any given period, and they will consequently be beaten in the race for life by the modified descendants of the commoner species.

From these several considerations I think it inevitably follows, that as new species in the course of time are formed through natural selection, others will become rarer and rarer, and finally extinct. The forms which stand in closest competition with those undergoing modification and improvement, will naturally suffer most. And we have seen in the chapter on the Struggle for Existence that it is the most closely-allied forms,—varieties of the same species, and species of the same genus or of related genera,—which, from having nearly the same structure, constitution, and habits, generally come into the severest competition with each other. Consequently, each new variety or species, during the progress of its formation, will generally press hardest on its nearest kindred, and tend to exterminate them. We see the same process of extermination amongst our domesticated productions, through the selection of improved forms by man. Many curious instances could be given showing how quickly new breeds of cattle, sheep, and other animals, and varieties of flowers, take the place of older and inferior kinds. In Yorkshire, it is historically known that the ancient black cattle were displaced by the long-horns, and that these "were swept away by the short-horns" (I quote the words of an agricultural writer) "as if by some murderous pestilence."

Divergence of Character.—The principle, which I have designated by this term, is of high importance on my theory, and explains, as I believe, several important facts. In the first place, varieties, even strongly-marked ones, though having somewhat of the character of species—as is shown by the hopeless doubts in many cases how to rank them—yet certainly differ from each other far less than do good and distinct species. Nevertheless, according to my view, varieties are species in the process of formation, or are, as I have called them, incipient species. How, then, does the lesser difference between varieties become augmented into the greater difference between species? That this does habitually happen, we must infer from most of the innumerable species throughout nature presenting well-marked differences; whereas varieties, the supposed prototypes and parents of future well-marked species, present slight and ill-defined differences. Mere chance, as we may call it, might cause one variety to differ in some character from its parents, and the offspring of this variety again to differ from its parent in the very same character and in a greater degree; but this alone would never account for so habitual and large an amount of difference as that between varieties of the same species and species of the same genus.

As has always been my practice, let us seek light on this head from our domestic productions. We shall here find something analogous. A fancier is struck by a pigeon having a slightly shorter beak; another fancier is struck by a pigeon having a rather longer beak; and on the acknowledged principle that "fanciers do not and will not admire a medium standard, but like extremes," they both go on (as has actually occurred with tumbler-pigeons) choosing and breeding from birds with longer and longer beaks, or with shorter and shorter beaks. Again, we may suppose that at an early period one man preferred swifter horses; another stronger and more bulky horses. The early differences would be very slight; in the course of time, from the continued selection of swifter horses by some breeders, and of stronger ones by others, the differences would become greater, and would be noted as forming two sub-breeds; finally, after the lapse of centuries, the sub-breeds would become converted into two well-established and distinct breeds. As the differences slowly become greater, the inferior animals with intermediate characters, being neither very swift nor very strong, will have been neglected, and will have tended to disappear. Here, then, we see in man's productions the action of what may be called the principle of divergence, causing differences, at first barely appreciable, steadily to increase, and the breeds to diverge in character both from each other and from their common parent.

But how, it may be asked, can any analogous principle apply in nature? I believe it can and does apply most efficiently, from the simple circumstance that the more diversified the descendants from any one species become in structure, constitution, and habits, by so much will they be better enabled to seize on many and widely diversified places in the polity of nature, and so be enabled to increase in numbers.

We can clearly see this in the case of animals with simple habits. Take the case of a carnivorous quadruped, of which the number that can be supported in any country has long ago arrived at its full average. If its natural powers of increase be allowed to act, it can succeed in increasing (the country not undergoing any change in its conditions) only by its varying descendants seizing on places at present occupied by other animals: some of them, for instance, being enabled to feed on new kinds of prey, either dead or alive; some inhabiting new stations, climbing trees, frequenting water, and some perhaps becoming less carnivorous. The more diversified in habits and structure the descendants of our carnivorous animal became, the more places they would be enabled to occupy. What applies to one animal will apply through-out all time to all animals—that is, if they vary—for otherwise natural se-lection can do nothing. So it will be with plants. It has been experimentally proved, that if a plot of ground be sown with one species of grass, and a similar plot be sown with several distinct genera of grasses, a greater num-ber of plants and a greater weight of dry herbage can thus be raised. The same has been found to hold good when first one variety and then several mixed varieties of wheat have been sown on equal spaces of ground. Hence, if any one species of grass were to go on varying, and those varieties were continually selected which differed from each other in at all the same man-ner as distinct species and genera of grasses differ from each other, a greater number of individual plants of this species of grass, including its modified descendants, would succeed in living on the same piece of ground. And we well know that each species and each variety of grass is annually sowing al-most countless seeds; and thus, as it may be said, is striving its utmost to increase its numbers. Consequently, I cannot doubt that in the course of many thousands of generations, the most distinct varieties of any one species of grass would always have the best chance of succeeding and of increasing in numbers, and thus of supplanting the less distinct varieties; and varieties, when rendered very distinct from each other, take the rank of species.

The truth of the principle, that the greatest amount of life can be sup-ported by great diversification of structure, is seen under many natural cir-cumstances. In an extremely small area, especially if freely open to immi-gration, and where the contest between individual and individual must be severe, we always find great diversity in its inhabitants. For instance, I found that a piece of turf, three feet by four in size, which had been exposed for many years to exactly the same conditions, supported twenty species of plants, and these belonged to eighteen genera and to eight orders, which shows how much these plants differed from each other. So it is with the plants and insects on small and uniform islets; and so in small ponds of fresh water. Farmers find that they can raise most food by a rotation of plants belonging to the most different orders: nature follows what may be called a simulta-neous rotation. Most of the animals and plants which live close round any small piece of ground, could live on it (supposing it not to be in any way peculiar in its nature), and may be said to be striving to the utmost to live there; but, it is seen, that where they come into the closest competition with each other, the advantages of diversification of structure, with the accom-panying differences of habit and constitution, determine that the inhabit-ants, which thus jostle each other most closely, shall, as a general rule, be-long to what we call different genera and orders.

The same principle is seen in the naturalisation of plants through man's agency in foreign lands. It might have been expected that the plants which have succeeded in becoming naturalised in any land would generally have been closely allied to the indigenes; for these are commonly looked at as specially created and adapted for their own country. It might, also, perhaps have been expected that naturalised plants would have belonged to a few groups more especially adapted to certain stations in their new homes. But the case is very different; and Alph. De Candolle has well remarked in his great and admirable work, that floras gain by naturalisation, proportionally with the number of the native genera and species, far more in new genera than in new species. To give a single instance: in the last edition of Dr. Asa Gray's 'Manual of the Flora of the Northern United States,' 260 naturalised plants are enumerated, and these belong to 162 genera. We thus see that these naturalised plants are of a highly diversified nature. They differ, moreover, to a large extent from the indigenes, for out of the 162 genera, no less than 100 genera are not there indigenous, and thus a large proportional addition is made to the genera of these States.

By considering the nature of the plants or animals which have struggled successfully with the indigenes of any country, and have there become naturalised, we can gain some crude idea in what manner some of the natives would have had to be modified, in order to have gained an advantage over the other natives; and we may, I think, at least safely infer that diversification of structure, amounting to new generic differences, would have been profitable to them.

The advantage of diversification in the inhabitants of the same region is, in fact, the same as that of the physiological division of labour in the organs of the same individual body—a subject so well elucidated by Milne Edwards. No physiologist doubts that a stomach by being adapted to digest vegetable matter alone, or flesh alone, draws most nutriment from these substances. So in the general economy of any land, the more widely and perfectly the animals and plants are diversified for different habits of life, so will a greater number of individuals be capable of there supporting themselves. A set of animals, with their organisation but little diversified, could hardly compete with a set more perfectly diversified in structure. It may be doubted, for instance, whether the Australian marsupials, which are divided into groups differing but little from each other, and feebly representing, as Mr. Waterhouse and others have remarked, our carnivorous, ruminant, and rodent mammals, could successfully compete with these well-pronounced orders. In the Australian mammals, we see the process of diversification in an early and incomplete stage of development.

After the foregoing discussion, which ought to have been much amplified, we may, I think, assume that the modified descendants of any one species will succeed by so much the better as they become more diversified in structure, and are thus enabled to encroach on places occupied by other beings. Now let us see how this principle of great benefit being derived from divergence of character, combined with the principles of natural selection and of extinction, will tend to act.

The accompanying diagram will aid us in understanding this rather perplexing subject. Let A to L represent the species of a genus large in its own country; these species are supposed to resemble each other in unequal degrees, as is so generally the case in nature, and as is represented in the diagram by the letters standing at unequal distances. I have said a large genus, because we have seen in the second chapter, that on an average more of the species of large genera vary than of small genera; and the varying species of the large genera present a greater number of varieties. We have, also, seen that the species, which are the commonest and the most widely-diffused, vary more than rare species with restricted ranges. Let (A) be a common, widely-diffused, and varying species, belonging to a genus large in its own country. The little fan of diverging dotted lines of unequal lengths proceeding from (A), may represent its varying offspring. The variations are supposed to be extremely slight, but of the most diversified nature; they are not supposed all to appear simultaneously, but often after long intervals of time; nor are they all supposed to endure for equal periods. Only those variations which are in some way profitable will be preserved or naturally selected. And here the importance of the principle of benefit being derived from divergence of character comes in; for this will generally lead to the most different or divergent variations (represented by the outer dotted lines) being preserved and accumulated by natural selection. When a dotted line reaches one of the horizontal lines, and is there marked by a small numbered letter, a sufficient amount of variation is supposed to have been accumulated to have formed a fairly well-marked variety, such as would be thought worthy of record in a systematic work.

The intervals between the horizontal lines in the diagram, may represent each a thousand generations; but it would have been better if each had represented ten thousand generations. After a thousand generations, species (A) is supposed to have produced two fairly well-marked varieties, namely a^1 and m^1. These two varieties will generally continue to be exposed to the same conditions which made their parents variable, and the tendency to variability is in itself hereditary, consequently they will tend to vary, and generally to vary in nearly the same manner as their parents varied. Moreover, these two varieties, being only slightly modified forms, will tend to inherit those advantages which made their common parent (A) more numerous than most of the other inhabitants of the same country; they will likewise partake of those more general advantages which made the genus to which the parent-species belonged, a large genus in its own country. And these circumstances we know to be favourable to the production of new varieties.

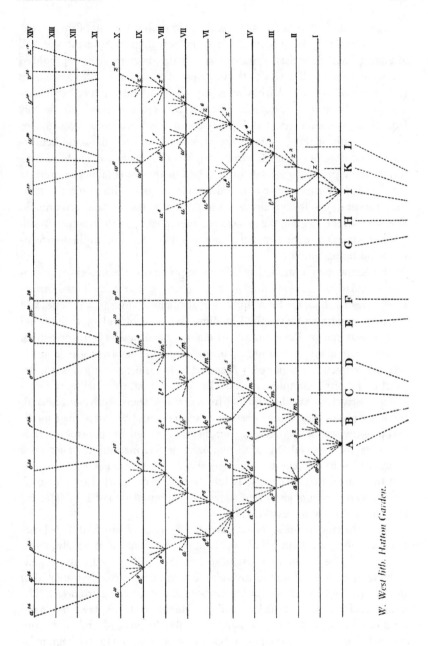

W. West lith. Hatton Garden.

If, then, these two varieties be variable, the most divergent of their variations will generally be preserved during the next thousand generations. And after this interval, variety a^1 is supposed in the diagram to have produced variety a^2, which will, owing to the principle of divergence, differ more from (A) than did variety a^1. Variety m^1 is supposed to have produced two varieties, namely m^2 *and* s^2, differing from each other, and more considerably from their common parent (A). We may continue the process by similar steps for any length of time; some of the varieties, after each thousand generations, producing only a single variety, but in a more and more modified condition, some producing two or three varieties, and some failing to produce any. Thus the varieties or modified descendants, proceeding from the common parent (A), will generally go on increasing in number and diverging in character. In the diagram the process is represented up to the ten-thousandth generation, and under a condensed and simplified form up to the fourteen-thousandth generation.

But I must here remark that I do not suppose that the process ever goes on so regularly as is represented in the diagram, though in itself made somewhat irregular. I am far from thinking that the most divergent varieties will invariably prevail and multiply: a medium form may often long endure, and may or may not produce more than one modified descendant; for natural selection will always act according to the nature of the places which are either unoccupied or not perfectly occupied by other beings; and this will depend on infinitely complex relations. But as a general rule, the more diversified in structure the descendants from any one species can be rendered, the more places they will be enabled to seize on, and the more their modified progeny will be increased. In our diagram the line of succession is broken at regular intervals by small numbered letters marking the successive forms which have become sufficiently distinct to be recorded as varieties. But these breaks are imaginary, and might have been inserted anywhere, after intervals long enough to have allowed the accumulation of a considerable amount of divergent variation.

As all the modified descendants from a common and widely-diffused species, belonging to a large genus, will tend to partake of the same advantages which made their parent successful in life, they will generally go on multiplying in number as well as diverging in character: this is represented in the diagram by the several divergent branches proceeding from (A). The modified offspring from the later and more highly improved branches in the lines of descent, will, it is probable, often take the place of, and so destroy, the earlier and less improved branches: this is represented in the diagram by some of the lower branches not reaching to the upper horizontal lines. In some cases I do not doubt that the process of modification will be confined to a single line of descent, and the number of the descendants will not be increased; although the amount of divergent modification may have been increased in the successive generations. This case would be represented in the diagram, if all the lines proceeding from (A) were removed, excepting that from a^1 to a^{10}. In the same way, for instance, the English race-horse and English pointer have apparently both gone on slowly diverging in character from their original stocks, without either having given off any fresh branches or races.

After ten thousand generations, species (A) is supposed to have produced three forms, a^{10}, f^{10}, and m^{10}, which, from having diverged in character during the successive generations, will have come to differ largely, but perhaps unequally, from each other and from their common parent. If we suppose the amount of change between each horizontal line in our diagram to be excessively small, these three forms may still be only well-marked varieties; or they may have arrived at the doubtful category of sub-species; but we have only to suppose the steps in the process of modification to be more numerous or greater in amount, to convert these three forms into well-defined species: thus the diagram illustrates the steps by which the small differences distinguishing varieties are increased into the larger differences distinguishing species. By continuing the same process for a greater number of generations (as shown in the diagram in a condensed and simplified manner), we get eight species, marked by the letters between a^{14} and m^{14}, all descended from (A). Thus, as I believe, species are multiplied and genera are formed.

In a large genus it is probable that more than one species would vary. In the diagram I have assumed that a second species (I) has produced, by analogous steps, after ten thousand generations, either two well-marked varieties (w^{10} and z^{10}) or two species, according to the amount of change supposed to be represented between the horizontal lines. After fourteen thousand generations, six new species, marked by the letters n^{14} to z^{14}, are supposed to have been produced. In each genus, the species, which are already extremely different in character, will generally tend to produce the greatest number of modified descendants; for these will have the best chance of filling new and widely different places in the polity of nature: hence in the diagram I have chosen the extreme species (A), and the nearly extreme species (I), as those which have largely varied, and have given rise to new varieties and species. The other nine species (marked by capital letters) of our original genus, may for a long period continue transmitting unaltered descendants; and this is shown in the diagram by the dotted lines not prolonged far upwards from want of space.

But during the process of modification, represented in the diagram, another of our principles, namely that of extinction, will have played an important part. As in each fully stocked country natural selection necessarily acts by the selected form having some advantage in the struggle for life over other forms, there will be a constant tendency in the improved descendants of any one species to supplant and exterminate in each stage of descent their predecessors and their original parent. For it should be remembered that the competition will generally be most severe between those forms which are most nearly related to each other in habits, constitution, and structure. Hence all the intermediate forms between the earlier and later states, that is between the less and more improved state of a species, as well as the original parent-species itself, will generally tend to become extinct. So it probably will be with many whole collateral lines of descent, which will be conquered by later and improved lines of descent. If, however, the modified offspring of a species get into some distinct country, or become quickly adapted to some quite new station, in which child and parent do not come into competition, both may continue to exist.

If then our diagram be assumed to represent a considerable amount of modification, species (A) and all the earlier varieties will have become extinct, having been replaced by eight new species (a^{14} to m^{14}); and (I) will have been replaced by six (n^{14} to z^{14}) new species.

But we may go further than this. The original species of our genus were supposed to resemble each other in unequal degrees, as is so generally the case in nature; species (A) being more nearly related to B, C, and D, than to the other species; and species (I) more to G, H, K, L, than to the others. These two species (A) and (I), were also supposed to be very common and widely diffused species, so that they must originally have had some advantage over most of the other species of the genus. Their modified descendants, fourteen in number at the fourteen-thousandth generation, will probably have inherited some of the same advantages: they have also been modified and improved in a diversified manner at each stage of descent, so as to have become adapted to many related places in the natural economy of their country. It seems, therefore, to me extremely probable that they will have taken the places of, and thus exterminated, not only their parents (A) and (I), but likewise some of the original species which were most nearly related to their parents. Hence very few of the original species will have transmitted offspring to the fourteen-thousandth generation. We may suppose that only one (F), of the two species which were least closely related to the other nine original species, has transmitted descendants to this late stage of descent.

The new species in our diagram descended from the original eleven species, will now be fifteen in number. Owing to the divergent tendency of natural selection, the extreme amount of difference in character between species a^{14} and z^{14} will be much greater than that between the most different of the original eleven species. The new species, moreover, will be allied to each other in a widely different manner. Of the eight descendants from (A) the three marked a^{14}, q^{14}, p^{14}, will be nearly related from having recently branched off from a^{10}; b^{14} and f^{14}, from having diverged at an earlier period from a^5, will be in some degree distinct from the three first-named species; and lastly, o^{14}, e^{14}, and m^{14}, will be nearly related one to the other, but from having diverged at the first commencement of the process of modification, will be widely different from the other five species, and may constitute a sub-genus or even a distinct genus.

The six descendants from (I) will form two sub-genera or even genera. But as the original species (I) differed largely from (A), standing nearly at the extreme points of the original genus, the six descendants from (I) will, owing to inheritance, differ considerably from the eight descendants from (A); the two groups, moreover, are supposed to have gone on diverging in different directions. The intermediate species, also (and this is a very important consideration), which connected the original species (A) and (I), have all become, excepting (F), extinct, and have left no descendants. Hence the six new species descended from (I), and the eight descended from (A), will have to be ranked as very distinct genera, or even as distinct sub-families.

Thus it is, as I believe, that two or more genera are produced by descent, with modification, from two or more species of the same genus. And the two or more parent-species are supposed to have descended from some one species of an earlier genus. In our diagram, this is indicated by the broken lines, beneath the capital letters, converging in sub-branches downwards towards a single point; this point representing a single species, the supposed single parent of our several new sub-genera and genera.

It is worth while to reflect for a moment on the character of the new species F^{14}, which is supposed not to have diverged much in character, but to have retained the form of (F), either unaltered or altered only in a slight degree. In this case, its affinities to the other fourteen new species will be of a curious and circuitous nature. Having descended from a form which stood between the two parent-species (A) and (I), now supposed to be extinct and unknown, it will be in some degree intermediate in character between the two groups descended from these species. But as these two groups have gone on diverging in character from the type of their parents, the new species (F^{14}) will not be directly intermediate between them, but rather between types of the two groups; and every naturalist will be able to bring some such case before his mind.

In the diagram, each horizontal line has hitherto been supposed to represent a thousand generations, but each may represent a million or hundred million generations, and likewise a section of the successive strata of the earth's crust including extinct remains. We shall, when we come to our chapter on Geology, have to refer again to this subject, and I think we shall then see that the diagram throws light on the affinities of extinct beings, which, though generally belonging to the same orders, or families, or genera, with those now living, yet are often, in some degree, intermediate in character between existing groups; and we can understand this fact, for the extinct species lived at very ancient epochs when the branching lines of descent had diverged less.

I see no reason to limit the process of modification, as now explained, to the formation of genera alone. If, in our diagram, we suppose the amount of change represented by each successive group of diverging dotted lines to be very great, the forms marked a^{14} to p^{14}, those marked b^{14} and f^{14}, and those marked o^{14} to m^{14}, will form three very distinct genera. We shall also have two very distinct genera descended from (I); and as these latter two genera, both from continued divergence of character and from inheritance from a different parent, will differ widely from the three genera descended from (A), the two little groups of genera will form two distinct families, or even orders, according to the amount of divergent modification supposed to be represented in the diagram. And the two new families, or orders, will have descended from two species of the original genus; and these two species are supposed to have descended from one species of a still more ancient and unknown genus.

We have seen that in each country it is the species of the larger genera which oftenest present varieties or incipient species. This, indeed, might have been expected; for as natural selection acts through one form having some advantage over other forms in the struggle for existence, it will chiefly act on those which already have some advantage; and the largeness of any group shows that its species have inherited from a common ancestor some advantage in common. Hence, the struggle for the production of new and modified descendants, will mainly lie between the larger groups, which are all trying to increase in number. One large group will slowly conquer another large group, reduce its numbers, and thus lessen its chance of further variation and improvement. Within the same large group, the later and more highly perfected sub-groups, from branching out and seizing on many new places in the polity of Nature, will constantly tend to supplant and destroy the earlier and less improved sub-groups. Small and broken groups and sub-groups will finally tend to disappear. Looking to the future, we can predict that the groups of organic beings which are now large and triumphant, and which are least broken up, that is, which as yet have suffered least extinction, will for a long period continue to increase. But which groups will ultimately prevail, no man can predict; for we well know that many groups, formerly most extensively developed, have now become extinct. Looking still more remotely to the future, we may predict that, owing to the continued and steady increase of the larger groups, a multitude of smaller groups will become utterly extinct, and leave no modified descendants; and consequently that of the species living at any one period, extremely few will transmit descendants to a remote futurity. I shall have to return to this subject in the chapter on Classification, but I may add that on this view of extremely few of the more ancient species having transmitted descendants, and on the view of all the descendants of the same species making a class, we can understand how it is that there exist but very few classes in each main division of the animal and vegetable kingdoms. Although extremely few of the most ancient species may now have living and modified descendants, yet at the most remote geological period, the earth may have been as well peopled with many species of many genera, families, orders, and classes, as at the present day.

Summary of Chapter.—If during the long course of ages and under varying conditions of life, organic beings vary at all in the several parts of their organisation, and I think this cannot be disputed; if there be, owing to the high geometrical powers of increase of each species, at some age, season, or year, a severe struggle for life, and this certainly cannot be disputed; then, considering the infinite complexity of the relations of all organic beings to each other and to their conditions of existence, causing an infinite diversity in structure, constitution, and habits, to be advantageous to them, I think it would be a most extraordinary fact if no variation ever had occurred useful to each being's own welfare, in the same way as so many variations have occurred useful to man. But if variations useful to any organic being do occur, assuredly individuals thus characterised will have the best chance of being preserved in the struggle for life; and from the strong principle of inheritance they will tend to produce offspring similarly characterised. This principle of preservation, I have called, for the sake of brevity, Natural Selection. Natural selection, on the principle of qualities being inherited at corresponding ages, can modify the egg, seed, or young, as easily as the adult. Amongst many animals, sexual selection will give its aid to ordinary selection, by assuring to the most vigorous and best adapted males the greatest number of offspring. Sexual selection will also give characters useful to the males alone, in their struggles with other males.

Whether natural selection has really thus acted in nature, in modifying and adapting the various forms of life to their several conditions and stations, must be judged of by the general tenour and balance of evidence given in the following chapters. But we already see how it entails extinction; and how largely extinction has acted in the world's history, geology plainly declares. Natural selection, also, leads to divergence of character; for more living beings can be supported on the same area the more they diverge in structure, habits, and constitution, of which we see proof by looking at the inhabitants of any small spot or at naturalised productions. Therefore during the modification of the descendants of any one species, and during the incessant struggle of all species to increase in numbers, the more diversified these descendants become, the better will be their chance of succeeding in the battle of life. Thus the small differences distinguishing varieties of the same species, will steadily tend to increase till they come to equal the greater differences between species of the same genus, or even of distinct genera.

We have seen that it is the common, the widely-diffused, and widely-ranging species, belonging to the larger genera, which vary most; and these will tend to transmit to their modified offspring that superiority which now makes them dominant in their own countries. Natural selection, as has just been remarked, leads to divergence of character and to much extinction of the less improved and intermediate forms of life. On these principles, I believe, the nature of the affinities of all organic beings may be explained. It is a truly wonderful fact—the wonder of which we are apt to overlook from familiarity—that all animals and all plants throughout all time and space should be related to each other in group subordinate to group, in the manner which we everywhere behold—namely, varieties of the same species most closely related together, species of the same genus less closely and unequally related together, forming sections and sub-genera, species of distinct genera much less closely related, and genera related in different degrees, forming sub-families, families, orders, sub-classes, and classes. The several subordinate groups in any class cannot be ranked in a single file, but seem rather to be clustered round points, and these round other points, and so on in almost endless cycles. On the view that each species has been independently created, I can see no explanation of this great fact in the classification of all organic beings; but, to the best of my judgment, it is explained through inheritance and the complex action of natural selection, entailing extinction and divergence of character, as we have seen illustrated in the diagram.

The affinities of all the beings of the same class have sometimes been represented by a great tree. I believe this simile largely speaks the truth. The green and budding twigs may represent existing species; and those produced during each former year may represent the long succession of extinct species. At each period of growth all the growing twigs have tried to branch out on all sides, and to overtop and kill the surrounding twigs and branches, in the same manner as species and groups of species have tried to overmaster other species in the great battle for life. The limbs divided into great branches, and these into lesser and lesser branches, were themselves once, when the tree was small, budding twigs; and this connexion of the former and present buds by ramifying branches may well represent the classification of all extinct and living species in groups subordinate to groups. Of the many twigs which flourished when the tree was a mere bush, only two or three, now grown into great branches, yet survive and bear all the other branches; so with the species which lived during long-past geological periods, very few now have living and modified descendants. From the first growth of the tree, many a limb and branch has decayed and dropped off; and these lost branches of various sizes may represent those whole orders, families, and genera which have now no living representatives, and which are known to us only from having been found in a fossil state. As we here and there see a thin straggling branch springing from a fork low down in a tree, and which by some chance has been favoured and is still alive on its summit, so we occasionally see an animal like the Ornithorhynchus or Lepidosiren, which in some small degree connects by its affinities two large branches of life, and which has apparently been saved from fatal competition by having inhabited a protected station. As buds give rise by growth to fresh buds, and these, if vigorous, branch out and overtop on all sides many a feebler branch, so by generation I believe it has been with the great Tree of Life, which fills with its dead and broken branches the crust of the earth, and covers the surface with its ever branching and beautiful ramifications.

DIFFICULTIES ON THEORY
WHAT NATURAL SELECTION CAN DO

The foregoing remarks lead me to say a few words on the protest lately made by some naturalists, against the utilitarian doctrine that every detail of structure has been produced for the good of its possessor. They believe that very many structures have been created for beauty in the eyes of man, or for mere variety. This doctrine, if true, would be absolutely fatal to my theory. Yet I fully admit that many structures are of no direct use to their possessors. Physical conditions probably have had some little effect on structure, quite independently of any good thus gained. Correlation of growth has no doubt played a most important part, and a useful modification of one part will often have entailed on other parts diversified changes of no direct use. So again characters which formerly were useful, or which formerly had arisen from correlation of growth, or from other unknown cause, may reappear from the law of reversion, though now of no direct use. The effects of sexual selection, when displayed in beauty to charm the females, can be called useful only in rather a forced sense. But by far the most important consideration is that the chief part of the organisation of every being is simply due to inheritance; and consequently, though each being assuredly is well fitted for its place in nature, many structures now have no direct relation to the habits of life of each species. Thus, we can hardly believe that the webbed feet of the upland goose or of the frigate-bird are of special use to these birds; we cannot believe that the same bones in the arm of the monkey, in the fore leg of the horse, in the wing of the bat, and in the flipper of the seal, are of special use to these animals. We may safely attribute these structures to inheritance. But to the progenitor of the up-land goose and of the frigate-bird, webbed feet no doubt were as useful as they now are to the most aquatic of existing birds. So we may believe that the progenitor of the seal had not a flipper, but a foot with five toes fitted for walking or grasping; and we may further venture to believe that the several bones in the limbs of the monkey, horse, and bat, which have been inherited from a common progenitor, were formerly of more special use to that progenitor, or its progenitors, than they now are to these animals hav-ing such widely diversified habits. Therefore we may infer that these sev-eral bones might have been acquired through natural selection, subjected formerly, as now, to the several laws of inheritance, reversion, correlation of growth, &c. Hence every detail of structure in every living creature (mak-ing some little allowance for the direct action of physical conditions) may be viewed, either as having been of special use to some ancestral form, or as being now of special use to the descendants of this form—either di-rectly, or indirectly through the complex laws of growth.

Natural selection cannot possibly produce any modification in any one species exclusively for the good of another species; though throughout na-ture one species incessantly takes advantage of, and profits by, the structure of another. But natural selection can and does often produce structures for the direct injury of other species, as we see in the fang of the adder, and in the ovipositor of the ichneumon, by which its eggs are deposited in the liv-ing bodies of other insects. If it could be proved that any part of the struc-ture of any one species had been formed for the exclusive good of another species, it would annihilate my theory, for such could not have been pro-duced through natural selection. Although many statements may be found in works on natural history to this effect, I cannot find even one which seems to me of any weight. It is admitted that the rattlesnake has a poison-fang for its own defence and for the destruction of its prey; but some authors sup-pose that at the same time this snake is furnished with a rattle for its own injury, namely, to warn its prey to escape. I would almost as soon believe that the cat curls the end of its tail when preparing to spring, in order to warn the doomed mouse. But I have not space here to enter on this and other such cases.

Natural selection will never produce in a being anything injurious to it-self, for natural selection acts solely by and for the good of each. No organ will be formed, as Paley has remarked, for the purpose of causing pain or for doing an injury to its possessor. If a fair balance be struck between the good and evil caused by each part, each will be found on the whole advantageous. After the lapse of time, under changing conditions of life, if any part comes to be injurious, it will be modified; or if it be not so, the being will become extinct, as myriads have become extinct.

Natural selection tends only to make each organic being as perfect as, or slightly more perfect than, the other inhabitants of the same country with which it has to struggle for existence. And we see that this is the degree of perfection attained under nature. The endemic productions of New Zealand, for instance, are perfect one compared with another; but they are now rapidly yielding before the advancing legions of plants and animals introduced from Europe. Natural selection will not produce absolute perfection, nor do we always meet, as far as we can judge, with this high standard under nature. The correction for the aberration of light is said, on high authority, not to be perfect even in that most perfect organ, the eye. If our reason leads us to admire with enthusiasm a multitude of inimitable contrivances in nature, this same reason tells us, though we may easily err on both sides, that some other contrivances are less perfect. Can we consider the sting of the wasp or of the bee as perfect, which, when used against many attacking animals, cannot be withdrawn, owing to the backward serratures, and so inevitably causes the death of the insect by tearing out its viscera?

If we look at the sting of the bee, as having originally existed in a remote progenitor as a boring and serrated instrument, like that in so many members of the same great order, and which has been modified but not perfected for its present purpose, with the poison originally adapted to cause galls subsequently intensified, we can perhaps understand how it is that the use of the sting should so often cause the insect's own death: for if on the whole the power of stinging be useful to the community, it will fulfil all the requirements of natural selection, though it may cause the death of some few members. If we admire the truly wonderful power of scent by which the males of many insects find their females, can we admire the production for this single purpose of thousands of drones, which are utterly useless to the community for any other end, and which are ultimately slaughtered by their industrious and sterile sisters? It may be difficult, but we ought to admire the savage instinctive hatred of the queen-bee, which urges her instantly to destroy the young queens her daughters as soon as born, or to perish herself in the combat; for undoubtedly this is for the good of the community; and maternal love or maternal hatred, though the latter fortunately is most rare, is all the same to the inexorable principle of natural selection. If we admire the several ingenious contrivances, by which the flowers of the orchis and of many other plants are fertilised through insect agency, can we consider as equally perfect the elaboration by our fir-trees of dense clouds of pollen, in order that a few granules may be wafted by a chance breeze on to the ovules?

INSTINCT

NEUTER INSECTS

No doubt many instincts of very difficult explanation could be opposed to the theory of natural selection,—cases, in which we cannot see how an instinct could possibly have originated; cases, in which no intermediate gradations are known to exist; cases of instinct of apparently such trifling importance, that they could hardly have been acted on by natural selection; cases of instincts almost identically the same in animals so remote in the scale of nature, that we cannot account for their similarity by inheritance from a common parent, and must therefore believe that they have been acquired by independent acts of natural selection. I will not here enter on these several cases, but will confine myself to one special difficulty, which at first appeared to me insuperable, and actually fatal to my whole theory. I allude to the neuters or sterile females in insect-communities: for these neuters often differ widely in instinct and in structure from both the males and fertile females, and yet, from being sterile, they cannot propagate their kind.

The subject well deserves to be discussed at great length, but I will here take only a single case, that of working or sterile ants. How the workers have been rendered sterile is a difficulty; but not much greater than that of any other striking modification of structure; for it can be shown that some insects and other articulate animals in a state of nature occasionally become sterile; and if such insects had been social, and it had been profitable to the community that a number should have been annually born capable of work, but incapable of procreation, I can see no very great difficulty in this being effected by natural selection. But I must pass over this preliminary difficulty. The great difficulty lies in the working ants differing widely from both the males and the fertile females in structure, as in the shape of the thorax and in being destitute of wings and sometimes of eyes, and in instinct. As far as instinct alone is concerned, the prodigious difference in this respect between the workers and the perfect females, would have been far better exemplified by the hive-bee. If a working ant or other neuter insect had been an animal in the ordinary state, I should have unhesitatingly assumed that all its characters had been slowly acquired through natural selection; namely, by an individual having been born with some slight profitable modification of structure, this being inherited by its offspring, which again varied and were again selected, and so onwards. But with the working ant we have an insect differing greatly from its parents, yet absolutely sterile; so that it could never have transmitted successively acquired modifications of structure or instinct to its progeny. It may well be asked how is it possible to reconcile this case with the theory of natural selection?

First, let it be remembered that we have innumerable instances, both in our domestic productions and in those in a state of nature, of all sorts of differences of structure which have become correlated to certain ages, and to either sex. We have differences correlated not only to one sex, but to that short period alone when the reproductive system is active, as in the nuptial plumage of many birds, and in the hooked jaws of the male salmon. We have even slight differences in the horns of different breeds of cattle in relation to an artificially imperfect state of the male sex; for oxen of certain breeds have longer horns than in other breeds, in comparison with the horns of the bulls or cows of these same breeds. Hence I can see no real difficulty in any character having become correlated with the sterile condition of certain members of insect-communities: the difficulty lies in understanding how such correlated modifications of structure could have been slowly accumulated by natural selection.

This difficulty, though appearing insuperable, is lessened, or, as I believe, disappears, when it is remembered that selection may be applied to the family, as well as to the individual, and may thus gain the desired end. Thus, a well-flavoured vegetable is cooked, and the individual is destroyed; but the horticulturist sows seeds of the same stock, and confidently expects to get nearly the same variety; breeders of cattle wish the flesh and fat to be well marbled together; the animal has been slaughtered, but the breeder goes with confidence to the same family. I have such faith in the powers of selection, that I do not doubt that a breed of cattle, always yielding oxen with extraordinarily long horns, could be slowly formed by carefully watching which individual bulls and cows, when matched, produced oxen with the longest horns; and yet no one ox could ever have propagated its kind. Thus I believe it has been with social insects: a slight modification of structure, or instinct, correlated with the sterile condition of certain members of the community, has been advantageous to the community: consequently the fertile males and females of the same community flourished, and transmitted to their fertile offspring a tendency to produce sterile members having the same modification. And I believe that this process has been repeated, until that prodigious amount of difference between the fertile and sterile females of the same species has been produced, which we see in many social insects.

CONCLUSION

When the views entertained in this volume on the origin of species, or when analogous views are generally admitted, we can dimly foresee that there will be a considerable revolution in natural history. Systematists will be able to pursue their labors as at present; but they will not be incessantly haunted by the shadowy doubt whether this or that form be in essence a species. This

I feel sure, and I speak after experience, will be no slight relief. The endless disputes whether or not some fifty species of British brambles are true species will cease. Systematists will have only to decide (not that this will be easy) whether any forms be sufficiently constant and distinct from other forms, to be capable of definition; and if definable, whether the differences be sufficiently important to deserve a specific name. This latter point will become a far more essential consideration than it is at present; for differences, however slight, between any two forms, if not blended by intermediate gradations, are looked at by most naturalists as sufficient to raise both forms to the rank of species. Hereafter we shall be compelled to acknowledge that the only distinction between species and well-marked varieties is, that the latter are known, or believed, to be connected at the present day by intermediate gradations, whereas species were formerly thus connected. Hence, without quite rejecting the consideration of the present existence of intermediate gradations between any two forms, we shall be led to weigh more carefully and to value higher the actual amount of difference between them. It is quite possible that forms now generally acknowledged to be merely varieties may hereafter be thought worthy of specific names, as with primrose and cowslip; and in this case scientific and common language will come into accordance. In short, we shall have to treat species in the same manner as those naturalists treat genera, who admit that genera are merely artificial combinations made for convenience. This may not be a cheering prospect; but we shall at least be freed from the vain search for the undiscovered and undiscoverable essence of the term species.

The other and more general departments of natural history will rise greatly in interest. The terms used by naturalists of affinity, relationship, community of type, paternity, morphology, adaptive characters, rudimentary and aborted organs, &c., will cease to be metaphorical, and will have a plain signification. When we no longer look at an organic being as a savage looks at a ship, as at something wholly beyond his comprehension; when we regard every production of nature as one which has had a history; when we contemplate every complex structure and instinct as the summing up of many contrivances, each useful to the possessor, nearly in the same way as when we look at any great mechanical invention as the summing up of the labour, the experience, the reason, and even the blunders of numerous workmen; when we thus view each organic being, how far more interesting, I speak from experience, will the study of natural history become!

A grand and almost untrodden field of inquiry will be opened, on the causes and laws of variation, on correlation of growth, on the effects of use and disuse, on the direct action of external conditions, and so forth. The study of domestic productions will rise immensely in value. A new variety raised by man will be a far more important and interesting subject for study than one more species added to the infinitude of already recorded species. Our classifications will come to be, as far as they can be so made, genealogies; and will then truly give what may be called the plan of creation. The rules for classifying will no doubt become simpler when we have a definite object in view. We possess no pedigrees or armorial bearings; and we have to discover and trace the many diverging lines of descent in our natural genealogies, by characters of any kind which have long been inherited. Rudimentary organs will speak infallibly with respect to the nature of long-lost structures. Species and groups of species, which are called aberrant, and which may fancifully be called living fossils, will aid us in forming a picture of the ancient forms of life. Embryology will reveal to us the structure, in some degree obscured, of the prototypes of each great class.

When we can feel assured that all the individuals of the same species, and all the closely allied species of most genera, have within a not very remote period descended from one parent, and have migrated from some one birthplace; and when we better know the many means of migration, then, by the light which geology now throws, and will continue to throw, on former changes of climate and of the level of the land, we shall surely be enabled to trace in an admirable manner the former migrations of the inhabitants of the whole world. Even at present, by comparing the differences of the inhabitants of the sea on the opposite sides of a continent, and the nature of the various inhabitants of that continent in relation to their apparent means of immigration, some light can be thrown on ancient geography.

The noble science of Geology loses glory from the extreme imperfection of the record. The crust of the earth with its embedded remains must not be looked at as a well-filled museum, but as a poor collection made at hazard and at rare intervals. The accumulation of each great fossiliferous formation will be recognised as having depended on an unusual concurrence of circumstances, and the blank intervals between the successive stages as having been of vast duration. But we shall be able to gauge with some security the duration of these intervals by a comparison of the preceding and succeeding organic forms. We must be cautious in attempting to correlate as strictly contemporaneous two formations, which include few identical species, by the general succession of their forms of life. As species are produced and exterminated by slowly acting and still existing causes, and not by miraculous acts of creation and by catastrophes; and as the most important of all causes of organic change is one which is almost independent of altered and perhaps suddenly altered physical conditions, namely, the mutual relation of organism to organism,—the improvement of one being entailing the improvement or the extermination of others; it follows, that the amount of organic change in the fossils of consecutive formations probably serves as a fair measure of the lapse of actual time. A number of species, however, keeping in a body might remain for a long period unchanged, whilst within this same period, several of these species, by migrating into new countries and coming into competition with foreign associates, might become modified; so that we must not overrate the accuracy of organic change as a measure of time. During early periods of the earth's history, when the forms of life were probably fewer and simpler, the rate of change was probably slower; and at the first dawn of life, when very few forms of the simplest structure existed, the rate of change may have been slow in an extreme degree. The whole history of the world, as at present known, although of a length quite incomprehensible by us, will hereafter be recognised as a mere fragment of time, compared with the ages which have elapsed since the first creature, the progenitor of innumerable extinct and living descendants, was created.

In the distant future I see open fields for far more important researches. Psychology will be based on a new foundation, that of the necessary acquirement of each mental power and capacity by gradation. Light will be thrown on the origin of man and his history.

Authors of the highest eminence seem to be fully satisfied with the view that each species has been independently created. To my mind it accords better with what we know of the laws impressed on matter by the Creator, that the production and extinction of the past and present inhabitants of the world should have been due to secondary causes, like those determining the birth and death of the individual. When I view all beings not as special creations, but as the lineal descendants of some few beings which lived long before the first bed of the Silurian system was deposited, they seem to me to become ennobled. Judging from the past, we may safely infer that not one living species will transmit its unaltered likeness to a distant futurity. And of the species now living very few will transmit progeny of any kind to a far distant futurity; for the manner in which all organic beings are grouped, shows that the greater number of species of each genus, and all the species of many genera, have left no descendants, but have become utterly extinct. We can so far take a prophetic glance into futurity as to foretel that it will be the common and widely-spread species, belonging to the larger and dominant groups, which will ultimately prevail and procreate new and dominant species. As all the living forms of life are the lineal descendants of those which lived long before the Silurian epoch, we may feel certain that the ordinary succession by generation has never once been broken, and that no cataclysm has desolated the whole world. Hence we may look with some confidence to a secure future of equally inappreciable length. And as natural selection works solely by and for the good of each being, all corporeal and mental endowments will tend to progress towards perfection.

It is interesting to contemplate an entangled bank, clothed with many plants of many kinds, with birds singing on the bushes, with various insects flitting about, and with worms crawling through the damp earth, and to reflect that these elaborately constructed forms, so different from each other, and dependent on each other in so complex a manner, have all been produced by laws acting around us. These laws, taken in the largest sense, being Growth and Reproduction; Inheritance which is almost implied by reproduction; Variability from the indirect and direct action of the external conditions of life, and from use and disuse; a Ratio of Increase so high as to lead to a Struggle for Life, and as a consequence to Natural Selection, entailing Divergence of Character and the Extinction of less-improved forms. Thus, from the war of nature, from famine and death, the most exalted object which we are capable of conceiving, namely, the production of the higher animals, directly follows. There is grandeur in this view of life, with its several powers, having been originally breathed into a few forms or into one; and that, whilst this planet has gone cycling on according to the fixed law of gravity, from so simple a beginning endless forms most beautiful and most wonderful have been, and are being, evolved.

THE SUCCESS OF *ON THE ORIGIN OF SPECIES*

The *Origin* was a great commercial success, the first edition selling out immediately.[1] Just how great a popular success, and how surprising this was to Darwin, was revealed in a letter to Lyell:

> What a grand, immense benefit you conferred on me by getting Murray to publish my book. I never till-today realised that it was getting widely distributed; for in a letter from a lady today to E., she says she heard a man enquiring for it *at the Railway Station!!!* at Waterloo Bridge; and the bookseller said that he had none till the new edition was out. The bookseller said he had not read it, but had heard it was a very remarkable book!!![2]

Darwin was also interested in his theory's reception in Europe:

> . . . I get letters occasionally, which show me that Natural Selection is making *great* progress in Germany, and some amongst the young in France. I have just received a pamphlet from Germany, with the complimentary title of "Darwinische Arten-Enstehung-Humbug"![3]

---

1. Darwin "heard that the stock of 1250 copies was oversubscribed when the book went on sale to the trade, 22 November," Adrian Desmond and James Moore, *Darwin: The Life of a Tormented Evolutionist* (New York, Time Warner, 1991), 477.

2. Darwin to Lyell, 14 January 1860, *Correspondence*, 8: 35.

3. Darwin to Hooker, 26 November 1864, *Life and Letters*, 3: 306, where *Entstehung* is misspelled.

Chapter 9

VARIATION OF
PLANTS AND ANIMALS
UNDER DOMESTICATION
(1868)

INTRODUCTION

In 1798, Darwin's grandfather Erasmus published *Zoonomia, or the Laws of Life*. *Zoonomia*'s final chapter, "Generation," was devoted to a wide range of topics in reproductive biology. When Charles began Notebook B in 1837, the first word he wrote was *Zoonomia* and his first topic was the distinction drawn by Erasmus between asexual and sexual reproduction.[1] Thus, it was fitting that "Tentative Hypothesis of Pangenesis" should be the last chapter of Charles Darwin's *Variation of Plants and Animals Under Domestication* (1868, 1875), for pangenesis was Darwin's attempt to offer a unified explanation for generation. Pangenesis was intended to be a single law of life that determined the commonality underlying the many faces of generation, from asexual and sexual reproduction to regeneration, hybridism, development, and inheritance. Pangenesis was the culmination of Darwin's nearly thirty-year struggle with the last chapter of *Zoonomia*.[2]

As the term implies, pangenesis is the common genesis of all forms of genesis. Simply put, Darwin proposed that cells are not only able to grow by means of cell division, but also are capable of "throwing off" minute replicas of themselves, which he called gemmules. The process was supposed to be analogous to budding, or gemmation, hence the name. Darwin's gem-

---

1. See "Darwin's Notebooks," chapter 2, 51.

2. See M.J.S. Hodge, "Darwin as a Lifelong Generation Theorist," in *The Darwinian Heritage*, ed. D. Kohn (Princeton, Princeton University Press, 1985), 207–243; see also Robert Olby, *Origins of Mendelism*, 2nd ed. (Chicago, University of Chicago Press, 1985), esp. pp. 84–85.

mules have many roles to play. A principal one is to accumulate, from all parts of the body, into the germ cells of the reproductive organs, where they constitute the formative material of sexual reproduction—the replicating essence of spermatozoa and ovum. At fertilization the gemmules from each parent combine. Gemmules from each part of one parent's body have what Darwin called an elective affinity, something like a chemical attraction, for gemmules from the corresponding part in the other parent.

This may sound inside out and backward to the modern student of biology, who is not used to thinking that somatic cells contribute to the formation of germ cells. But Darwin's idea that there are minute cell replicas, capable of self-recognition and organization by means of chemical affinity, and capable of replicating the entire individual, is both simple and profound. Once we get past the inversion of our "commonsense" modern notion and go beyond dismissing Darwin's hypothesis as a failure compared to Mendel's theory, it becomes apparent that Darwin was trying to grapple with a fundamental question: How is biological information replicated? At a time when the cell theory had become a key organizing principle but little knowledge existed of subcellular structure, function, and chemistry, Darwin conceived of particulate gemmules and asserted that a chemical attraction must be involved in their replication. It would be a long way from the elective affinity of gemmules to base pairing and hydrogen bonds, but given where Darwin stood on the landscape of biological knowledge, his vision was remarkable.

Let us consider just some of the detailed workings of pangenesis as they apply to sexual reproduction. Consider three properties Darwin ascribed to gemmules: coordinated replication, variation, and dormancy. In sexual reproduction, every part of the body obviously changes as it undergoes development. Darwin proposed that gemmules were thrown off at every developmental stage. Therefore, by the stage of sexual maturity, the accumulated gemmules formed a dynamic trace of the entire unfolding life of the organism. Thus, pangenesis accounted for the faithful replication and development of a complex organism. Furthermore, if the cells of one part of the body underwent change as the result of environmental change, these changes were likewise recorded by the gemmules produced by that body part. Thus, pangenesis accounted for the origin of hereditary variation. In addition, gemmules, like seeds or spores, were capable of dormancy. A character could lie dormant only to be expressed after several generations. This accounted for the phenomenon of hereditary reversion to ancestral characters. Together, these three properties of gemmules account for what Darwin took to be the three basic laws of inheritance: (1) like produces like (gemmule replication and elective affinity), (2) like produces novelty (gemmule variation), and

(3) grandchildren can revert to grandparents (gemmule dormancy). This boils down to two laws: (1) like produces like and (2) like produces unlike.

When studying Darwin's pangenesis, we should bear in mind that Darwin was not only a "lifelong generation theorist" (Hodge, 1985). Generation theory, starting with the Red Notebook (see chapter 2), was always coordinated with Darwin's attempt to understand evolution. This meant that Darwin was not trying to understand each of the various phenomena of reproduction in its own right. Rather, he was trying to understand how they all fit into the evolutionary scheme of things. In his later years, facing criticism of natural selection, Darwin came to rely on Lamarckian notions of the inheritance of acquired characters, but he never abandoned natural selection. From Darwin's evolutionist perspective, pangenesis was a plausible hypothesis. It covered what his theory of evolution needed to cover: the fact of heredity and the fact of deviation from hereditary stability. That is true whether or not natural selection is the key agent of evolution. Perhaps this is why Darwin, for all his study of generation, including many studies of breeding and hybridization, did not proceed down Mendel's path to systematic analysis of the transmission of hereditary characters over a number of generations. The theory of evolution, which regulated all his theorizing, did not seem to Darwin to require such analysis. Subsequent generations of Darwinists found otherwise.

There is an unresolved debate about when Darwin formulated the theory of pangenesis. A manuscript version has been dated to 1865, three years before the publication of *Variation Under Domestication*, and, ironically, the same year in which Mendel published his paper. Darwin claimed in an 1867 letter to Lyell to have formulated the hypothesis in the early 1840s. Hodge (1985) and Olby (1985) both believe pangenesis came out of Darwin's early generational thinking, which Hodge has very carefully reconstructed. This seems highly plausible; however, no 1840s manuscript has been found which posits the main idea in pangenesis: that cells generate minute replicas (gemmules). As Olby (1985, 85) points out, there are rather suggestive marginalia in Darwin's copy of Johannes Müller's *Physiology* (1838–1842). But these are scraps. Perhaps, like a dormant gemmule, the idea of pangenesis was "thrown off" in 1841 or 1842, only to germinate into a full-blown hypothesis in 1865 when Darwin reread his notes on Müller.

PANGENESIS

I have now enumerated the chief facts which everyone would desire to see connected by some intelligible bond. This can be done, if we make the following assumptions, and much may be advanced in favour of the chief one. The secondary assumptions can likewise be supported by various physiological considerations. It is universally admitted that the cells or units of the body increase by self-division or proliferation, retaining the same nature, and that they ultimately become converted into the various tissues and substances of the body. But besides this means of increase I assume that the units throw off minute granules which are dispersed throughout the whole system; that these, when supplied with proper nutriment, multiply by self-division, and are ultimately developed into units like those from which they were originally derived. These granules may be called gemmules. They are collected from all parts of the system to constitute the sexual elements, and their development in the next generation forms a new being; but they are likewise capable of transmission in a dormant state to future generations and may then be developed. Their development depends on their union with other partially developed or nascent cells which precede them in the regular course of growth. Why I use the term union, will be seen when we discuss the direct action of pollen on the tissues of the mother plant. Gemmules are supposed to be thrown off by every unit, not only during the adult state, but during each stage of development of every organism; but not necessarily during the continued existence of the same unit. Lastly, I assume that the gemmules in their dormant state have a mutual affinity for each other, leading to their aggregation into buds or into the sexual elements. Hence, it is not the reproductive organs or buds which generate new organisms, but the units of which each individual is composed. These assumptions constitute the provisional hypothesis which I have called pangenesis. Views in many respects similar have been propounded by various authors.

Before proceeding to show, firstly, how far these assumptions are in themselves probable, and secondly, how far they connect and explain the various groups of facts with which we are concerned, it may be useful to give an illustration, as simple as possible, of the hypothesis. If one of the Protozoa be formed, as it appears under the microscope, of a small mass of homogeneous gelatinous matter, a minute particle or gemmule thrown off from any part and nourished under favourable circumstances would reproduce the whole; but if the upper and lower surfaces were to differ in texture from each other and from the central portion, then all three parts would have to throw off gemmules, which when aggregated by mutual affinity would form either buds or the sexual elements, and would ultimately be developed into a similar organism. Precisely the same view may be extended to one of the higher animals; although in this case many thousand gemmules must be thrown off from the various parts of the body at each stage of development; these gemmules being developed in union with pre-existing nascent cells in due order of succession.

Physiologists maintain, as we have seen, that each unit of the body, though to a large extent dependent on others, is likewise to a certain extent independent or autonomous, and has the power of increasing by self-division. I go one step further, and assume that each unit casts off free gemmules which are dispersed throughout the system, and are capable under proper conditions of being developed into similar units. Nor can this assumption be considered as gratuitous and improbable. It is manifest that the sexual elements and buds include formative matter of some kind, capable of development; and we now know from the production of graft-hybrids that similar matter is dispersed throughout the tissues of plants, and can combine with that of another and distinct plant, giving rise to a new being, intermediate in character. We know also that the male element can act directly on the partially developed tissues of the mother plant, and on the future progeny of female animals. The formative matter which is thus dispersed throughout the tissues of plants, and which is capable of being developed into each unit or part, must be generated there by some means; and my chief assumption is that this matter consists of minute particles or gemmules cast off from each unit or cell.

But I have further to assume that the gemmules in their undeveloped state are capable of largely multiplying themselves by self-division, like independent organisms. Delpino insists that to 'admit of multiplication by fissiparity in corpuscles, analogous to seeds or buds . . . is repugnant to all analogy'. But this seems a strange objection, as Thuret has seen the zoospore of an alga divide itself, and each half germinated. Haeckel divided the segmented ovum of a siphonophora into many pieces, and these were developed. Nor does the extreme minuteness of the gemmules, which can hardly differ much in nature from the lowest and simplest organisms, render it improbable that they should grow and multiply. A great authority, Dr Beale, says 'that minute yeast cells are capable of throwing off buds or gemmules, much less than the $\frac{1}{100000}$ of an inch in diameter'; and these he thinks are 'capable of subdivision practically ad infinitum'.

A particle of smallpox matter, so minute as to be borne by the wind, must multiply itself many thousandfold in a person thus inoculated; and so with the contagious matter of scarlet fever. It has recently been ascertained that a minute portion of the mucous discharge from an animal affected with rinderpest, if placed in the blood of a healthy ox, increases so fast that in a short space of time 'the whole mass of blood, weighing many pounds, is infected, and every small particle of that blood contains enough poison to give, within less than forty-eight hours, the disease to another animal'.

The retention of free and undeveloped gemmules in the same body from early youth to old age will appear improbable, but we should remember how long seeds lie dormant in the earth and buds in the bark of a tree. Their transmission from generation to generation will appear still more improbable; but here again we should remember that many rudimentary and use-

less organs have been transmitted during an indefinite number of generations. We shall presently see how well the long-continued transmission of
undeveloped gemmules explains many facts.

As each unit, or group of similar units, throughout the body, casts off its
gemmules, and as all are contained within the smallest ovule, and within
each spermatozoon or pollen grain, and as some animals and plants produce
an astonishing number of pollen grains and ovules, the number and minuteness of the gemmules must be something inconceivable. But considering how minute the molecules are, and how many go to the formation of the
smallest granule of any ordinary substance, this difficulty with respect to the
gemmules is not insuperable. From the data arrived at by Sir W. Thomson,
my son George finds that a cube of $\frac{1}{10000}$ of an inch of glass or water must
consist of between 16 million millions, and 131 thousand million million
molecules. No doubt the molecules of which an organism is formed are larger,
from being more complex, than those of an inorganic substance, and probably many molecules go to the formation of a gemmule; but when we bear
in mind that a cube of $\frac{1}{10000}$ of an inch is much smaller than any pollen
grain, ovule or bud, we can see what a vast number of gemmules one of
these bodies might contain.

The gemmules derived from each part or organ must be thoroughly dispersed throughout the whole system. We know, for instance, that even a minute
fragment of a leaf of a begonia will reproduce the whole plant; and that if
a freshwater worm is chopped into small pieces, each will reproduce the
whole animal. Considering also the minuteness of the gemmules and the
permeability of all organic tissues, the thorough dispersion of the gemmules
is not surprising. That matter may be readily transferred without the aid of
vessels from part to part of the body, we have a good instance in a case recorded by Sir J. Paget of a lady, whose hair lost its colour at each successive
attack of neuralgia and recovered it again in the course of a few days. With
plants, however, and probably with compound animals, such as corals, the
gemmules do not ordinarily spread from bud to bud, but are confined to the
parts developed from each separate bud; and of this fact no explanation can
be given.

The assumed elective affinity of each gemmule for that particular cell
which precedes it in due order of development is supported by many analogies. In all ordinary cases of sexual reproduction, the male and female elements certainly have a mutual affinity for each other; thus, it is believed that
about ten thousand species of Compositae exist, and there can be no doubt
that if the pollen of all these species could be simultaneously or successively
placed on the stigma of any one species, this one would elect with unerring
certainty its own pollen. This elective capacity is all the more wonderful, as
it must have been acquired since the many species of this great group of

plants branched off from a common progenitor. On any view of the nature of sexual reproduction, the formative matter of each part contained within the ovules and the male element act on each other by some law of special affinity, so that corresponding parts affect one another; thus, a calf produced from a shorthorned cow by a longhorned bull has its horns affected by the union of the two forms, and the offspring from two birds with differently coloured tails have their tails affected.

The various tissues of the body plainly show, as many physiologists have insisted, an affinity for special organic substances, whether natural or foreign to the body. We see this in the cells of the kidneys attracting urea from the blood; in curare affecting certain nerves; *Lytta vesicatoria* the kidneys; and the poisonous matter of various diseases, as small-pox, scarlet fever, hooping-cough, glanders, and hydrophobia, affecting certain definite parts of the body.

It has also been assumed that the development of each gemmule depends on its union with another cell or unit which has just commenced its development, and which precedes it in due order of growth. That the formative matter within the pollen of plants, which by our hypothesis consists of gemmules, can unite with and modify the partially developed cells of the mother plant, we have clearly seen in the section devoted to this subject. As the tissues of plants are formed, as far as is known, only by the proliferation of pre-existing cells, we must conclude that the gemmules derived from the foreign pollen do not become developed into new and separate cells, but penetrate and modify the nascent cells of the mother plant. This process may be compared with what takes place in the act of ordinary fertilization, during which the contents of the pollen tubes penetrate the closed embryonic sac within the ovule, and determine the development of the embryo. According to this view, the cells of the mother plant may almost literally be said to be fertilized by the gemmules derived from the foreign pollen. In this case and in all others the proper gemmules must combine in due order with pre-existing nascent cells, owing to their elective affinities. A slight difference in nature between the gemmules and the nascent cells would be far from interfering with their mutual union and development, for we well know in the case of ordinary reproduction that such slight differentiation in the sexual elements favours in a marked manner their union and subsequent development, as well as the vigour of the offspring thus produced.

Thus far we have been able by the aid of our hypothesis to throw some obscure light on the problems which have come before us; but it must be confessed that many points remain altogether doubtful. Thus it is useless to speculate at what period of development each unit of the body casts off its gemmules, as the whole subject of the development of the various tissues is as yet far from clear. We do not know whether the gemmules are merely collected by some unknown means at certain seasons within the reproductive organs, or whether after being thus collected they rapidly multiply there, as the flow of blood to these organs at each breeding season seems to render probable. Nor do we know why the gemmules collect to form buds in certain definite places, leading to the symmetrical growth of trees and corals. We have no means of deciding whether the ordinary wear and tear of the tissues is made good by means of gemmules, or merely by the proliferation of pre-existing cells. If the gemmules are thus consumed, as seems probable from the intimate connection between the repair of waste, regrowth, and development, and more especially from the periodical changes which many male animals undergo in colour and structure, then some light would be thrown on the phenomena of old age, with its lessened power of reproduction and of the repair of injuries, and on the obscure subject of longevity. The fact of castrated animals, which do not cast off innumerable gemmules in the act of reproduction, not being longer-lived than perfect males, seems opposed to the belief that gemmules are consumed in the ordinary repair of wasted tissues; unless indeed the gemmules after being collected in small numbers within the reproductive organs are there largely multiplied.

That the same cells or units may live for a long period and continue multiplying without being modified by their union with free gemmules of any kind, is probable from such cases as that of the spur of a cock which grew to an enormous size when grafted into the ear of an ox. How far units are modified during their normal growth by absorbing peculiar nutriment from the surrounding tissues, independently of their union with gemmules of a distinct nature, is another doubtful point. We shall appreciate this difficulty by calling to mind what complex yet symmetrical growths the cells of plants yield when inoculated by the poison of a gall-insect. With animals various polypoid excrescences and tumours are generally admitted to be the direct product, through proliferation, of normal cells which have become abnormal. In the regular growth and repair of bones, the tissues undergo, as Virchow remarks, a whole series of permutations and substitutions. 'The cartilage cells may be converted by a direct transformation into marrow cells, and continue as such; or they may first be converted into osseous and then into medullary tissue; or lastly, they may first be converted into marrow and then into bone. So variable are the permutations of these tissues, in themselves so nearly allied, and yet in their external appearance so completely distinct'. But as these tissues thus change their nature at any age, without any obvious change in their nutrition, we must suppose in accordance with our hypothesis that gemmules derived from one kind of tissue combine with the cells of another kind, and cause the successive modifications.

We have good reason to believe that several gemmules are requisite for the development of one and the same unit or cell; for we cannot otherwise understand the insufficiency of a single or even of two or three pollen grains or spermatozoa. But we are far from knowing whether the gemmules of all the units are free and separate from one another, or whether some are from the first united into small aggregates. A feather, for instance, is a complex structure, and, as each separate part is liable to inherited variations, I conclude that each feather generates a large number of gemmules; but it is possible that these may be aggregated into a compound gemmule. The same remark applies to the petals of flowers, which are sometimes highly complex structures, with each ridge and hollow contrived for a special purpose, so that each part must have been separately modified, and the modifications transmitted; consequently, separate gemmules, according to our hypothesis, must have been thrown off from each cell or unit. But, as we sometimes see half an anther or a small portion of a filament becoming petali-form, or parts or mere stripes of the calyx assuming the colour and texture of the corolla, it is probable that with petals the gemmules of each cell are not aggregated together into a compound gemmule, but are free and separate. Even in so simple a case as that of a perfect cell, with its protoplasmic contents, nucleus, nucleolus, and walls, we do not know whether or not its development depends on a compound gemmule derived from each part.

Having now endeavoured to show that the several foregoing assumptions are to a certain extent supported by analogous facts, and having alluded to some of the most doubtful points, we will consider how far the hypothesis brings under a single point of view the various cases enumerated in the First Part. All the forms of reproduction graduate into one another and agree in their product; for it is impossible to distinguish between organisms produced from buds, from self-division, or from fertilized germs; such organisms are liable to variations of the same nature and to reversions of the same kind; and as, according to our hypothesis, all the forms of reproduction depend on the aggregation of gemmules derived from the whole body, we can understand this remarkable agreement. Parthenogenesis is no longer wonderful, and if we did not know that great good followed from the union of the sexual elements derived from two distinct individuals, the wonder would be that parthenogenesis did not occur much oftener than it does. On any ordinary theory of reproduction the formation of graft-hybrids, and the action of the male element on the tissues of the mother plant, as well as on the future progeny of female animals, are great anomalies; but they are intelligible on our hypothesis. The reproductive organs do not actually create the sexual elements; they merely determine the aggregation and perhaps the multiplication of the gemmules in a special manner. These organs, however, together with their accessory parts, have high functions to perform. They adapt one or both elements for independent temporary existence, and for mutual union. The stigmatic secretion acts on the pollen of a plant of the same species in a wholly different manner to what it does on the pollen of one belonging to a distinct genus or family. The spermatophores of the Cephalopoda are wonderfully complex structures, which were formerly mistaken for parasitic worms; and the spermatozoa of some animals possess attributes which, if observed in an independent animal, would be put down to instinct guided by sense-organs — as when the spermatozoa of an insect find their way into the minute micropyle of the egg.

The antagonism which has long been observed, with certain exceptions, between growth and the power of sexual reproduction — between the repair of injuries and gemmation — and with plants, between rapid increase by buds, rhizomes, etc., and the production of seed, is partly explained by the gemmules not existing in sufficient numbers for these processes to be carried on simultaneously.

Hardly any fact in physiology is more wonderful than the power of regrowth; for instance, that a snail should be able to reproduce its head, or a salamander its eyes, tail, and legs, exactly at the points where they have been cut off. Such cases are explained by the presence of gemmules derived from each part, and disseminated throughout the body. I have heard the process compared with that of the repair of the broken angles of a crystal by recrystallization; and the two processes have much in common, that in the one case the polarity of the molecules is the efficient cause, and in the other the affinity of the gemmules for particular nascent cells. But we have here to encounter two objections which apply not only to the regrowth of a part, or of a bisected individual, but to fissiparous generation and budding. The first objection is that the part which is reproduced is in the same stage of development as that of the being which has been operated on or bisected; and in the case of buds, that the new beings thus producd are in the same stage as that of the budding parent. Thus a mature salamander, of which the tail has been cut off, does not reproduce a larval tail; and a crab does not reproduce a larval leg. In the case of budding it was shown in the first part of this chapter that the new being thus produced does not retrograde in development — that is, does not pass through those earlier stages, which the fertilized germ has to pass through. Nevertheless, the organisms operated on or multiplying themselves by buds must, by our hypothesis, include innumerable gemmules derived from every part or unit of the earlier stages of development; and why do not such gemmules reproduce the amputated part or the whole body at a corresponding early stage of development?

The second objection, which has been insisted on by Delpino, is that the tissues, for instance, of a mature salamander or crab, of which a limb has been removed, are already differentiated and have passed through their whole course of development; and how can such tissues in accordance with our hypothesis attract and combine with the gemmules of the part which is to be reproduced? In answer to these two objections we must bear in mind the evidence which has been advanced, showing that at least in a large number of cases the power of regrowth is a localized faculty, acquired for the sake of repairing special injuries to which each particular creature is liable; and in the case of buds or fissiparous generation, for the sake of quickly multiplying the organism at a period of life when it can be supported in large numbers. These considerations lead us to believe that in all such cases a stock of nascent cells or of partially developed gemmules are retained for this special purpose either locally or throughout the body, ready to combine with the gemmules derived from the cells which come next in due succession. If this be admitted we have a sufficient answer to the above two objections. Anyhow, pangenesis seems to throw a considerable amount of light on the wonderful power of regrowth.

It follows, also, from the view just given, that the sexual elements differ from buds in not including nascent cells or gemmules in a somewhat advanced stage of development, so that only the gemmules belonging to the earliest stages are first developed. As young animals and those which stand low in the scale generally have a much greater capacity for regrowth than older and higher animals, it would also appear that they retain cells in a nascent state, or partially developed gemmules, more readily than do animals which have already passed through a long series of developmental changes. I may here add that although ovules can be detected in most or all female animals at an extremely early age, there is no reason to doubt that gemmules derived from parts modified during maturity can pass into the ovules.

With respect to hybridism, pangenesis agrees well with most of the ascertained facts. We must believe, as previously shown, that several gemmules are requisite for the development of each cell or unit. But from the occurrence of parthenogenesis, more especially from those cases in which an embryo is only partially formed, we may infer that the female element generally includes gemmules in nearly sufficient number for independent development, so that when united with the male element the gemmules are superabundant. Now, when two species or races are crossed reciprocally, the offspring do not commonly differ, and this shows that the sexual elements agree in power, in accordance with the view that both include the same gemmules. Hybrids and mongrels are also generally intermediate in character between the two parent forms, yet occasionally they closely resemble one parent in one part and the other parent in another part, or even in their whole structure: nor is this difficult to understand on the admission that the gemmules in the fertilized germ are superabundant in number, and that those derived from one parent may have some advantage in number, affinity, or vigour over those derived from the other parent. Crossed forms sometimes exhibit the colour or other characters of either parent in stripes or blotches; and this occurs in the first generation, or through reversion in succeeding bud and seminal generations, of which fact several instances were given in the eleventh chapter. In these cases we must follow Naudin, and admit that the 'essence' or 'element' of the two species — terms which I should translate into the gemmules — have an affinity for their own kind, and thus separate themselves into distinct stripes or blotches; and reasons were given, when discussing in the fifteenth chapter the incompatibility of certain characters to unite, for believing in such mutual affinity. When two forms are crossed, one is not rarely found to be prepotent in the transmission of its characters over the other; and this we can explain by again assuming that the one form has some advantage over the other in the number, vigour, or affinity of its gemmules. In some cases, however, certain characters are present in the one form and latent in the other; for instance, there is a latent tendency in all pigeons to become blue, and, when a blue pigeon is crossed with one of any other colour, the blue tint is generally prepotent. The explanation of this form of prepotency will be obvious when we come to the consideration of reversion.

When two distinct species are crossed, it is notorious that they do not yield the full or proper number of offspring; and we can only say on this head that, as the development of each organism depends on such nicely balanced affinities between a host of gemmules and nascent cells, we need not feel at all surprised that the commixture of gemmules derived from two distinct species should lead to partial or complete failure of development. With respect to the sterility of hybrids produced from the union of two distinct species, it was shown in the nineteenth chapter that this depends exclusively on the reproductive organs being specially affected; but why these organs should be thus affected we do not know, any more than why unnatural conditions of life, though compatible with health, should cause sterility; or why continued close interbreeding, or the illegitimate unions of heterostyled plants, induce the same result. The conclusion that the reproductive organs alone are affected, and not the whole organization, agrees perfectly with the unimpaired or even increased capacity in hybrid plants for propagation by buds; for this implies, according to our hypothesis, that the cells of the hybrids throw off hybridized gemmules, which become aggregated into buds, but fail to become aggregated within the reproductive organs, so as to form the sexual elements. In a similar manner many plants, when placed under unnatural conditions, fail to produce seed, but can readily be propagated by buds. We shall presently see that pangenesis agrees well with the strong tendency to reversion exhibited by all crossed animals and plants.

Each organism reaches maturity through a longer or shorter course of growth and development: the former term being confined to mere increase of size, and development to changed structure. The changes may be small and insensibly slow, as when a child grows into a man, or many, abrupt, and slight, as in the metamorphoses of certain ephemerous insects, or, again, few and strongly marked, as with most other insects. Each newly formed part may be moulded within a previously existing and corresponding part, and in this case it will appear, falsely as I believe, to be developed from the old part; or it may be formed within a distinct part of the body, as in the extreme cases of metagenesis. An eye, for instance, may be developed at a spot where no eye previously existed. We have also seen that allied organic beings in the course of their metamorphoses sometimes attain nearly the same structure after passing through widely different forms; or conversely, after passing through nearly the same early forms, arrive at widely different mature forms. In these cases it is very difficult to accept the common view that the first-formed cells or units possess the inherent power, independently of any external agency, of producing new structures wholly different in form, position, and function. But all these cases become plain on the hypothesis of pangenesis. The units, during each stage of development, throw off gemmules, which, multiplying, are transmitted to the offspring. In the offspring, as soon as any particular cell or unit becomes partially developed, it unites with (or, to speak metaphorically, is fertilized by) the gemmule of the next succeeding cell, and so onwards. But organisms have often been subjected to changed conditions of life at a certain stage of their development, and in consequence have been slightly modified; and the gemmules cast off from such modified parts will tend to reproduce parts modified in the same manner. This process may be repeated until the structure of the part becomes greatly changed at one particular stage of development, but this will not necessarily affect other parts, whether previously or subsequently formed. In this manner we can understand the remarkable independence of structure in the successive metamorphoses, and especially in the successive metageneses of many animals. In the case, however, of diseases which supervene during old age, subsequently to the ordinary period of procreation, and which, nevertheless, are sometimes inherited, as occurs with brain and heart complaints, we must suppose that the organs were affected at an early age and threw off at this period affected gemmules; but that the affection became visible or injurious only after the prolonged growth, in the strict sense of the word, of the part. In all the changes of structure which regularly supervene during old age, we probably see the effects of deteriorated growth, and not of true development.

The principle of the independent formation of each part, owing to the union of the proper gemmules with certain nascent cells, together with the superabundance of the gemmules derived from both parents, and the subsequent self-multiplication of the gemmules, throws light on a widely different group of facts, which on any ordinary view of development appears very strange. I allude to organs which are abnormally transposed or multiplied. For instance, a curious case has been recorded by Dr. Elliott Coues of a monstrous chicken with a perfect additional *right* leg articulated to the *left* side of the pelvis. Gold-fish often have supernumerary fins placed on various parts of their bodies. When the tail of a lizard is broken off, a double tail is sometimes reproduced; and when the foot of the salamander was divided longitudinally by Bonnet, additional digits were occasionally formed. Valentin injured the caudal extremity of an embryo, and three days afterwards it produced rudiments of a double pelvis and of double hind-limbs. When frogs, toads, etc., are born with their limbs doubled, as sometimes happens, the doubling, as Gervais remarks, cannot be due to the complete fusion of two embryos, with the exception of the limbs, for the larvae are limbless. The same argument is applicable to certain insects produced with multiple legs or antennae, for these are metamorphosed from apodal or antennae-less larvae. Alphonse Milne-Edwards has described the curious case of a crustacean in which one eye peduncle supported, instead of a complete eye, only an imperfect cornea, and out of the centre of this a portion of an antenna was developed. A case has been recorded of a man who had during both dentitions a double tooth in place of the left second incisor, and he inherited this peculiarity from his paternal grandfather. Several cases are known of additional teeth having been developed in the orbit of the eye, and, more especially with horses, in the palate. Hairs occasionally appear in strange situations, as 'within the substance of the brain'. Certain breeds of sheep bear a whole crowd of horns on their foreheads. As many as five spurs have been seen on both legs of certain game-fowls. In the Polish fowl the male is ornamented with a top-knot of hackles like those on his neck, whilst the female has a top-knot formed of common feathers. In feather-footed pigeons and fowls, feathers like those on the wing arise from the outer side of the legs and toes. Even the elemental parts of the same feather may be transposed; for in the Sebastopol goose, barbules are developed on the divided filaments of the shaft. Imperfect nails sometimes appear on the stumps of the amputated fingers of man; and it is an interesting fact that with the snake-like Saurians, which present a series with more and more imperfect limbs, the terminations of the phalanges first disappear, 'the nails becoming transferred to their proximal remnants, or even to parts which are not phalanges'.

Analogous cases are of such frequent occurrence with plants that they do not strike us with sufficient surprise. Supernumerary petals, stamens, and pistils, are often produced. I have seen a leaflet low down in the compound leaf of *Vicia sativa* replaced by a tendril; and a tendril possesses many peculiar properties, such as spontaneous movement and irritability. The calyx sometimes assumes, either wholly or by stripes, the colour and texture of the corolla. Stamens are so frequently converted into petals, more or less completely, that such cases are passed over as not deserving notice; but as petals have special functions to perform, namely, tó protect the included organs, to attract insects, and in not a few cases to guide their entrance by well-adapted contrivances, we can hardly account for the conversion of stamens into petals merely by unnatural or superfluous nourishment. Again, the edge of a petal may occasionally be found including one of the highest products of the plant, namely, pollen; for instance, I have seen the pollen mass of an Ophrys, which is a very complex structure, developed in the edge of an upper petal. The segments of the calyx of the common pea have been observed partially converted into carpels, including ovules, and with their tips converted into stigmas. Mr Salter and Dr Maxwell Masters have found pollen within the ovules of the passion-flower and of the rose. Buds may be developed in the most unnatural positions, as on the petal of a flower. Numerous analogous facts could be given.

I do not know how physiologists look at such facts as the foregoing. According to the doctrine of pangenesis, the gemmules of the transposed organs become developed in the wrong place, from uniting with wrong cells or aggregates of cells during their nascent state; and this would follow from a slight modification in their elective affinities. Nor ought we to feel much surprise at the affinities of cells and gemmules varying, when we remember the many curious cases given in the seventeenth chapter, of plants which absolutely refuse to be fertilized by their own pollen, though abundantly fertile with that of any other individual of the same species, and in some cases only with that of a distinct species. It is manifest that the sexual elective affinities of such plants — to use the term employed by Gärtner — have been modified. As the cells of adjoining or homologous parts will have nearly the same nature, they will be particularly liable to acquire by variation each other's elective affinities; and we can thus understand to a certain extent such cases as a crowd of horns on the heads of certain sheep, of several spurs on the legs of fowls, hackle-like feathers on the heads of the males of other fowls, and with the pigeon wing-like feathers on their legs and membrane between their toes, for the leg is the homologue of the wing. As all the organs of plants are homologous and spring from a common axis, it is natural that they should be eminently liable to transposition. It ought to be observed that when any compound part, such as an additional limb or an antenna, springs from a false position, it is only necessary that the few first gemmules should be wrongly attached; for these whilst developing would attract other gemmules in due succession, as in the regrowth of an amputated limb. When parts which are homologous and similar in structure, as the vertebrae of snakes or the stamens of polyandrous flowers, etc., are repeated many times in the same organism, closely allied gemmules must be extremely numerous, as well as the points to which they ought to become united; and, in accordance with the foregoing views, we can to a certain extent understand Isid. Geoffroy Saint-Hilaire's law, that parts, which are already multiple, are extremely liable to vary in number.

Variability often depends, as I have attempted to show, on the reproductive organs being injuriously affected by changed conditions; and in this case the gemmules derived from the various parts of the body are probably aggregated in an irregular manner, some superfluous and others deficient. Whether a superabundance of gemmules would lead to the increased size of any part cannot be told; but we can see that their partial deficiency, without necessarily leading to the entire abortion of the part, might cause considerable modifications; for in the same manner as plants, if their own pollen be excluded, are easily hybridized, so, in the case of cells, if the properly succeeding gemmules were absent, they would probably combine easily with other and allied gemmules, as we have just seen with transposed parts.

In variations caused by the direct action of changed conditions, of which several instances have been given, certain parts of the body are directly affected by the new conditions, and consequently throw off modified gemmules, which are transmitted to the offspring. On any ordinary view it is unintelligible how changed conditions, whether acting on the embryo, the young, or the adult, can cause inherited modifications. It is equally or even more unintelligible on any ordinary view, how the effects of the long-continued use or disuse of a part, or of changed habits of body or mind, can be inherited. A more perplexing problem can hardly be proposed; but on our view we have only to suppose that certain cells become at last structurally modified; and that these throw off similarly modified gemmules. This may occur at any period of development, and the modification will be inherited at a corresponding period; for the modified gemmules will unite in all ordinary cases with the proper preceding cells, and will consequently be developed at the same period at which the modification first arose. With respect to mental habits or instincts, we are so profoundly ignorant of the relation between the brain and the power of thought that we do not know positively whether a fixed habit induces any change in the nervous system, though this seems highly probable; but when such habit or other mental attribute, or insanity, is inherited, we must believe that some actual modification is transmitted; and this implies, according to our hypothesis, that gemmules derived from modified nerve-cells are transmitted to the offspring.

It is generally necessary than an organism should be exposed during several generations to changed conditions or habits, in order that any modification thus acquired should appear in the offspring. This may be partly due to the changes not being at first marked enough to catch attention, but this explanation is insufficient; and I can account for the fact only by the assumption, which we shall see under the head of reversion is strongly supported, that gemmules derived from each unmodified unit or part are transmitted in large numbers to successive generations, and that the gemmules derived from the same unit after it has been modified go on multiplying under the same favourable conditions which first caused the modification, until at last they become sufficiently numerous to overpower and supplant the old gemmules.

A difficulty may be here noticed; we have seen that there is an important difference in the frequency, though not in the nature, of the variations in plants propagated by sexual and asexual generation. As far as variability depends on the imperfect action of the reproductive organs under changed conditions, we can at once see why plants propagated asexually should be far less variable than those propagated sexually. With respect to the direct action of changed conditions, we know that organisms produced from buds do not pass through the earlier phases of development; they will therefore not be exposed at that period of life when structure is most readily modified, to the various causes inducing variability in the same manner as are embryos and young larval forms; but whether this is a sufficient explanation I know not.

With respect to variations due to reversion, there is a similar difference between plants propagated from buds and seeds. Many varieties can be propagated securely by buds, but generally or invariably revert to their parent forms by seed. So, also, hybridized plants can be multiplied to any extent by buds, but are continually liable to reversion by seed — that is, to the loss of their hybrid or intermediate character. I can offer no satisfactory explanation of these facts. Plants with variegated leaves, phloxes with striped flowers, barberries with seedless fruit, can all be securely propagated by buds taken from the stem or branches; but buds from the roots of these plants almost invariably lose their character and revert to their former condition. This latter fact is also inexplicable, unless buds developed from the roots are as distinct from those on the stem, as is one bud on the stem from another, and we know that these latter behave like independent organisms.

Finally, we see that on the hypothesis of pangenesis variability depends on at least two distinct groups of causes. Firstly, the deficiency, superabundance, and transposition of gemmules, and the redevelopment of those which have long been dormant: the gemmules themselves not having undergone any modification; and such changes will amply account for much fluctuating variability. Secondly, the direct action of changed conditions on the organization, and of the increased use or disuse of parts; and in this case the gemmules from the modified units will be themselves modified, and, when sufficiently multiplied, will supplant the old gemmules and be developed into new structures.

Turning now to the laws of inheritance. If we suppose a homogeneous gelatinous protozoon to vary and assume a reddish colour, a minute separated particle would naturally, as it grew to full size, retain the same colour; and we should have the simplest form of inheritance. Precisely the same view may be extended to the infinitely numerous and diversified units of which the whole body of one of the higher animals is composed; the separated particles being our gemmules. We have already sufficiently discussed by implication, the important principle of inheritance at corresponding ages. Inheritance as limited by sex and by the season of the year (for instance with animals becoming white in winter) is intelligible if we may believe that the elective affinities of the units of the body are slightly different in the two sexes, especially at maturity, and in one or both sexes at different seasons, so that they unite with different gemmules. It should be remembered that, in the discussion on the abnormal transposition of organs, we have seen reason to believe that such elective affinities are readily modified. But I shall soon have to recur to sexual and seasonal inheritance. These several laws are therefore explicable to a large extent through pangenesis, and on no other hypothesis which has as yet been advanced.

But it appears at first sight a fatal objection to our hypothesis that a part or organ may be removed during several successive generations, and if the operation be not followed by disease, the lost part reappears in the offspring. Dogs and horses formerly had their tails docked during many generations without any inherited effect; although, as we have seen, there is some reason to believe that the tailless condition of certain sheep-dogs is due to such inheritance. Circumcision has been practised by the Jews from a remote period, and in most cases the effects of the operation are not visible in the offspring; though some maintain that an inherited effect does occasionally appear. If inheritance depends on the presence of disseminated gemmules derived from all the units of the body, why does not the amputation or mutilation of a part, especially if effected on both sexes, invariably affect the offspring? The answer in accordance with our hypothesis probably is that gemmules multiply and are transmitted during a long series of generations — as we see in the reappearance of zebrine stripes on the horse — in the reappearance of muscles and other structures in man which are proper to his lowly organized progenitors, and in many other such cases. Therefore the long-continued inheritance of a part which has been removed during many generations is no real anomaly, for gemmules formerly derived from the part are multiplied and transmitted from generation to generation.

We have as yet spoken only of the removal of parts, when not followed by morbid action: but when the operation is thus followed, it is certain that the deficiency is sometimes inherited. In a former chapter instances were given, as of a cow, the loss of whose horn was followed by suppuration, and her calves were destitute of a horn on the same side of their heads. But the evidence which admits of no doubt is that given by Brown-Séquard with respect to guinea-pigs, which after their sciatic nerves had been divided, gnawed off their own gangrenous toes, and the toes of their offspring were deficient in at least thirteen instances on the corresponding feet. The inheritance of the lost part in several of these cases is all the more remarkable as only one parent was affected; but we know that a congenital deficiency is often transmitted from one parent alone — for instance, the offspring of hornless cattle of either sex, when crossed with perfect animals, are often hornless. How, then, in accordance with our hypothesis can we account for mutilations being sometimes strongly inherited, if they are followed by diseased action? The answer probably is that all the gemmules of the mutilated or amputated part are gradually attracted to the diseased surface during the reparative process, and are there destroyed by the morbid action.

A few words must be added on the complete abortion of organs. When a part becomes diminished by disuse prolonged during many generations, the principle of economy of growth, together with intercrossing, will tend to reduce it still further as previously explained, but this will not account for the complete or almost complete obliteration of, for instance, a minute papilla of cellular tissue representing a pistil, or of a microscopically minute nodule of bone representing a tooth. In certain cases of suppression not yet completed, in which a rudiment occasionally reappears through reversion, dispersed gemmules derived from this part must, according to our view, still exist; we must therefore suppose that the cells, in union with which the rudiment was formerly developed, fail in their affinity for such gemmules, except in the occasional cases of reversion. But when the abortion is complete and final, the gemmules themselves no doubt perish; nor is this in any way improbable, for, though a vast number of active and long-dormant gemmules are nourished in each living creature, yet there must be some limit to their number; and it appears natural that gemmules derived from reduced and useless parts would be more liable to perish than those freshly derived from other parts which are still in full functional activity.

The last subject that needed be discussed, namely, reversion, rests on the principle that transmission and development, though generally acting in conjunction, are distinct powers; and the transmission of gemmules with their subsequent development shows us how this is possible. We plainly see the

distinction in the many cases in which a grandfather transmits to his grand-
son, through his daughter, characters which she does not, or cannot, possess. But before proceeding, it will be advisable to say a few words about
latent or dormant characters. Most, or perhaps all, of the secondary characters, which appertain to one sex, lie dormant in the other sex; that is gemmules capable of development into the secondary male sexual characters are
included within the female; and conversely female characters in the male:
we have evidence of this in certain masculine characters, both corporeal and
mental, appearing in the female, when her ovaria are diseased or when they
fail to act from old age. In like manner female characters appear in castrated
males, as in the shape of the horns of the ox, and in the absence of horns in
castrated stags. Even a slight change in the conditions of life due to confinement sometimes suffices to prevent the development of masculine characters in male animals, although their reproductive organs are not permanently injured. In the many cases in which masculine characters are
periodically renewed, these are latent at other seasons; inheritance as limited
by sex and season being here combined. Again, masculine characters generally lie dormant in male animals until they arrive at the proper age for
reproduction. The curious case formerly given of a hen which assumed the
masculine characters, not of her own breed but of a remote progenitor, illustrates the close connection between latent sexual characters and ordinary
reversion.

With those animals and plants which habitually produce several forms, as
with certain butterflies described by Mr Wallace, in which three female forms
and one male form co-exist, or, as with the trimorphic species of Lythrum
and Oxalis, gemmules capable of reproducing these different forms must be
latent in each individual.

Insects are occasionally produced with one side or one-quarter of their
bodies like that of the male, with the other half or three-quarters like that of
the female. In such cases the two sides are sometimes wonderfully different
in structure, and are separated from each other by a sharp line. As gemmules derived from every part are present in each individual of both sexes,
it must be the elective affinities of the nascent cells which in these cases
differ abnormally on the two sides of the body. Almost the same principle
comes into play with those animals, for instance, certain gasteropods and
Verruca among cirripedes, which normally have the two sides of the body
constructed on a very different plan; and yet a nearly equal number of individuals have either side modified in the same remarkable manner.

Reversion, in the ordinary sense of the word, acts so incessantly, that it
evidently forms an essential part of the general law of inheritance. It occurs
with beings, however propagated, whether by buds or seminal generation,
and sometimes may be observed with advancing age even in the same indi-

vidual. The tendency to reversion is often induced by a change of conditions, and in the plainest manner by crossing. Crossed forms of the first generation are generally nearly intermediate in character between their two parents; but in the next generation the offspring commonly revert to one or both of their grandparents, and occasionally to more remote ancestors. How can we account for these facts? Each unit in a hybrid must throw off, according to the doctrine of pangenesis, an abundance of hybridized gemmules, for crossed plants can be readily and largely propagated by buds; but by the same hypothesis dormant gemmules derived from both pure parent forms are likewise present; and as these gemmules retain their normal condition, they would, it is probable, be enabled to multiply largely during the lifetime of each hybrid. Consequently the sexual elements of a hybrid will include both pure and hybridized gemmules; and when two hybrids pair, the combination of pure gemmules derived from the one hybrid with the pure gemmules of the same parts derived from the other, would necessarily lead to complete reversion of character; and it is, perhaps, not too bold a supposition that unmodified and undeteriorated gemmules of the same nature would be especially apt to combine. Pure gemmules in combination with hybridized gemmules would lead to partial reversion. And lastly, hybridized gemmules derived from both parent hybrids would simply reproduce the original hybrid form. All these cases and degrees of reversion incessantly occur.

It was shown in the fifteenth chapter that certain characters are antagonistic to each other or do not readily blend; hence, when two animals with antagonistic characters are crossed, it might well happen that a sufficiency of gemmules in the male alone for the reproduction of his peculiar characters, and in the female alone for the reproduction of her peculiar characters, would not be present; and in this case dormant gemmules derived from the same part in some remote progenitor might easily gain the ascendancy, and cause the reappearance of the long-lost character. For instance, when black and white pigeons, or black and white fowls, are crossed — colours which do not readily blend — blue plumage in the one case, evidently derived from the rock-pigeon, and red plumage in the other case, derived from the wild jungle-cock, occasionally reappear. With uncrossed breeds the same result follows, under conditions which favour the multiplication and development of certain dormant gemmules, as when animals become feral and revert to their pristine character. A certain number of gemmules being requisite for the development of each character, as is known to be the case from several spermatozoa or pollen grains being necessary for fertilization, and time favouring their multiplication, will perhaps account for the curious cases, insisted on by Mr Sedgwick, of certain diseases which regularly appear in alternate generations. This likewise holds good, more or less strictly, with other weakly inherited modifications. Hence, as I have heard it remarked, certain diseases appear to gain strength by the intermission of a generation. The transmission of dormant gemmules during many successive generations is hardly in itself more improbable, as previously remarked, than the retention during many ages of rudimentary organs, or even only of a tendency to the production of a rudiment; but there is no reason to suppose that dormant gemmules can be transmitted and propagated for ever. Excessively minute and numerous as they are believed to be, an infinite number derived, during a long course of modification and descent, from each unit of each progenitor, could not be supported or nourished by the organism. But it does not seem improbable that certain gemmules, under favourable conditions, should be retained and go on multiplying for a much longer period than others. Finally, on the view here given, we certainly gain some insight into the wonderful fact that the child may depart from the type of both its parents, and resemble its grandparents, or ancestors removed by many hundreds of generations.

CONCLUSION

The hypothesis of pangenesis, as applied to the several great classes of facts just discussed, no doubt is extremely complex, but so are the facts. The chief assumption is that all the units of the body, besides having the universally admitted power of growing by self-division, throw off minute gemmules which are dispersed through the system. Nor can this assumption be considered as too bold, for we know from the cases of graft-hybridization that formative matter of some kind is present in the tissues of plants, which is capable of combining with that included in another individual, and of reproducing every unit of the whole organism. But we have further to assume that the gemmules grow, multiply, and aggregate themselves into buds and the sexual elements; their development depending on their union with other nascent cells or units. They are also believed to be capable of transmission in a dormant state, like seeds in the ground, to successive generations.

In a highly organized animal, the gemmules thrown off from each different unit throughout the body must be inconceivably numerous and minute. Each unit of each part, as it changes during development, and we know that some insects undergo at least twenty metamorphoses, must throw off its gemmules. But the same cells may long continue to increase by self-division, and even become modified by absorbing peculiar nutriment, without necessarily throwing off modified gemmules. All organic beings, moreover, include many dormant gemmules derived from their grandparents and more remote progenitors, but not from all their progenitors. These almost infinitely numerous and minute gemmules are contained within each bud, ovule, spermatozoon, and pollen grain. Such an admission will be declared impossible;

but number and size are only relative difficulties. Independent organisms exist which are barely visible under the highest powers of the microscope, and their germs must be excessively minute. Particles of infectious matter, so small as to be wafted by the wind or to adhere to smooth paper, will multiply so rapidly as to infect within a short time the whole body of a large animal. We should also reflect on the admitted number and minuteness of the molecules composing a particle of ordinary matter. The difficulty, therefore, which at first appears insurmountable, of believing in the existence of gemmules so numerous and small as they must be according to our hypothesis, has no great weight.

The units of the body are generally admitted by physiologists to be autonomous. I go one step further and assume that they throw off reproductive gemmules. Thus an organism does not generate its kind as a whole, but each separate unit generates its kind. It has often been said by naturalists that each cell of a plant has the potential capacity of reproducing the whole plant; but it has this power only in virtue of containing gemmules derived from every part. When a cell or unit is from some cause modified, the gemmules derived from it will be in like manner modified. If our hypothesis be provisionally accepted, we must look at all the forms of asexual reproduction, whether occurring at maturity or during youth, as fundamentally the same, and dependent on the mutual aggregation and multiplication of the gemmules. The regrowth of an amputated limb and the healing of a wound is the same process partially carried out. Buds apparently include nascent cells, belonging to that stage of development at which the budding occurs, and these cells are ready to unite with the gemmules derived from the next succeeding cells. The sexual elements, on the other hand, do not include such nascent cells; and the male and female elements taken separately do not contain a sufficient number of gemmules for independent development, except in the cases of parthenogenesis. The development of each being, including all the forms of metamorphosis and metagenesis, depends on the presence of gemmules thrown off at each period of life, and on their development, at a corresponding period, in union with preceding cells. Such cells may be said to be fertilized by the gemmules which come next in due order of development. Thus the act of ordinary impregnation and the development of each part in each being are closely analogous processes. The child, strictly speaking, does not grow into the man, but includes germs which slowly and successively become developed and form the man. In the child, as well as in the adult, each part generates the same part. Inheritance must be looked at as merely a form of growth, like the self-division of a lowly organized unicellular organism. Reversion depends on the transmission from the forefather to his descendants of dormant gemmules, which occasionally become developed under certain known or unknown conditions. Each animal and plant may be compared with a bed of soil full of seeds, some of which soon germinate, some lie dormant for a period, whilst others perish. When we hear it said that a man carries in his constitution the seeds of an inherited disease, there is much truth in the expression. No other attempt, as far as I am aware, has been made, imperfect as this confessedly is, to connect under one point of view these several grand classes of facts. An organic being is a microcosm — a little universe, formed of a host of self-propagating organisms, inconceivably minute and numerous as the stars in heaven.

TROUBLES WITH PANGENESIS

Of all of Darwin's hypotheses, it was pangenesis that caused him the most problems in the scientific community, especially among those who favored evolution and natural selection. He reviewed some of this reaction in a letter to Hooker:

> I fear Pangenesis is stillborn; [Henry] Bates says he has read it twice, and is not sure that he understands it. H. Spencer says the view is quite different from his (and this is a great relief to me, as I feared to be accused of plagiarism, but utterly failed to be sure what he meant, so thought it safest to give my view as almost the same as his), and he says he is not sure he understands it. . . . Am I not a poor devil? yet I took such pains, I must think that I expressed myself clearly. Old Sir H. Holland says he has read it twice, and thinks it very tough; but believes that sooner or later, "some view akin to it" will be accepted.
>
> You will think me very self-sufficient, when I declare that I feel *sure* if Pangenesis is not stillborn it will, thank God, at some future time appear, begotten by some other father, and christened by some other name.
>
> Have you ever met with any tangible and clear view of what takes place in generation, whether by seeds or buds, or how a long-lost character can possibly reappear; or how the male element can possibly affect the other plant, or the mother animal, so that her future progeny are affected? Now all these points and many others are connected together, whether truely or falsely is another question, by Pangenesis. You see I die hard, and stick up for my poor child.[1]

1. Darwin to Hooker, 23 February [1860], *Life and Letters*, 3: 78.

Chapter 10

THE DESCENT OF MAN (SECOND EDITION, 1874)

INTRODUCTION: MORAL FACULTIES

In the famous discussion of "moral faculties" in *The Descent of Man*, Darwin sought to address human culture from an evolutionary perspective and, in so doing, to probe the limits of natural selection. The chapter was enormously influential because it provided an obvious point of intersection between Darwin's biological discourse and the social, ethical, and theological concerns of many of his readers. In fact, there was little that was original about the chapter; it was a restatement of Enlightenment social theory, recapitulating without attribution widely known ideas culled from Hobbes, Locke, Hume, and Adam Smith. In social thought, at least, Darwin had much more in common with thinkers of the eighteenth century than with those of the nineteenth.

Darwin's discussion is an extended speculation on the evolution of early human society, expressed in the same format as and centering on the familiar themes of natural-man theory.[1] Darwin's primitive human was governed by instincts such as sympathy (which is precisely the "fellow feeling"

1. The influence of natural-man theory on Darwin has not been explicitly considered by historians. On the influence of Enlightenment social theory on Darwin see, in particular, Sylvan S. Schweber, "Darwin and the Political Economists: Divergence of Character," *Journal of the History of Biology* 13 (1980): 195–289; *idem*, "The Wider British Context in Darwin's Theorizing," in D. Kohn, ed., *The Darwinian Heritage* (Princeton, Princeton University Press, 1985), 35–69; and Robert M. Young, "Malthus and the Evolutionists: The Common Context and Biological and Social Theory," *Past and Present* 43 (May 1969), reprinted in Young, *Nature's Place in Victorian Society* (Cambridge, Cambridge University Press, 1985), 23–55.

of Adam Smith), a social quality acquired through natural selection. This quality develops out of self-interest from the individual's observation of the benefits accruing from group solidarity. In this case, natural selection acts to confer a competitive advantage not directly to the individual, but rather to the kinship group in the "never-ceasing wars of savages." The last image, as Darwin's nineteenth-century English readers must have instantly recognized, was an allusion to Hobbes's wars of "each against all" that a "natural" society, one with no laws, must inevitably suffer—hence, his equally Hobbesian conclusion that "any form of government is better than none." The human instinct for self-preservation then is transferred to the group, which provides a biological rationale for Locke's notion that society's primordial objective is the preservation of life.

Civilization further encouraged the development of sympathy, but through cultural rather than strictly biological mechanisms. Morality also developed through self-interest: Darwin, like Smith, believed that morality was the result of the individual's acting to enhance his own survival by using society's approval as a mirror by which to guide his own actions. Others will be more likely to help persons who act in socially acceptable ways than those who do not. Therefore, in Darwin's view, a person who acts "morally"—that is, in accordance with the approval of his social reference group—also acts to preserve himself and thereby gains a slight competitive advantage in the struggle for life.

In developing this discussion, Darwin was bound to form some judgment concerning the relationship between culture and natural selection, which raises the issue of to what extent he himself might be regarded a "Social Darwinist"—that is, to what extent did he think that natural selection, a law of nature, still, in spite of the rise of civilization, affected human society? We can see from the way his argument is formulated that Darwin realized that culture tended to mute the action of selection in that, most particularly, morality demanded that the strong help the weak to survive. The sorry state of the social world led him to observe that "the vicious members of society increase faster than the virtuous," a conclusion that tended to support the general evolutionary principle that progress was not, and could not be, an invariable rule. The reason for what his progress-minded Victorian readers must have experienced as a pessimistic conclusion followed from the tentative nature of natural selection's action. But even though muted, it still acts on human society, especially through cultural mechanisms: Thus, he noted the deleterious effects of celibacy in Spain, where many of the most talented members of that society were removed from the breeding pool, or the more drastic form of selection that the Inquisition practiced in executing talented converted Jews. A positive example was the self-selection whereby the "best people" in Europe, as defined by such qualities as energy and courage, had formed an exceptionally talented population in emigrating to the United States.

ON THE DEVELOPMENT OF THE
INTELLECTUAL AND MORAL FACULTIES
DURING PRIMEVAL AND CIVILISED TIMES

The subjects to be discussed in this chapter are of the highest interest, but are treated by me in a most imperfect and fragmentary manner. Mr. Wallace, in an admirable paper before referred to, argues that man after he had partially acquired those intellectual and moral faculties which distinguish him from the lower animals, would have been but little liable to have had his bodily structure modified through natural selection or any other means. For man is enabled through his mental faculties "to keep with an unchanged body in harmony with the changing universe." He has great power of adapting his habits to new conditions of life. He invents weapons, tools and various stratagems, by which he procures food and defends himself. When he migrates into a colder climate he uses clothes, builds sheds, and makes fires; and, by the aid of fire, cooks food otherwise indigestible. He aids his fellow-men in many ways, and anticipates future events. Even at a remote period he practised some subdivision of labour.

The lower animals, on the other hand, must have their bodily structure modified in order to survive under greatly changed conditions. They must be rendered stronger, or acquire more effective teeth or claws, in order to defend themselves from new enemies; or they must be reduced in size so as to escape detection and danger. When they migrate into a colder climate they must become clothed with thicker fur, or have their constitutions altered. If they fail to be thus modified, they will cease to exist.

The case, however, is widely different, as Mr. Wallace has with justice insisted, in relation to the intellectual and moral faculties of man. These faculties are variable; and we have every reason to believe that the variations tend to be inherited. Therefore, if they were formerly of high importance to primeval man and to his ape-like progenitors, they would have been perfected or advanced through natural selection. Of the high importance of the intellectual faculties there can be no doubt, for man mainly owes to them his predominant position in the world. We can see that, in the rudest state of society, the individuals who were the most sagacious, who invented and used the best weapons or traps, and who were best able to defend themselves, would rear the greatest number of offspring. The tribes which included the largest number of men thus endowed would increase in number and supplant other tribes. Numbers depend primarily on the means of subsistence, and this, depends partly on the physical nature of the country, but in a much higher degree on the arts which are there practised. As a tribe increases and is victorious, it is often still further increased by the absorption of other tribes. The stature and strength of the men of a tribe are likewise of some importance for its success, and these depend in part on the nature and amount of the food which can be obtained. In Europe the men of the Bronze period were supplanted by a more powerful and, judging from their sword-handles, larger-handed race; but their success was probably due in a much higher degree to their superiority in the arts.

All that we know about savages, or may infer from their traditions and from old monuments, the history of which is quite forgotten by the present inhabitants shew that from the remotest times successful tribes have supplanted other tribes. Relics of extinct or forgotten tribes have been discovered throughout the civilised regions of the earth, on the wild plains of America, and on the isolated islands in the Pacific Ocean. At the present day civilised nations are everywhere supplanting barbarous nations, excepting where the climate opposes a deadly barrier; and they succeed mainly, though not exclusively, through their arts, which are the products of the intellect. It is, therefore, highly probable that with mankind the intellectual faculties have been gradually perfected through natural selection; and this conclusion is sufficient for our purpose. Undoubtedly it would be interesting to have traced the development of each separate faculty from the state in which it exists in the lower animals to that in which it exists in man; but neither my ability nor knowledge permit the attempt.

It deserves notice that as soon as the progenitors of man became social (and this probably occurred at a very early period), the advancement of the intellectual faculties will have been aided and modified in an important manner, of which we see only traces in the lower animals, namely, through the principle of imitation, together with reason and experience. Apes are much given to imitation, as are the lowest savages; and the simple fact previously referred to, that after a time no animal can be caught in the same place by the same sort of trap, shews that animals learn by experience, and imitate each others' caution. Now, if some one man in a tribe, more sagacious than the others, invented a new snare or weapon, or other means of attack or defence, the plainest self-interest, without the assistance of much reasoning power, would prompt the other members to imitate him; and all would thus profit. The habitual practice of each new art must likewise in some slight degree strengthen the intellect. If the new invention were an important one, the tribe would increase in number, spread, and supplant other tribes. In a tribe thus rendered more numerous there would always be a rather better chance of the birth of other superior and inventive members. If such men left children to inherit their mental superiority, the chance of the birth of still more ingenious members would be somewhat better, and in a very small tribe decidedly better. Even if they left no children, the tribe would still include their blood-relations; and it has been ascertained by agriculturists that by preserving and breeding from the family of an animal, which when slaughtered was found to be valuable, the desired character has been obtained.

Turning now to the social and moral faculties. In order that primeval men, or the ape-like progenitors of man, should have become social, they must have acquired the same instinctive feelings which impel other animals to live in a body; and they no doubt exhibited the same general disposition. They would have felt uneasy when separated from their comrades, for whom they would have felt some degree of love; they would have warned each other of danger, and have given mutual aid in attack or defence. All this implies some degree of sympathy, fidelity, and courage. Such social qualities, the paramount importance of which to the lower animals is disputed by no one, were no doubt acquired by the progenitors of man in a similar manner, namely, through natural selection, aided by inherited habit. When two tribes of primeval man, living in the same country, came into competition, if the one tribe included (other circumstances being equal) a great number of courageous, sympathetic and faithful members, who were always ready to warn each other of danger, to aid and defend each other, this tribe would without doubt succeed best and conquer the other. Let it be borne in mind how all-important, in the never-ceasing wars of savages, fidelity and courage must be. The advantage which disciplined soldiers have over undisciplined hordes follows chiefly from the confidence which each man feels in his comrades. Obedience, as Mr. Bagehot has well shewn, is of the highest value, for any form of government is better than none. Selfish and contentious people will not cohere, and without coherence nothing can be effected. A tribe possessing the above qualities in a high degree would spread and be victorious over other tribes; but in the course of time it would, judging from all past history, be in its turn overcome by some other and still more highly endowed tribe. Thus the social and moral qualities would tend slowly to advance and be diffused throughout the world.

But it may be asked, how within the limits of the same tribe did a large number of members first become endowed with these social and moral qualities, and how was the standard of excellence raised? It is extremely doubtful whether the offspring of the more sympathetic and benevolent parents, or of those which were the most faithful to their comrades, would be reared in greater number than the children of selfish and treacherous parents of the same tribe. He who was ready to sacrifice his life, as many a savage has been, rather than betray his comrades, would often leave no offspring to inherit his noble nature. The bravest men, who were always willing to come to the front in war, and who freely risked their lives for others, would on an average perish in larger numbers than other men. Therefore, it seems scarcely possible (bearing in mind that we are not here speaking of one tribe being victorious over another) that the number of men gifted with such virtues, or that the standard of their excellence, could be increased through natural selection, that is, by the survival of the fittest.

Although the circumstances which lead to an increase in the number of men thus endowed within the same tribe are too complex to be clearly followed out, we can trace some of the probable steps. In the first place, as the reasoning powers and foresight of the members became improved, each man would soon learn from experience that if he aided his fellow-men, he would commonly receive aid in return. From this low motive he might acquire the habit of aiding his fellows; and the habit of performing benevolent actions certainly strengthens the feeling of sympathy, which gives the first impulse to benevolent actions. Habits, moreover, followed during many generations probably tend to be inherited.

But there is another and much more powerful stimulus to the development of the social virtues, namely, the praise and the blame of our fellow-men. The love of approbation and the dread of infamy, as well as the bestowal of praise or blame, are primarily due, as we have seen in the third chapter, to the instinct of sympathy; and this instinct no doubt was originally acquired, like all the other social instincts, through natural selection. At how early a period the progenitors of man, in the course of their development, became capable of feeling and being impelled by the praise or blame of their fellow-creatures, we cannot, of course say. But it appears that even dogs appreciate encouragement, praise, and blame. The rudest savages feel the sentiment of glory, as they clearly show by preserving the trophies of their prowess, by their habit of excessive boasting, and even by the extreme care which they take of their personal appearance and decorations; for unless they regarded the opinion of their comrades, such habits would be senseless.

They certainly feel shame at the breach of some of their lesser rules; but how far they experience remorse is doubtful. I was at first surprised that I could not recollect any recorded instances of this feeling in savages; and Sir J. Lubbock states that he knows of none. But if we banish from our minds all cases given in novels and plays and in death-bed confessions made to priests, I doubt whether many of us have actually witnessed remorse; though we may have often seen shame and contrition for smaller offences. Remorse is a deeply hidden feeling. It is incredible that a savage, who will sacrifice his life rather than betray his tribe, or one who will deliver himself up as a prisoner rather than break his parole, would not feel remorse in his inmost soul, though he might conceal it, if he had failed in a duty which he held sacred.

We may therefore conclude that primeval man, at a very remote period, would have been influenced by the praise and blame of his fellows. It is obvious, that the members of the same tribe would approve of conduct which appeared to them to be for the general good, and would reprobate that which appeared evil. To do good unto others—to do unto others as ye would they should do unto you,—is the foundation-stone of morality. It is, therefore, hardly possible to exaggerate the importance during rude times of the love of praise and the dread of blame. A man who was not impelled by any deep, instinctive feeling, to sacrifice his life for the good of others, yet was roused to such actions by a sense of glory, would by his example excite the same wish for glory in other men, and would strengthen by exercise the noble feeling of admiration. He might thus do far more good to his tribe than by begetting offspring with a tendency to inherit his own high character.

With increased experience and reason, man perceives the more remote consequences of his actions, and the self-regarding virtues, such as temperance, chastity, &c., which during early times are, as we have before seen, utterly disregarded, come to be highly esteemed or even held sacred. I need not, however, repeat what I have said on this head in the third chapter. Ultimately, a highly complex sentiment, having its first origin in the social instincts, largely guided by the approbation of our fellow-men, ruled by reason, self-interest, and in later times by deep religious feelings, and confirmed by instruction and habit, all combined, constitute our moral sense or conscience.

It must not be forgotten that although a high standard of morality gives but a slight or no advantage to each individual man and his children over the other men of the same tribe, yet that an advancement in the standard of morality and an increase in the number of well-endowed men will certainly give an immense advantage to one tribe over another. There can be no doubt that a tribe including many members who, from possessing in a high degree the spirit of patriotism, fidelity, obedience, courage, and sympathy, were always ready to give aid to each other and to sacrifice themselves for the common good, would be victorious over most other tribes; and this would be natural selection. At all times throughout the world tribes have supplanted other tribes; and as morality is one element in their success, the standard of morality and the number of well-endowed men will thus everywhere tend to rise and increase.

It is, however, very difficult to form any judgement why one particular tribe and not another has been successful and has risen in the scale of civilisation. Many savages are in the same condition as when first discovered several centuries ago. As Mr. Bagehot has remarked, we are apt to look at the progress as normal in human society; but history refutes this. The ancients did not even entertain the idea, nor do the oriental nations at the present day. According to another high authority, Mr. Maine, "the greatest part of mankind has never shewn a particle of desire that its civil institutions should be improved." Progress seems to depend on many concurrent favourable conditions, far too complex to be followed out. But it has often been remarked, that a cool climate, from leading to industry and to the various arts has been highly favourable, or even indispensable for this end. The Esquimaux, pressed by hard necessity, have succeeded in many ingenious inventions, but their climate has been too severe for continued progress. Nomadic habits, whether over wide plains, or through the dense forests of the tropics, or along the shores of the sea, have in every case been highly detrimental. Whilst observing the barbarous inhabitants of Tierra del Fuego, it struck me that the possession of some property, a fixed abode, and the union of many families under a chief, were the indispensable requisites for civilisation. Such habits almost necessitate the cultivation of the ground; and the first steps in cultivation would probably result, as I have elsewhere shewn,[1] from some such accident as the seeds of a fruit-tree falling on a heap of refuse, and producing an unusually fine variety. The problem, however, of the first advance of savages towards civilisation is at present much too difficult to be solved.

Natural Selection as affecting Civilised Nations.—In the last and present chapters I have considered the advancement of man from a former semi-human condition to his present state as a barbarian. But some remarks on the agency of natural selection on civilised nations may be here worth adding. This subject has been ably discussed by Mr. W. R. Greg, and previously by Mr. Wallace and Mr. Galton. Most of my remarks are taken from these three authors. With savages, the weak in body or mind are soon eliminated; and those that survive commonly exhibit a vigorous state of health. We civilised men, on the other hand, do our utmost to check the process of elimination; we build asylums for the imbecile, the maimed, and the sick; we institute poorlaws; and our medical men exert their utmost skill to save the life of every one to the last moment. There is reason to believe that vaccination has preserved thousands, who from a weak constitution would formerly have succumbed to small-pox. Thus the weak members of civilised societies propagate their kind. No one who has attended to the breeding of domestic animals will doubt that this must be highly injurious to the race of man. It is surprising how soon a want of care, or care wrongly directed, leads to the degeneration of a domestic race; but excepting in the case of man himself, hardly any one is so ignorant as to allow his worst animals to breed.

1. 'The Variation of Animals and Plants under Domestication,' vol. i. p. 309.

The aid which we feel impelled to give to the helpless is mainly an incidental result of the instinct of sympathy, which was originally acquired as part of the social instincts, but subsequently rendered, in the manner previously indicated, more tender and more widely diffused. Nor could we check our sympathy, if so urged by hard reason, without deterioration in the noblest part of our nature. The surgeon may harden himself whilst performing an operation, for he knows that he is acting for the good of his patient; but if we were intentionally to neglect the weak and helpless, it could only be for a contingent benefit, with a certain and great present evil. Hence we must bear without complaining the undoubtedly bad effects of the weak surviving and propagating their kind; but there appears to be at least one check in steady action, namely the weaker and inferior members of society not marrying so freely as the sound; and this check might be indefinitely increased, though this is more to be hoped for than expected, by the weak in body or mind refraining from marriage.

In all civilised countries man accumulates property and bequeaths it to his children. So that the children in the same country do not by any means start fair in the race for success. But this is far from an unmixed evil; for without the accumulation of capital the arts could not progress; and it is chiefly through their power that the civilised races have extended, and are now everywhere extending, their range, so as to take the place of the lower races. Nor does the moderate accumulation of wealth interfere with the process of selection. When a poor man becomes moderately rich, his children enter trades or professions in which there is struggle enough, so that the able in body and mind succeed best. The presence of a body of well-instructed men, who have not to labour for their daily bread, is important to a degree which cannot be over-estimated; as all high intellectual work is carried on by them, and on such work material progress of all kinds mainly depends, not to mention other and higher advantages. No doubt wealth when very great tends to convert men into useless drones, but their number is never large; and some degree of elimination here occurs, for we daily see rich men, who happen to be fools or profligate, squandering away their wealth.

Primogeniture with entailed estates is a more direct evil, though it may formerly have been a great advantage by the creation of a dominant class, and any government is better than anarchy. The eldest sons, though they may be weak in body or mind, generally marry, whilst the younger sons, however superior in these respects, do not so generally marry. Nor can worthless eldest sons with entailed estates squander their wealth. But here, as elsewhere, the relations of civilised life are so complex that some compensatory checks intervene. The men who are rich through primogeniture are able to select generation after generation the more beautiful and charming women;

and these must generally be healthy in body and active in mind. The evil consequences, such as they may be, of the continued preservation of the same line of descent, without any selection, are checked by men of rank always wishing to increase their wealth and power; and this they effect by marrying heiresses. But the daughters of parents who have produced single children, are themselves, as Mr. Galton has shewn, apt to be sterile; and thus noble families are continually cut off in the direct line, and their wealth flows into some side channel; but unfortunately this channel is not determined by superiority of any kind.

Although civilisation thus checks in many ways the action of natural selection, it apparently favours, by means of improved food and the freedom from occasional hardships, the better development of the body. This may be inferred from civilised men having been found, wherever compared, to be physically stronger than savages. They appear also to have equal powers of endurance, as has been proved in many adventurous expeditions. Even the great luxury of the rich can be but little detrimental; for the expectation of life of our aristocracy, at all ages and of both sexes, is very little inferior to that of healthy English lives in the lower classes.

We will now look to the intellectual faculties alone. If in each grade of society the members were divided into two equal bodies, the one including the intellectually superior and the other the inferior, there can be little doubt that the former would succeed best in all occupations and rear a greater number of children. Even in the lowest walks of life, skill and ability must be of some advantage, though in many occupations, owing to the great division of labour, a very small one. Hence in civilised nations there will be some tendency to an increase both in the number and in the standard of the intellectually able. But I do not wish to assert that this tendency may not be more than counterbalanced in other ways, as by the multiplication of the reckless and improvident; but even to such as these, ability must be some advantage.

It has often been objected to views like the foregoing, that the most eminent men who have ever lived have left no offspring to inherit their great intellect. Mr. Galton says, "I regret I am unable to solve the simple question whether, and how far, men and women who are prodigies of genius are infertile. I have, however, shewn that men of eminence are by no means so." Great lawgivers, the founders of beneficent religions, great philosophers and discoverers in science, aid the progress of mankind in a far higher degree by their works than by leaving a numerous progeny. In the case of corporeal structures, it is the selection of the slightly better-endowed and the elimination of the slightly less well-endowed individuals, and not the preservation of strongly-marked and rare anomalies, that leads to the advancement of a species.[2] So it will be with the intellectual faculties, namely from the somewhat more able men in each grade of society succeeding rather better than the less able, and consequently increasing in number, if not otherwise prevented. When in any nation the standard of intellect and the number of intellectual men have increased, we may expect from the law of the deviation from an average, as shown by Mr. Galton, that prodigies of genius will appear somewhat more frequently than before.

In regard to the moral qualities, some elimination of the worst dispositions is always in progress even in the most civilised nations. Malefactors are executed, or imprisoned for long periods, so that they cannot freely transmit their bad qualities. Melancholic and insane persons are confined, or commit suicide. Violent and quarrelsome men often come to a bloody end. Restless men who will not follow any steady occupation—and this relic of barbarism is a great check to civilisation—emigrate to newly-settled countries, where they prove useful pioneers. Intemperance is so highly destructive, that the expectation of life of the intemperate, at the age for instance, of thirty is only 13.8 years; whilst for the rural labourers of England at the same age it is 40.59 years. Profligate women bear few children, and profligate men rarely marry; both suffer from disease. In the breeding of domestic animals, the elimination of those individuals, though few in number, which are in any marked manner inferior, is by no means an unimportant element towards success. This especially holds good with injurious characters which tend to reappear through reversion, such as blackness in sheep; and with mankind some of the worst dispositions, which occasionally without any assignable cause make their appearance in families, may perhaps be reversions to a savage state, from which we are not removed by very many generations. This view seems indeed recognised in the common expression that such men are the black sheep of the family.

With civilised nations, as far as an advanced standard of morality, and an increased number of fairly well-endowed men are concerned, natural selection apparently effects but little; though the fundamental social instincts were originally thus gained. But I have already said enough, whilst treating of the lower races, on the causes which lead to the advance of morality, namely, the approbation of our fellow-men—the strengthening of our sympathies by habit—example and imitation—reason—experience, and even self-interest—instruction during youth, and religious feelings.

A most important obstacle in civilised countries to an increase in the number of men of a superior class has been strongly urged by Mr. Greg and Mr. Galton, namely, the fact that the very poor and reckless, who are often degraded by vice, almost invariably marry early, whilst the careful and frugal,

2. 'Origin of Species' (fifth edition, 1869), p. 104.

who are generally otherwise virtuous, marry late in life, so that they may be able to support themselves and their children in comfort. Those who marry early produce within a given period not only a greater number of generations, but, as shewn by Dr. Duncan, they produce many more children. The children, moreover, that are born by mothers during the prime of life are heavier and larger, and therefore probably more vigorous, than those born at other periods. Thus the reckless, degraded, and often vicious members of society, tend to increase at a quicker rate than the provident and generally virtuous members. Or as Mr. Greg puts the case: "The careless, squalid, unaspiring Irishman multiplies like rabbits: the frugal, foreseeing, self-respecting, ambitious Scot, stern in his morality, spiritual in his faith, sagacious and disciplined in his intelligence, passes his best years in struggle and in celibacy, marries late, and leaves few behind him. Given a land originally peopled by a thousand Saxons and a thousand Celts—and in a dozen generations five-sixths of the population would be Celts, but five-sixths of the property, of the power, of the intellect, would belong to the one-sixth of Saxons that remained. In the eternal 'struggle for existence,' it would be the inferior and *less* favoured race that had prevailed—and prevailed by virtue not of its good qualities but of its faults."

There are, however, some checks to this downward tendency. We have seen that the intemperate suffer from a high rate of mortality, and the extremely profligate leave few offspring. The poorest classes crowd into towns, and it has been proved by Dr. Stark from the statistics of ten years in Scotland, that at all ages the death-rate is higher in towns than in rural districts, "and during the first five years of life the town death-rate is almost exactly double that of the rural districts." As these returns include both the rich and the poor, no doubt more than twice the number of births would be requisite to keep up the number of the very poor inhabitants in the towns, relatively to those in the country. With women, marriage at too early an age is highly injurious; for it has been found in France that, "twice as many wives under twenty die in the year, as died out of the same number of the unmarried." The mortality, also, of husbands under twenty is "excessively high," but what the cause of this may be seems doubtful. Lastly, if the men who prudently delay marrying until they can bring up their families in comfort, were to select, as they often do, women in the prime of life, the rate of increase in the better class would be only slightly lessened.

It was established from an enormous body of statistics, taken during 1853, that the unmarried men throughout France, between the ages of twenty and eighty, die in a much larger proportion than the married: for instance, out of every 1000 unmarried men, between the ages of twenty and thirty, 11.3 annually died, whilst of the married, only 6.5 died. A similar law was proved to hold good, during the years 1863 and 1864, with the entire population above the age of twenty in Scotland: for instance, out of every 1000 unmarried men, between the ages of twenty and thirty, 14.97 annually died, whilst of the married only 7.24 died, that is less than half. Dr. Stark remarks on this, "Bachelorhood is more destructive to life than the most unwholesome trades, or than residence in an unwholesome house or district where there has never been the most distant attempt at sanitary improvement." He considers that the lessened mortality is the direct result of "marriage, and the more regular domestic habits which attend that state." He admits, however, that the intemperate, profligate, and criminal classes, whose duration of life is low, do not commonly marry; and it must likewise be admitted that men with a weak constitution, ill health, or any great infirmity in body or mind, will often not wish to marry, or will be rejected. Dr. Stark seems to have come to the conclusion that marriage in itself is a main cause of prolonged life, from finding that aged married men still have a considerable advantage in this respect over the unmarried of the same advanced age; but every one must have known instances of men, who with weak health during youth did not marry, and yet have survived to old age, though remaining weak, and therefore always with a lessened chance of life. There is another remarkable circumstance which seems to support Dr. Stark's conclusion, namely, that widows and widowers in France suffer in comparison with the married a very heavy rate of mortality; but Dr. Farr attributes this to the poverty and evil habits consequent on the disruption of the family, and to grief. On the whole we may conclude with Dr. Farr that the lesser mortality of married than of unmarried men, which seems to be a general law, "is mainly due to the constant elimination of imperfect types, and to the skilful selection of the finest individuals out of each successive generation"; the selection relating only to the marriage state, and acting on all corporeal, intellectual, and moral qualities. We may, therefore, infer that sound and good men who out of prudence remain for a time unmarried do not suffer a high rate of mortality.

If the various checks specified in the two last paragraphs, and perhaps others as yet unknown, do not prevent the reckless, the vicious and otherwise inferior members of society from increasing at a quicker rate than the better class of men, the nation will retrograde, as has occurred too often in the history of the world. We must remember that progress is no invariable rule. It is most difficult to say why one civilised nation rises, becomes more powerful, and spreads more widely, than another; or why the same nation progresses more at one time than at another. We can only say that it depends on an increase in the actual number of the population, on the number of men endowed with high intellectual and moral faculties, as well as on their standard of excellence. Corporeal structure, except so far as vigour of body leads to vigour of mind, appears to have little influence.

It has been urged by several writers that as high intellectual powers are advantageous to a nation, the old Greeks, who stood some grades higher in intellect than any race that has ever existed, ought to have risen, if the power of natural selection were real, still higher in the scale, increased in number, and stocked the whole of Europe. Here we have the tacit assumption, so often made with respect to corporeal structures, that there is some innate tendency towards continued development in mind and body. But development of all kinds depends on many concurrent favourable circumstances. Natural selection acts only in a tentative manner. Individuals and races may have acquired certain indisputable advantages, and yet have perished from failing in other characters. The Greeks may have retrograded from a want of coherence between the many small states, from the small size of their whole country, from the practice of slavery, or from extreme sensuality; for they did not succumb until "they were enervated and corrupt to the very core." The western nations of Europe, who now so immeasurably surpass their former savage progenitors and stand at the summit of civilisation, owe little or none of their superiority to direct inheritance from the old Greeks; though they owe much to the written works of this wonderful people.

Who can positively say why the Spanish nation, so dominant at one time, has been distanced in the race? The awakening of the nations of Europe from the dark ages is a still more perplexing problem. At this early period, as Mr. Galton has remarked, almost all the men of a gentle nature, those given to meditation or culture of the mind, had no refuge except in the bosom of a Church which demanded celibacy; and this could hardly fail to have had a deteriorating influence on each successive generation. During this same period the Holy Inquisition selected with extreme care the freest and boldest men in order to burn or imprison them. In Spain alone some of the best men—those who doubted and questioned, and without doubting there can be no progress—were eliminated during three centuries at the rate of a thousand a year. The evil which the Catholic Church has thus effected, though no doubt counterbalanced to a certain, perhaps large extent in other ways, is incalculable; nevertheless, Europe has progressed at an unparalleled rate.

The remarkable success of the English as colonists over other European nations, which is well illustrated by comparing the progress of the Canadians of English and French extraction, has been ascribed to their "daring and persistent energy"; but who can say how the English gained their energy? There is apparently much truth in the belief that the wonderful progress of the United States, as well as the character of the people, are the results of natural selection; the more energetic, restless, and courageous men from all parts of Europe have emigrated during the last ten or twelve generations to that great country, and having there succeeded best. Looking to the distant future, I do not think that the Rev. Mr. Zincke takes an exaggerated view when he says: "All other series of events—as that which resulted in the culture of mind in Greece, and that which resulted in the empire of Rome—only appear to have purpose and value when viewed in connection with, or rather as subsidiary to . . . the great stream of Anglo-Saxon emigration to the west." Obscure as is the problem of the advance of civilisation, we can at least see that a nation which produced during a lengthened period the greatest number of highly intellectual, energetic, brave, patriotic, and benevolent men, would generally prevail over less favoured nations.

Natural selection follows from the struggle for existence; and this from a rapid rate of increase. It is impossible not bitterly to regret, but whether wisely is another question, the rate at which man tends to increase; for this leads in barbarous tribes to infanticide and many other evils, and in civilised nations to abject poverty, celibacy, and to the late marriages of the prudent. But as man suffers from the same physical evils with the lower animals, he has no right to expect an immunity from the evils consequent on the struggle for existence. Had he not been subjected to natural selection, assuredly he would never have attained to the rank of manhood. When we see in many parts of the world enormous areas of the most fertile land peopled by a few wandering savages, but which are capable of supporting numerous happy homes, it might be argued that the struggle for existence had not been sufficiently severe to force man upwards to his highest standard. Judging from all that we know of man and the lower animals, there has always been sufficient variability in their intellectual and moral faculties, for their steady advancement through natural selection. No doubt such advance demands many favourable concurrent circumstances; but it may well be doubted whether the most favourable would have sufficed, had not the rate of increase been rapid, and the consequent struggle for existence severe to an extreme degree.

On the evidence that all civilised nations were once barbarous.—As we have had to consider the steps by which some semi-human creature has been gradually raised to the rank of man in his most perfect state, the present subject cannot be quite passed over. But it has been treated in so full and admirable a manner by Sir J. Lubbock, Mr. Tylor, Mr. M'Lennan, and others, that I need here give only the briefest summary of their results. The arguments recently advanced by the Duke of Argyll and formerly by Archbishop Whately, in favour of the belief that man came into the world as a civilised being, and that all savages have since undergone degradation, seem to me weak in comparison with those advanced on the other side. Many nations, no doubt, have fallen away in civilisation, and some may have lapsed into utter barbarism, though on this latter head I have not met with any evidence. The Fuegians were probably compelled by other conquering hordes to settle in their inhospitable country, and they may have become in consequence somewhat more degraded; but it would be difficult to prove that they have fallen much below the Botocudos, who inhabit the finest parts of Brazil.

The evidence that all civilised nations are the descendants of barbarians, consists, on the one side, of clear traces of their former low condition in still-existing customs, beliefs, language, &c.; and on the other side, of proofs that savages are independently able to raise themselves a few steps in the scale of civilisation, and have actually thus risen. The evidence on the first head is extremely curious, but cannot be here given: I refer to such cases as that, for instance, of the art of enumeration, which, as Mr. Tylor clearly shews by the words still used in some places, originated in counting the fingers, first of one hand and then of the other, and lastly of the toes. We have traces of this in our own decimal system, and in the Roman numerals, which after reaching to the number V., change into VI., &c., when the other hand no doubt was used. So again, "when we speak of three-score and ten, we are counting by the vigesimal system, each score thus ideally made, standing for 20—for 'one man' as a Mexican or Carib would put it." According to a large and increasing school of philologists, every language bears the marks of its slow and gradual evolution. So it is with the art of writing, for letters are rudiments of pictorial representations. It is hardly possible to read Mr. M'Lennan's work and not admit that almost all civilised nations still retain some traces of such rude habits as the forcible capture of wives. What ancient nation, as the same author asks, can be named that was originally monogamous? The primitive idea of justice, as shewn by the law of battle and other customs of which traces still remain, was likewise most rude. Many existing superstitions are the remnants of former false religious beliefs. The highest form of religion—the grand idea of God hating sin and loving righteousness—was unknown during primeval times.

Turning to the other kind of evidence: Sir J. Lubbock has shewn that some savages have recently improved a little in some of their simpler arts. From the extremely curious account which he gives of the weapons, tools, and arts, used or practised by savages in various parts of the world, it cannot be doubted that these have nearly all been independent discoveries, excepting perhaps the art of making fire. The Australian boomerang is a good instance of one such independent discovery. The Tahitians when first visited had advanced in many respects beyond the inhabitants of most of the other Polynesian islands. There are no just grounds for the belief that the high culture of the native Peruvians and Mexicans was derived from any foreign source; many native plants were there cultivated, and a few native animals domesticated. We should bear in mind that a wandering crew from some semi-civilised land, if washed to the shores of America, would not, judging from the small influence of most missionaries, have produced any marked effect on the natives, unless they had already become somewhat advanced. Looking to a very remote period in the history of the world, we find, to use Sir J. Lubbock's well-known terms, a paleolithic and neolithic period; and no one will pretend that the art of grinding rough flint tools was a borrowed one. In all parts of Europe, as far east as Greece, in Palestine, India, Japan, New Zealand, and Africa, including Egypt, flint tools have been discovered in abundance; and of their use the existing inhabitants retain no tradition. There is also indirect evidence of their former use by the Chinese and ancient Jews. Hence there can hardly be a doubt that the inhabitants of these countries, which include nearly the whole civilised world, were once in a barbarous condition. To believe that man was aboriginally civilised and then suffered utter degradation in so many regions, is to take a pitiably low view of human nature. It is apparently a truer and more cheerful view that progress has been much more general than retrogression; that man has risen, though by slow and interrupted steps, from a lowly condition to the highest standard as yet attained by him in knowledge, morals and religion.

INTRODUCTION: SEXUAL SELECTION

Darwin's interest in sexual selection predates his understanding of natural selection. In Notebook C, written in 1838 about six months before he read Malthus, Darwin asked the following suggestive question:

> Whether species may not be made by a little more vigour being given to the chance offspring who have any slight peculiarity of structure <<hence seals take victorious seals, hence deer victorious deer, hence males armed & pugnacious (all order; cocks all warlike)>> (C:61)

Thus, before he understood how Malthusian population pressure could lead to adaptations affecting all aspects of an organism's life, Darwin could already perceive how competition for mates would discriminate among advantageous variations in the sexual displays associated with charming or battling for mates.

For the subsequent three decades, Darwin collected the descriptive evidence on sexual selection throughout the animal kingdom that would fill 70 percent of *The Descent of Man*. And one unmistakable reason for Darwin's long march through sexual selection in the lower animals to sexual selection in humans was to prove once and for all that "Man still bears in his bodily frame the indelible stamp of his lowly origin" (*Descent* 2:405). But Darwin also thought that sexual selection had particular importance for human evolution. In passages spread throughout the *Descent*, a cumulative but rather disjointed argument is made that sexual selection is a key to the evolution of human difference—from the origin of gender differences to the origin of racial differences.

First we need to consider the principles of sexual selection, which "depends on the advantage certain individuals have over other individuals of the same sex and species, in exclusive relation to reproduction." Darwin attributes the evolution of many structures to this form of competition—notably, gender differences in size, strength, and pugnacity; weapons of defense and offense; and gaudy sexual displays. He is careful to note that many sexual characteristics are the products of natural selection rather than sexual selection; indeed, the characters he is most interested in have often been shaped by both forms of selection. While sexual selection occurs throughout the animal kingdom, it becomes more prevalent in more advanced groups. In most cases competition is among males. Females are the selectors, and males are selected. Thus, it is the males that show the effects of sexual selection: Peacocks are gaudy and peahens are plain. In some cases, however, notably among humans, males may become the selectors. Indeed, in humans, Darwin argues, we find double sexual selection. The males select women for their physical beauty, and the females select men for their physical strength and status.

Beyond these principles, we should also note an important difference between natural and sexual selection. The "selection" in natural selection is a metaphor for the adaptations that result from the survival of the fittest. Although Darwin used the term natural selection as an analogy to artificial selection by plant and animal breeders (see chapter 8, 165), there is no selector in natural selection. But an analogy between sexual selection and artificial selection would be much more accurate. For in sexual selection, one organism literally "selects" another on the basis of some explicit characteristic. This is quite like a pigeon breeder "fancying" one pigeon over another for its color or the shape of its crop. When we consider sexual selection as practiced by humans, we are witnessing the effects on humans of selection by humans. This leads to the interesting observation that human sexual selection is a natural form of human domestication.

Perhaps because of its early appearance in Darwin's thought, sexual selection is one of several evolutionary agents that Darwin relied on as alternates and supplements to natural selection. Besides sexual selection, these agents included the inheritance of acquired characteristics through the "direct and definite action of the surrounding conditions of life," the "inherited effects of the long-continued use or disuse of parts," and the principle of the correlation of parts. By the time Darwin published the *Descent*, his reliance on these alternatives, including sexual selection, seemed to mark a painful retreat from his strong claims for natural selection:

I now admit . . . that in the earlier editions of my 'Origin of Species' I probably attributed too much to the action of natural selection or the survival of the fittest. I have altered the fifth edition of the Origin so as to confine my remarks to adaptive changes of structure. (*Descent* 1:152)

SEXUAL SELECTION AND HUMAN EVOLUTION

Sexual Selection, according to Darwin, has influenced the evolution of the human body and mind. Men are larger and stronger than women because of "the strongest and boldest men having succeeded best in the general struggle for life, as well as in securing wives, and thus having left a large number of offspring." While civilization has diminished male combat for mates, the crucial human mental powers—intelligence and morality—continue to be shaped by sexual selection, as they have been since primeval times. Both sexes have intelligence and morality, but there are marked sexual differences in mental disposition. Women predominate in morality, whereas men have stronger intellects. Female superiority in sympathy for the plight of others, the key measure of the Darwinian moral sense (see the preceding section, "Moral Faculties"), is an extension of the maternal instinct, which produces a "greater tenderness and less selfishness" toward her infants that then flows "towards her fellow-creatures." But man "is the rival of other men; he delights in competition, and this leads to ambition which passes too easily into selfishness." Male rivalry, the very factor that limits male moral capacity, drives his intellectual superiority:

> The half-human male progenitors of man, and men in a savage state, have struggled together during many generations for the possession of the females. But mere bodily strength and size would do little for victory, unless associated with courage, perseverance, and determined energy. . . . But to avoid enemies, or to attack them with success, to capture wild animals, and to invent and fashion weapons, requires the aid of the higher mental faculties, namely, observation, reason, invention, or imagination. These various faculties will thus have been continually put to the test, and selected during manhood. . . . Now, when two men are put into competition, or a man with a woman, who possess every mental quality in the same perfection, with the exception that the one has higher energy, perseverance, and courage, this one will generally become more eminent. . . . But [patience . . . unflinching, undaunted perseverance] and [the higher powers of the imagination and reason] will have been developed in man, partly through sexual selection,—that is, through the contest of rival males. . . . Thus man has ultimately become superior to woman. (*Descent* 2:327–328)

Although Darwin grants female strength in the intellectual "powers of intuition, of rapid perception, and perhaps imitation," he rates these as characteristics of "the lower races, and therefore of a past and lower state of civilisation." Whatever the importance of sexual selection in human cognitive evolution, the modern reader will recognize a male bias in Darwin's conclusion of male superiority, and for the student familiar with Victorian attitudes about sharp gender-role separation, it will be difficult not to see Darwin projecting the values of a nineteenth-century gentleman onto his "half-human male progenitors."

Darwin sees at least one other role for sexual selection in human evolution: It was one of the forces that simultaneously shaped the origin of cultural attainments in the arts and the formation of distinct human races. Although Darwin was definitely a biological determinist, he was nevertheless a cultural relativist. For example, what he calls the "musical powers," the ability to make and respond to music, derived from sexual selection in our early ancestors:

> . . . we may assume that musical tones and rhythm were used by the half-human progenitors of man, during the season of courtship, when animals of all kinds are excited by the strongest passions . . . the suspicion does not appear improbable that the progenitors of man, either the males or females, or both sexes, before they had acquired the power of expressing their mutual love in articulate language, endeavoured to charm each other with musical notes and rhythm. (*Descent* 2:336–337)

Since the musical powers are ancestral, predating language, they are also universal. But musical taste is different for different human cultures. Darwin took taste, which, after all, is a form of selectivity, to be the key to this and many other differences between human races—today, we would say cultures. Relative isolation and taste driven by sexual selection result in the elaboration of distinctive cultural styles. The same process is seen most powerfully at work in "the influence of beauty in determining the marriages of mankind." Surveying extensive anthropological reports on bodily ornamentation, from tattooing to knocking out teeth to painting the body with red ochre, Darwin again uses the argument that different groups have different tastes. Through sexual selection, these tastes can contribute to hereditary racial differences. This line of thought culminates in the primordial-horde hypothesis for the formation of human races:

> Let us suppose the members of a tribe, in which some form of marriage was practised, to spread over an unoccupied continent; they would soon split up into distinct hordes, which would be separated from each other by various barriers, and still more effectually by the incessant wars between barbarous nations. The hordes would thus be exposed to slightly different conditions and habits of life, and would sooner or later come to differ in some small degree. As soon as this occurred, each isolated tribe would form for itself a slightly different standard of beauty; and then unconscious selection would come into action through the more powerful and leading savages preferring certain women to others. Thus the differences be-

tween the tribes, at first very slight, would gradually and inevitably be increased to a greater and greater degree. (*Descent* 2:370–371)

SEXUAL SELECTION
CHAPTER 8
PRINCIPLES OF SEXUAL SELECTION.

With animals which have their sexes separated, the males necessarily differ from the females in their organs of reproduction; and these afford the primary sexual characters. But the sexes often differ in what Hunter has called secondary sexual characters, which are not directly connected with the act of reproduction; for instance, the male possessing certain organs of sense or locomotion, of which the female is quite destitute, or having them more highly-developed, in order that he may readily find or reach her; or again, in the male having special organs of prehension so as to hold her securely. These latter organs of infinitely diversified kinds graduate into, and in some cases can hardly be distinguished from, those which are commonly ranked as primary, such as the complex appendages at the apex of the abdomen in male insects.

. . . Numerous similar cases could be given, but they do not here concern us. There are, however, other sexual differences quite disconnected with the primary organs, with which we are more especially concerned— such as the greater size, strength, and pugnacity of the male, his weapons of offence or means of defence against rivals, his gaudy colouring and various ornaments, his power of song, and other such characters. . . .

We are . . . here concerned only with that kind of selection, which I have called sexual selection. This depends on the advantage which certain individuals have over other individuals of the same sex and species, in exclusive relation to reproduction. . . .

When the two sexes follow exactly the same habits of life, and the male has more highly developed sense or locomotive organs than the female, it may be that these in their perfected state are indispensable to the male for finding the female; but in the vast majority of cases, they serve only to give one male an advantage over another, for the less well-endowed males, if time were allowed them, would succeed in pairing with the females; and they would in all other respects, judging from the structure of the female, be equally well adapted for their ordinary habits of life. In such cases sexual selection must have come into action, for the males have acquired their present structure, not from being better fitted to survive in the struggle for existence, but from having gained an advantage over other males, and from having transmitted this advantage to their male offspring alone. It was the importance of this distinction which led me to designate this form of selection as sexual selection. So again, if the chief service rendered to the male by his prehensile organs is to prevent the escape of the female before the arrival of other males, or when assaulted by them, these organs will have been perfected through sexual selection, that is by the advantage acquired by certain males over their rivals. But in most cases it is scarcely possible to distinguish between the effects of natural and sexual selection. . . .

There are many other structures and instincts which must have been developed through sexual selection—such as the weapons of offence and the means of defence are eminently variable. In the same manner as man can give beauty, according to his standard of taste, to his male poultry—can give to the Sebright bantam a new and elegant plumage, an erect and peculiar carriage—so it appears that in a state of nature female birds, by having long selected the more attractive males, have added to their beauty. No doubt this implies powers of discrimination and taste on the part of the female which will at first appear extremely improbable; but I hope hereafter to shew that this is not the case.

From our ignorance on several points, the precise manner in which sexual selection acts is to a certain extent uncertain. Nevertheless if those naturalists who already believe in the mutability of species, will read the following chapters, they will, I think, agree with me that sexual selection has played an important part in the history of the organic world. It is certain that with almost all animals there is a struggle between the males for the possession of the female. This fact is so notorious that it would be superfluous to give instances. Hence the females, supposing that their mental capacity sufficed for the exertion of a choice, could select one out of several males. But in numerous cases it appears as if it had been specially arranged that there should be a struggle between many males. Thus with migratory birds, the males generally arrive before the females at their place of breeding, so that many males are ready to contend for each female. The bird-catchers assert that this is invariably the case with the nightingale and blackcap, as I am informed by Mr. Jenner Weir, who confirms the statement with respect to the latter species. . . .

Our difficulty in regard to sexual selection lies in understanding how it is that the males which conquer other males, or those which prove the most attractive to the females, leave a greater number of offspring to inherit their superiority than the beaten and less attractive rivals. Unless this result followed, the characters which give to certain males an advantage over others, could not be perfected and augmented through sexual selection. When the sexes exist in exactly equal numbers, the worst-endowed males will ultimately find females (excepting where polygamy prevails), and leave as many offspring, equally well fitted for their general habits of life, as the best-endowed males. From various facts and considerations, I formerly inferred that with most animals, in which secondary sexual characters were well developed, the males considerably exceeded the females in number; and this does hold good in some few cases. If the males were to the females as two to one, or as three to two, or even in a somewhat lower ratio, the whole affair would be simple; for the better-armed or more attractive males would leave the largest number of offspring. But after investigating, as far as possible, the numerical proportion of the sexes, I do not believe that any great inequality in number commonly exists. In most cases sexual selection appears to have been effective in the following manner.

Let us take any species, a bird for instance, and divide the females inhabiting a district into two equal bodies: the one consisting of the more vigorous and better-nourished individuals, and the other of the less vigorous and healthy. The former, there can be little doubt, would be ready to breed in the spring before the others; and this is the opinion of Mr. Jenner Weir, who has during many years carefully attended to the habits of birds. There can also be no doubt that the most vigorous, healthy, and best-nourished females would on an average succeed in rearing the largest number of offspring. The males, as we have seen, are generally ready to breed before the females; of the males the strongest, and with some species the best armed, drive away the weaker males; and the former would then unite with the more vigorous and better-nourished females, as these are the first to breed. Such vigorous pairs would surely rear a larger number of offspring than the retarded females, which would be compelled, supposing the sexes to be numerically equal, to unite with the conquered and less powerful males; and this is all that is wanted to add, in the course of successive generations, to the size, strength and courage of the males, or to improve their weapons.

But in a multitude of cases the males which conquer other males, do not obtain possession of the females, independently of the choice on the part of the latter. The courtship of animals is by no means so simple and short an affair as might be thought. The females are most excited by, or prefer pairing with, the more ornamented males, or those which are the best songsters, or play the best antics; but it is obviously probable, as has been actually observed in some cases, that they would at the same time prefer the more vigorous and lively males. Thus the more vigorous females, which are the first to breed, will have the choice of many males; and though they may not always select the strongest or best armed, they will select those which are vigorous and well armed, and in other respects the most attractive. Such early pairs would have the same advantage in rearing offspring on the female side as above explained, and nearly the same advantage on the male side. And this apparently has sufficed during a long course of generations to add not only to the strength and fighting-powers of the males, but likewise to their various ornaments or other attractions.

Numerical Proportion of the Two Sexes.—I have remarked that sexual selection would be a simple affair if the males considerably exceeded in number the females. Hence I was led to investigate, as far as I could, the proportions between the two sexes of as many animals as possible; but the materials are scanty. . . .

For our present purpose we are concerned with the proportions of the sexes, not at birth, but at maturity, and this adds another element of doubt; for it is a well ascertained fact that with man a considerably larger proportion of males than of females die before or during birth, and during the first few years of infancy. . . . Hence, with animals in a state of nature, in order to judge of the proportions of the sexes at maturity, we must rely on mere estimation; and this, except perhaps when the inequality is strongly marked, is but little trustworthy. Nevertheless, as far as a judgement can be formed, we may conclude from the facts given in the supplement, that the males of some few mammals, of many birds, of some fish and insects, considerably exceed in number the females.

Polygamy.—The practice of polygamy leads to the same results as would follow from an actual inequality in the number of the sexes; for if each male secures two or more females, many males will not be able to pair; and the latter assuredly will be the weaker or less attractive individuals. Many mammals and some few birds are polygamous, but with animals belonging to the lower classes I have found no evidence of this habit. The intellectual powers of such animals are, perhaps, not sufficient to lead them to collect and guard a harem of females. That some relation exists between polygamy and the development of secondary sexual characters, appears nearly certain; and this supports the view that a numerical preponderance of males would be eminently favourable to the action of sexual selection. Nevertheless many animals, especially birds, which are strictly monogamous, display strongly-marked secondary sexual characters; whilst some few animals, which are polygamous, are not thus characterised.

The Male generally more modified than the Female.—Throughout the animal kingdom, when the sexes differ from each other in external appearance, it is the male which, with rare exceptions, has been chiefly modified; for the female still remains more like the young of her own species, and more like the other members of the same group. The cause of this seems to lie in the males of almost all animals having stronger passions than the females. Hence it is the males that fight together and sedulously display their charms before the females; and those which are victorious transmit their superiority to their male offspring. Why the males do not transmit their characters to both sexes will hereafter be considered. That the males of all mammals eagerly pursue the females is notorious to every one. So it is with birds; but many male birds do not so much pursue the female, as display their plumage, perform strange antics, and pour forth their song, in her presence. With the few fish which have been observed, the male seems much more eager than the female; and so it is with alligators, and apparently with Batrachians. Throughout the enormous class of insects, as Kirby remarks, "the law is, that the male shall seek the female." With spiders and crustaceans, as I hear from two great authorities, Mr. Blackwall and Mr. C. Spence Bate, the males are more active and more erratic in their habits than the females. With insects and crustaceans, when the organs of sense or locomotion are present in the one sex and absent in the other, or when, as is frequently the case, they are more highly developed in the one than the other, it is almost invariably the male, as far as I can discover, which retains such organs, or has them most developed; and this shews that the male is the more active member in the courtship of the sexes.

The female, on the other hand, with the rarest exceptions, is less eager than the male. As the illustrious Hunter long ago observed, she generally "requires to be courted"; she is coy, and may often be seen endeavouring for a long time to escape from the male. Every one who has attended to the habits of animals will be able to call to mind instances of this kind. Judging from various facts, hereafter to be given, and from the results which may fairly be attributed to sexual selection, the female, though comparatively passive, generally exerts some choice and accepts one male in preference to others. Or she may accept, as appearances would sometimes lead us to believe, not the male which is the most attractive to her, but the one which is the least distasteful. The exertion of some choice on the part of the female seems almost as general a law as the eagerness of the male. . . .

In various classes of animals a few exceptional cases occur, in which the female instead of the male has acquired well pronounced secondary sexual characters, such as brighter colours, greater size, strength, or pugnacity. With birds, as we shall hereafter see, there has sometimes been a complete transposition of the ordinary characters proper to each sex; the females having become the more eager in courtship, the males remaining comparatively passive, but apparently selecting, as we may infer from the results, the more attractive females. Certain female birds have thus been rendered more highly coloured or otherwise ornamented, as well as more powerful and pugnacious than the males, these characters being transmitted to the female offspring alone.

It may be suggested that in some cases a double process of selection has been carried on; that the males have selected the more attractive females, and the latter the more attractive males. This process however, though it might lead to the modification of both sexes, would not make the one sex different from the other, unless indeed their tastes for the beautiful differed; but this is a supposition too improbable in the case of any animal, excepting man, to be worth considering. . . .

Sexual selection acts in a less rigorous manner than natural selection. The latter produces its effects by the life or death at all ages of the more or less successful individuals. Death, indeed, not rarely ensues from the conflicts of rival males. But generally the less successful male merely fails to obtain a female, or obtains later in the season a retarded and less vigorous female, or, if polygamous, obtains fewer females; so that they leave fewer, less vigorous, or no offspring. In regard to structures acquired through ordinary or natural selection, there is in most cases, as long as the conditions of life remain the same, a limit to the amount of advantageous modification in relation to certain special ends; but in regard to structures adapted to make one male victorious over another, either in fighting or in charming the female, there is no definite limit to the amount of advantageous modification; so that as long as the proper variations arise the work of sexual selection will go on. This circumstance may partly account for the frequent and extraordinary amount of variability presented by secondary sexual characters. Nevertheless, natural selection will determine that characters of this kind shall not be acquired by the victorious males, which would be highly injurious to them in any high degree, either by expending too much of their vital powers, or by exposing them to any great danger. The development, however, of certain structures— of the horns, for instance, in certain stags—has been carried to a wonderful extreme; and in some cases to an extreme which, as far as the general conditions of life are concerned, must be slightly injurious to the male. From this fact we learn that the advantages which favoured males have derived from conquering other males in battle or courtship, and thus leaving a numerous progeny, have been in the long run greater than those derived from rather more perfect adaptation to the external conditions of life. We shall further see, and this could never have been anticipated, that the power to charm the female has been in some few instances more important than the power to conquer other males in battle.

CHAPTER 19
SECONDARY SEXUAL CHARACTERS OF MAN.

With mankind the differences between the sexes are greater than in most species of Quadrumana, but not so great as in some, for instance, the mandrill. Man on an average is considerably taller, heavier, and stronger than woman, with squarer shoulders and more plainly-pronounced muscles. Owing to the relation which exists between muscular development and the projection of the brows, the superciliary ridge is generally more strongly marked in man than in woman. His body, and especially his face, is more hairy, and his voice has a different and more powerful tone. In certain races the women are said, whether truly I know not, to differ slightly in tint from the men; and with Europeans, the women are perhaps the more brightly coloured of the two, as may be seen when both sexes have been equally exposed to the weather.

Man is more courageous, pugnacious and energetic than woman, and has a more inventive genius. His brain is absolutely larger, but whether relatively to the larger size of his body, in comparison with that of woman, has not, I believe been fully ascertained. In woman the face is rounder; the jaws and the base of the skull smaller; the outlines of the body rounder, in parts more prominent; and her pelvis is broader than in man; but this latter character may perhaps be considered rather as a primary than a secondary sexual character. She comes to maturity at an earlier age than man. . . .

I have specified the foregoing differences between the male and female sex in mankind, because they are curiously the same as in the Quadrumana. With these animals the female is mature at an earlier age than the male; at least this is certainly the case with the *Cebus azarae*. With most of the species the males are larger and stronger than the females, of which fact the gorilla affords a well-known instance. Even in so trifling a character as the greater prominence of the superciliary ridge, the males of certain monkeys differ from the females, and agree in this respect with mankind. In the gorilla and certain other monkeys, the cranium of the adult male presents a strongly-marked sagittal crest, which is absent in the female; and Ecker found a trace of a similar difference between the two sexes in the Australians. With monkeys when there is any difference in the voice, that of the male is the more powerful. We have seen that certain male monkeys have a well-developed beard, which is quite deficient, or much less developed in the female. No instance is known of the beard, whiskers, or moustache being larger in the female than in the male monkey.

Law of Battle.—With barbarous nations, for instance with the Australians, the women are the constant cause of war both between individuals of the same tribe and between distinct tribes. So no doubt it was in ancient times; "nam fuit ante Helenam mulier teterrima belli causa." With the North American Indians, the contest is reduced to a system.

Other similar facts could be given; but even if we had no evidence on this head, we might feel almost sure, from the analogy of the higher Quadrumana, that the law of battle had prevailed with man during the early stages of his development. The occasional appearance at the present day of canine teeth which project above the others, with traces of diastema or open space for the reception of the opposite canines, is in all probability a case of reversion to a former state, when the progenitors of man were provided with

Darwin here summarizes a longer passage in chapter 4 of the *Descent* (pp. 143–44), in which he explains the evolution of erect posture:

As the progenitors of man became more and more erect, with their hands and arms more and more modified for prehension and other purposes, with their feet and legs at the same time modified for firm support and progression, endless other changes of structure would have been necessary. The pelvis would have had to be made broader, the spine peculiarly curved and the head fixed in an altered position, and all these changes have been attained by man. Prof. Schaaffhausen maintains that "the powerful mastoid processes of the human skull are the result of his erect position;" and these processes are absent in the orang, chimpanzee, &c., and are smaller in the gorilla than in man. Various other structures might here have been specified, which appear connected with man's erect position. It is very difficult to decide how far all these correlated modifications are the result of natural selection, and how far of the inherited effects of the increased use of certain parts, or of the action of one part on another. No doubt these means of change act and react on each other: thus when certain muscles, and the crests of bone to which they are attached, become enlarged by habitual use, this shews that certain actions are habitually performed and must be serviceable. Hence the individuals which performed them best, would tend to survive in greater numbers.

The free use of the arms and hands, partly the cause and partly the result of man's erect position, appears to have led in an indirect manner to other modifications of structure. The early male progenitors of man were, as previously stated, probably furnished with great canine teeth; but as they gradually acquired the habit of using stones, clubs, or other weapons, for fighting with their enemies, they would have used their jaws and teeth less and less. In this case, the jaws, together with the teeth, would have become reduced in size, as we may feel sure from innumerable analogous cases. In a future chapter we shall meet with a closely-parallel case, in the reduction or complete disappearance of the canine teeth in male ruminants, apparently in relation with the development of their horns; and in horses, in relation with their habit of fighting with their incisor teeth and hoofs.

these weapons, like so many existing male Quadrumana. It was remarked in a former chapter that as man gradually became erect, and continually used his hands and arms for fighting with sticks and stones, as well as for the other purposes of life, he would have used his jaws and teeth less and less. The jaws, together with their muscles, would then have been reduced through disuse, as would the teeth through the not well understood principles of correlation and the economy of growth; for we everywhere see that parts which are no longer of service are reduced in size. By such steps the original inequality between the jaws and teeth in the two sexes of mankind would ultimately have been obliterated. The case is almost parallel with that of many male Ruminants, in which the canine teeth have been reduced to mere rudiments, or have disappeared, apparently in consequence of the development of horns. As the prodigious difference between the skulls of the two sexes in the Gorilla and Orang, stands in close relation with the development of the immense canine teeth in the males, we may infer that the reduction of the jaws and teeth in the early male progenitors of man led to a most striking and favourable change in his appearance.

There can be little doubt that the greater size and strength of man, in comparison with woman, together with his broader shoulders, more developed muscles, rugged outline of body, his greater courage and pugnacity, are all due in chief part to inheritance from some early male progenitor, who, like the existing anthropoid apes, was thus characterised. These characters will, however, have been preserved or even augmented during the long ages whilst man was still in a barbarous condition, by the strongest and boldest men having succeeded best in the general struggle for life, as well as in securing wives, and thus having left a large number of offspring. It is not probable that the greater strength of man was primarily acquired through the inherited effects of his having worked harder than woman for his own subsistence and that of his family; for the women in all barbarous nations are compelled to work at least as hard as the men. With civilised people the arbitrament of battle for the possession of the women has long ceased; on the other hand, the men, as a general rule, have to work harder than the women for their mutual subsistence; and thus their greater strength will have been kept up.

Difference in the Mental Powers of the two Sexes.—With respect to differences of this nature between man and woman, it is probable that sexual selection has played a very important part. I am aware that some writers doubt whether there is any inherent difference; but this is at least probable from the analogy of the lower animals which present other secondary sexual characters. No one will dispute that the bull differs in disposition from the cow, the wild-boar from the sow, the stallion from the mare, and, as is well known to the keepers of menageries, the males of the larger apes from the females. Woman seems to differ from man in mental disposition, chiefly in her greater tenderness and less selfishness; and this holds good even with savages, as shewn by a well-known passage in Mungo Park's Travels, and by statements made by many other travellers. Woman, owing to her maternal instincts, displays these qualities towards her infants in an eminent degree; therefore it is likely that she would often extend them towards her fellow-creatures. Man is the rival of other men; he delights in competition, and this leads to ambition which passes too easily into selfishness. These latter qualities seem to be his natural and unfortunate birthright. It is generally admitted that with woman the powers of intuition, of rapid perception, and perhaps of imitation, are more strongly marked than in man; but some, at least, of these faculties are characteristic of the lower races, and therefore of a past and lower state of civilisation.

The chief distinction in the intellectual powers of the two sexes is shewn by man's attaining to a higher eminence, in whatever he takes up, than woman can attain—whether requiring deep thought, reason, or imagination, or merely the use of the senses and hands. If two lists were made of the most eminent men and women in poetry, painting, sculpture, music,—comprising composition and performance, history, science, and philosophy, with half-a-dozen names under each subject, the two lists would not bear comparison. We may also infer, from the law of the deviation of averages, so well illustrated by Mr. Galton, in his work on 'Hereditary Genius,' that if men are capable of decided eminence over women in many subjects, the average standard of mental power in man must be above that of woman.

The half-human male progenitors of man, and men in a savage state, have struggled together during many generations for the possession of the females. But mere bodily strength and size would do little for victory, unless associated with courage, perseverance, and determined energy. With social animals, the young males have to pass through many a contest before they win a female, and the older males have to retain their females by renewed battles. They have, also, in the case of mankind, to defend their females, as well as their young, from enemies of all kinds, and to hunt for their joint subsistence. But to avoid enemies, or to attack them with success, to capture wild animals, and to invent and fashion weapons, requires the aid of the higher mental faculties, namely, observation, reason, invention, or imagination. These various faculties will thus have been continually put to the test, and selected during manhood; they will, moreover, have been strengthened by use during this same period of life. Consequently in accordance with the principle often alluded to, we might expect that they would at least tend to be transmitted chiefly to the male offspring at the corresponding period of manhood.

Now, when two men are put into competition, or a man with a woman, who possesses every mental quality in the same perfection, with the exception that the one has higher energy, perseverance, and courage, this one will generally become more eminent, whatever the object may be, and will gain the victory.[1] He may be said to possess genius—for genius has been declared by a great authority to be patience; and patience, in this sense, means un-flinching, undaunted perseverance. But this view of genius is perhaps de-ficient; for without the higher powers of the imagination and reason, no emi-nent success in many subjects can be gained. But these latter as well as the former faculties will have been developed in man, partly through sexual selection,—that is, through the contest of rival males, and partly through natural selection,—that is, from success in the general struggle for life; and as in both cases the struggle will have been during maturity, the characters thus gained will have been transmitted more fully to the male than to the female offspring. Thus man has ultimately become superior to woman. It is, indeed, fortunate that the law of the equal transmission of characters to both sexes has commonly prevailed throughout the whole class of mammals; oth-erwise it is probable that man would have become as superior in mental en-dowment to woman, as the peacock is in ornamental plumage to the peahen.

It must be borne in mind that the tendency in characters acquired at a late period of life by either sex, to be transmitted to the same sex at the same age, and of characters acquired at an early age to be transmitted to both sexes, are rules which, though general, do not always hold good. If they al-ways held good, we might conclude (but I am here wandering beyond my proper bounds) that the inherited effects of the early education of boys and girls would be transmitted equally to both sexes; so that the present in-equality between the sexes in mental power could not be effaced by a similar course of early training; nor can it have been caused by their dissimilar early training. In order that woman should reach the same standard as man, she ought, when nearly adult, to be trained to energy and perseverance, and to have her reason and imagination exercised to the highest point; and then she would probably transmit these qualities chiefly to her adult daughters. The whole body of women, however, could not be thus raised, unless during many generations the women who excelled in the above robust virtues were mar-ried, and produced offspring in larger numbers than other women. As be-fore remarked with respect to bodily strength, although men do not now fight for the sake of obtaining wives, and this form of selection has passed away, yet they generally have to undergo, during manhood, a severe struggle

1. J. Stuart Mill remarks ("The Subjection of Women," 1869, p. 122), "the things in which man most excels woman are those which require most plodding, and long hammering at single thoughts." What is this but energy and perseverance?

in order to maintain themselves and their families; and this will tend to keep up or even increase their mental powers, and, as a consequence, the present inequality between the sexes.

Voice and Musical Powers.—In some species of Quadrumana there is a great difference between the adult sexes, in the power of the voice and in the development of the vocal organs; and man appears to have inherited this difference from his early progenitors. His vocal cords are about one-third longer than in woman, or than in boys; and emasculation produces the same effect on him as on the lower animals, for it "arrests that prominent growth of the thyroid, &c., which accompanies the elongation of the cords.". . .

The capacity and love for singing or music, though not a sexual character in man, must not here be passed over. Although the sounds emitted by animals of all kinds serve many purposes, a strong case can be made out, that the vocal organs were primarily used and perfected in relation to the propagation of the species. Insects and some few spiders are the lowest animals which voluntarily produce any sound; and this is generally effected by the aid of beautifully constructed stridulating organs, which are often confined to the males alone. The sounds thus produced consist, I believe in all cases, of the same note, repeated rhythmically; and this is sometimes pleasing even to the ears of man. The chief, and in some cases exclusive use appears to be either to call or to charm the opposite sex. . . .

With all those animals, namely insects, amphibians, and birds, the males of which during the season of courtship incessantly produce musical notes or mere rhythmical sounds, we msut believe that the females are able to appreciate them, and are thus excited or charmed; otherwise the incessant efforts of the males and the complex structures often possessed exclusively by them would be useless.

With man song is generally admitted to be the basis or origin of instrumental music. As neither the enjoyment nor the capacity of producing musical notes are faculties of the least direct use to man in reference to his ordinary habits of life, they must be ranked amongst the most mysterious with which he is endowed. They are present, though in a very rude and as it appears almost latent condition, in men of all races, even the most savage; but so different is the taste of the different races, that our music gives not the least pleasure to savages, and their music is to us hideous and unmeaning. Dr. Seemann, in some interesting remarks on this subject, "doubts whether even amongst the nations of Western Europe, intimately connected as they are by close and frequent intercourse, the music of the one is interpreted in the same sense by the others. By travelling eastwards we find that there is certainly a different language of music. Songs of joy and dance-accompaniments are no longer, as with us, in the major keys, but always in the minor." Whether or not the half-human progenitors of man possessed, like the before-mentioned gibbon, the capacity of producing, and no doubt of appreciating, musical notes, we have every reason to believe that man possessed these faculties at a very remote period, for singing and music are extremely ancient arts. Poetry, which may be considered as the offspring of song, is likewise so ancient that many persons have felt astonishment that it should have arisen during the earliest ages of which we have any record. . . .

Music affects every emotion, but does not by itself excite in us the more terrible emotions of horror, rage, &c. It awakens the gentler feelings of tenderness and love, which readily pass into devotion. It likewise stirs up in us the sensation of triumph and the glorious ardour for war. These powerful and mingled feelings may well give rise to the sense of sublimity. We can concentrate, as Dr. Seemann observes, greater intensity of feeling in a single musical note than in pages of writing. Nearly the same emotions, but much weaker and less complex, are probably felt by birds when the male pours forth his full volume of song, in rivalry with other males, for the sake of captivating the female. Love is still the commonest theme of our songs. As Herbert Spencer remarks, music "arouses dormant sentiments of which we had not conceived the possibility, and do not know the meaning; or, as Richter says, tells us of things we have not seen and shall not see."

All these facts with respect to music become to a certain extent intelligible if we may assume that musical tones and rhythm were used by the half-human progenitors of man, during the season of courtship, when animals of all kinds are excited by the strongest passions. In this case, from the deeply-laid principle of inherited associations, musical tones would be likely to excite in us, in a vague and indefinite manner, the strong emotions of a long-past age. Bearing in mind that the males of some quadrumanous animals have their vocal organs much more developed than in the females, and that one anthropomorphous species pours forth a whole octave of musical notes and may be said to sing, the suspicion does not appear improbable that the progenitors of man, either the males or females, or both sexes, before they had acquired the power of expressing their mutual love in articulate language, endeavoured to charm each other with musical notes and rhythm. . . .

The impassioned orator, bard, or musician, when with his varied tones and cadences he excites the strongest emotions in his hearers, little suspects that he uses the same means by which, at an extremely remote period, his half-human ancestors aroused each other's ardent passions, during their mutual courtship and rivalry.

On the influence of beauty in determining the marriages of mankind.—In civilised life man is largely, but by no means exclusively, influenced in the choice of his wife by external appearance; but we are chiefly concerned with primeval times, and our only means of forming a judgement on this subject is to study the habits of existing semi-civilised and savage nations. If it can be shewn that the men of different races prefer women having certain characteristics, or conversely that the women prefer certain men, we have then to enquire whether such choice, continued during many generations, would produce any sensible effect on the race, either on one sex or both sexes; this latter circumstance depending on the form of inheritance which prevails.

It will be well first to shew in some detail that savages pay the greatest attention to their personal appearance. That they have a passion for ornament is notorious; and an English philosopher goes so far as to maintain that clothes were first made for ornament and not for warmth. As Professor Waitz remarks, "however poor and miserable man is, he finds a pleasure in adorning himself." The extravagance of the naked Indians of South America in decorating themselves is shewn "by a man of large stature gaining with difficulty enough by the labour of a fortnight to procure in exchange the *chica* necessary to paint himself red." The ancient barbarians of Europe during the Reindeer period brought to their caves any brilliant or singular objects which they happened to find. Savages at the present day everywhere deck themselves with plumes, necklaces, armlets, earrings, &c. They paint themselves in the most diversified manner. "If painted nations," as Humboldt observes, "had been examined with the same attention as clothed nations, it would have been perceived that the most fertile imagination and the most mutable caprice have created the fashions of painting, as well as those of garments."

The hair is treated with especial care in various countries; it is allowed to grow to full length, so as to reach to the ground, or is combed into "a compact frizzled mop, which is the Papuan's pride and glory." In Northern Africa "a man requires a period of from eight to ten years to perfect his coiffure." With other nations the head is shaved, and in parts of South America and Africa even the eyebrows and eyelashes are eradicated. The natives of the Upper Nile knock out the four front teeth, saying that they do not wish to resemble brutes.

As the face with us is chiefly admired for its beauty, so with savages it is the chief seat of mutilation. In all quarters of the world the septum, and more rarely the wings of the nose are pierced, with rings, sticks, feathers, and other ornaments being inserted into the holes. The ears are everywhere pierced and similarly ornamented, and with the Botocudos and Lenguas of South America the hole is gradually so much enlarged that the lower edge touches the shoulder. In North and South America and in Africa either the upper or lower lip is pierced; and with the Botocudos the hole in the lower lip is so large that a disc of wood, four inches in diameter, is placed in it.

Hardly any part of the body, which can be unnaturally modified, has escaped. The amount of suffering thus caused must have been wonderfully great, for many of the operations require several years for their completion, so that the idea of their necessity must be imperative. The motives are various; the men paint their bodies to make themselves appear terrible in battle; certain mutilations are connected with religious rites, or they mark the age of puberty, or the rank of the man, or they serve to distinguish the tribes.

Having made these preliminary remarks on the admiration felt by savages for various ornaments, and for deformities most unsightly in our eyes, let us see how far the men are attracted by the appearance of their women, and what are their ideas of beauty. As I have heard it maintained that savages are quite indifferent about the beauty of their women, valuing them solely as slaves; it may be well to observe that this conclusion does not at all agree with the care which the women take in ornamenting themselves, or with their vanity. Burchell gives an amusing account of a Bush-woman who used so much grease, red ochre, and shining powder "as would have ruined any but a very rich husband." She displayed also "much vanity and too evident a consciousness of her superiority." Mr. Winwood Reade informs me that the negroes of the West Coast often discuss the beauty of their women. Some competent observers have attributed the fearfully common practice of infanticide partly to the desire felt by the women to retain their good looks. In several regions the women wear charms and use love-philters to gain the affections of the men; and Mr. Brown enumerates four plants used for this purpose by the women of North-Western America.

We thus see how widely the different races of man differ in their taste for the beautiful. In every nation sufficiently advanced to have made effigies of their gods or of their deified rulers, the sculptors no doubt have endeavoured to express their highest ideal of beauty and grandeur. Under this point of view it is well to compare in our mind the Jupiter or Apollo of the Greeks with the Egyptian or Assyrian statues; and these with the hideous bas-reliefs on the ruined buildings of Central America. . . .

The truth of the principle, long ago insisted on by Humboldt, that man admires and often tries to exaggerate whatever characters nature may have given him, is shown in many ways. The practice of beardless races extirpating every trace of a beard, and generally all the hairs on the body, offers one illustration. The skull has been greatly modified during ancient and modern times by many nations; and there can be little doubt that this has been practised, especially in N. and S. America, in order to exaggerate some natural and admired peculiarity. . . . It is certainly not true that there is in the mind of man any universal standard of beauty with respect to the human body. It is, however, possible that certain tastes may in the course of time become inherited, though there is no evidence in favour of this belief: and if so, each race would possess its own innate ideal standard of beauty.

CHAPTER 20
SECONDARY SEXUAL CHARACTERS OF MAN, CONT.

On the Manner of Action of Sexual Selection with mankind.—With primeval men under the favourable conditions just stated, and with those savages who at the present time enter into any marriage tie (but subject to greater or less interference according as the habits of female infanticide, early betrothals, &c., are more or less practised), sexual selection will probably have acted in the following manner. The strongest and most vigorous men,—those who could best defend and hunt for their families, and during later times the chiefs or heads-men,—those who were provided with the best weapons and possessed the most property, such as a large number of dogs or other animals, would have succeeded in rearing a greater average number of offspring, than would the weaker, poorer and lower members of the same tribes. There can, also, be no doubt that such men would generally have been able to select the more attractive women. At present the chiefs of nearly every tribe throughout the world succeed in obtaining more than one wife. Until recently, as I hear from Mr. Mantell, almost every girl in New Zealand, who was pretty, or promised to be pretty, was *tapu* to some chief. With the Kafirs, as Mr. C. Hamilton states, "the chiefs generally have the pick of the women for many miles round, and are most persevering in establishing or confirming their privilege." We have seen that each race has its own style of beauty, and we know that it is natural to man to admire each characteristic point in his domestic animals, dress, ornaments, and personal appearance, when carried a little beyond the common standard. If then the several foregoing propositions be admitted, and I cannot see that they are doubtful, it would be an inexplicable circumstance, if the selection of the more attractive women by the more powerful men of each tribe, who would rear on an average a greater number of children, did not after the lapse of many generations modify to a certain extent the character of the tribe.

With our domestic animals, when a foreign breed is introduced into a new country, or when a native breed is long and carefully attended to, either for use or ornament, it is found after several generations to have undergone, whenever the means of comparison exist, a greater or less amount of change. This follows from unconscious selection during a long series of generations—that is, the preservation of the most approved individuals—without any wish or expectation of such a result on the part of the breeder. So again, if two careful breeders rear during many years animals of the same family, and do not compare them together or with a common standard, the animals are found after a time to have become to the surprise of their owners slightly different.

Each breeder has impressed, as Von Nathusius well expresses it, the character of his own mind—his own taste and judgement—on his animals. What reason, then, can be assigned why similar results should not follow from the long-continued selection of the most admired women by those men of each tribe who were able to rear to maturity the greatest number of children? This would be unconscious selection, for an effect would be produced, independently of any wish or expectation on the part of the men who preferred certain women to others.

Let us suppose the members of a tribe, in which some form of marriage was practised, to spread over an unoccupied continent, they would soon split up into distinct hordes, which would be separated from each other by various barriers, and still more effectually by the incessant wars between all barbarous nations. The hordes would thus be exposed to slightly different conditions and habits of life, and would sooner or later come to differ in some small degree. As soon as this occurred, each isolated tribe would form for itself a slightly different standard of beauty; and then unconscious selection would come into action through the more powerful and leading savages preferring certain women to others. Thus the differences between the tribes, at first very slight, would gradually and inevitably be increased to a greater and greater degree.

WALLACE ON THE DESCENT OF MAN

Darwin counted on Wallace for critical but sympathetic readings of his books by a colleague so favorable to natural selection. They differed on sexual selection and on the role played by natural selection in the evolution of humankind specifically. Wallace touched upon both of these issues in a letter written after he had finished reading the first of *Descent*'s two volumes:

> Many thanks for your first volume, which I have just finished reading through with the greatest pleasure and interest; and I have also to thank you for the great tenderness with which you have treated me and my heresies.
>
> On the subject of "sexual selection" and "protection," you do not yet convince me that I am wrong; but I expect your heaviest artillery will be brought up in your second volume, and I may have to capitulate. You seem, however, to have somewhat misunderstood my exact meaning, and I do not think the difference between us is quite so great as you seem to think it. There are a number of passages in which you argue against the view that the female has in any large number of cases been "specially modified" for protection, or that colour has generally been obtained by either sex for purposes of protection. But my view is, as I thought I had made it clear, that the female has (in most cases) been simply prevented from acquiring the gay tints of the male (even when there was a tendency for her to inherit it), because it was hurtful; and that, when protection is not needed, gay colours are so generally inherited by both sexes as to show that inheritance by both

sexes of colour variations is the most usual, when not prevented from acting by Natural Selection. The colour itself may be acquired either by sexual selection or by other unknown causes.

There are, however, difficulties in the very wide application you give to sexual selection which at present stagger me, though no one was or is more ready than myself to admit the perfect truth of the principle or the immense importance and great variety of its applications.

Your chapters on "Man" are of intense interest—but as touching my special heresy, not as yet altogether convincing, though, of course, I fully agree with every word and every argument which goes to prove the "evolution" or "development" of man out of a lower form. My *only* difficulties are, as to whether you have accounted for *every step* of the development by ascertained laws.

I feel sure the book will keep up and increase your high reputation, and be immensely successful, as it deserves to be.[1]

1. Wallace to Darwin, 27 January 1871, *More Letters*, 2: 92–93.

Chapter 11

FLOWERS AND ADAPTATION

INTRODUCTION

Darwin's three great studies of reproductive botany, *The Various Contrivances by Which Orchids are Fertilised by Insects* (1862; 2nd ed. 1877); *The Effects of Cross and Self-Fertilisation in the Vegetable Kingdom* (1876; 2nd ed. 1878), and *The Different Forms of Flowers in Plants of the Same Species* (1877; 2nd ed. 1884) are variations on the same theme of adaptation that was so central to Darwin's conception of how natural selection works. Here we combine sections from the three volumes in one chapter because Darwin not only stressed the same themes in each, but also, by inserting pertinent cross-references (particularly in the second editions of the first two titles), created a distinct impression of their interrelationship.[1] The themes that are stressed in all three volumes are, first, the various and complex ways in which the same plant organs have been adapted to serve different purposes, and second, the ways in which plants, whether hermaphroditic or divided into separate sexes, are designed to promote the kind of fertilization mechanisms (usually cross-fertilization) that ensure the production of the most vigorous and viable offspring. A subtext of the second theme has to do with the rationale behind the origin of sex in the first place, although this subplot is secondary in Darwin's presentation of the adaptive nature of sex differentiation in plants.

Orchids were an early interest of Darwin's. The title of his book on them draws attention to the notion of "contrivance," a term connoting artfulness

---

1. The sixth edition of *On the Origin of Species*, which appeared in 1872, naturally reflected Darwin's heightened concern with reproductive adaptations of plants at this stage of his scientific career.

and ingenuity which was much favored by natural theologian William Paley, whose doctrine of "perfect adaptation" impressed Darwin when he read Paley's book at Cambridge. He was drawn to the orchids because, as a group, they displayed a dazzling array of structures—contrivances—that served the same purpose. Moreover, the organs of a plant that originally served one purpose later came, through the action of natural selection, to serve another. This peculiar and striking kind of adaptation has been titled "exaptation" by Gould and Vrba.[2] It was this phenomenon, in which characters evolved for a certain use were later "co-opted for another function," that attracted Darwin's attention in the first place, because many orchid species had developed ingenious contrivances to secure their efficient pollination.

In the first edition of *On the Origin of Species*, in the chapter (6) on difficulties of the theory of natural selection, Darwin introduces the orchid in a discussion of organs that appear to have changed their function over time. In plants, he notes,

> the very curious contrivance of a mass of pollen-grains, borne on a foot-stalk with a sticky gland at the end, is the same in Orchis and Asclepias,—genera almost as remote as possible among the flowering plants. In all these cases of two very distinct species furnished with apparently the same anomalous organ, it should be observed that, although the general appearance and function of the organ may be the same, yet some fundamental difference can generally be detected. I am inclined to believe that in nearly the same way as two men have sometimes independently hit on the very same invention, so natural selection, working for the good of each being and taking advantage of analogous variations, has sometimes modified in very nearly the same manner two parts in two organic beings, which owe but little of their structure in common to inheritance from the same ancestor.[3]

In the sixth edition (1872), Darwin includes a long passage on the orchid as an example of how natural selection works in adaptive ways. The question here is why certain structures or organs of plants have been adapted for no reason apparent at first glance. This leads to a discussion of an orchid, *Coryanthes*, as described by a German biologist, Dr. Crüger:

> This orchid has part of its labellum or lower lip hollowed out into a great bucket, into which drops of almost pure water continually fall from two secreting horns which stand above it; and when the bucket is half full, the water overflows by a spout on one side. The basal part of the labellum stands over the bucket, and is

2. Stephen Jay Gould and Elisabeth S. Vrba, "Exaptation—A Missing Term in the Science of Form," *Paleobiology* 8 (1982): 4–15.

3. *On the Origin of Species*, 1st ed., 193–194.

itself hollowed out into a sort of a chamber with two lateral entrances; within this chamber are two curious fleshy ridges. The most ingenious man, if he had not witnessed what takes place, could never have imagined what purpose all these parts serve. But Dr. Crüger saw crowds of large humble-bees visiting the gigantic flowers of this orchid, not in order to suck nectar, but to gnaw off the ridges within the chamber above the bucket, and their wings being thus wetted they could not fly away, but were compelled to crawl out through the passage formed by the spout or overflow. Dr. Crüger saw a 'continual' procession of bees thus crawling out of their involuntary bath. The passage is narrow, and is roofed over by the column, so that a bee, in forcing its way out, first rubs its back against the viscid stigma and then against the viscid glands of the pollen-masses. The pollen-masses are thus glued to the back of the bee which first happens to crawl out through the passage of a lately expanded flower, and are thus carried away.[4]

Darwin goes on, in the same passage, to discuss the equally curious adaptation for pollination of another orchid *Catasetum*, wherein the visitation of a bee sets off a kind of explosion in the flower whereby a mass of pollen shoots forth so that it adheres to the back of the visiting insect. *Catasetum* is an orchid with three flowers—one male, one female, the third hermaphroditic—each of which had, before Darwin's analysis, been considered separate species. When Darwin described his findings to Thomas Huxley, the latter replied, "Do you think that I can believe all that!"[5]

The key to the vast diversity of structures all designed to bring about efficient fertilization was the process of coadaptation. One organ—a stamen, a pistil, a petal—would become highly specialized in order to enhance some specific mode of fertilization, and as a result, other organs of the same plant would themselves have to adapt to the changed function of the former. In general, the adaptations or contrivances that Darwin discusses are those designed to enhance cross-fertilization, a process that produces, in Darwin's view, greater numbers of more viable progeny. The counter-case he gives would be a plant, formerly pollinated by a specific insect, which must, in order to survive, readapt itself for *self*-fertilization when, for whatever reason, that insect is no longer available to cross-pollinate.

Darwin himself was keenly aware of the relationship of his orchid observations to the core of his theory. As he wrote to Hooker in 1862, "I found the study of Orchids eminently useful in showing me how nearly all parts of the flower are co-adapted for fertilization by insects, and therefore the results of natural selection—even the most trifling details of structure."[6]

---

4. *On the Origin of Species*, 6th ed. (New York, Random House, 1993), 244–45.

5. Quoted in Adrian Desmond and James Moore, *Darwin: The Life of a Tormented Evolutionist* (New York, Time Warner, 1991), 510.

6. Darwin to Hooker, 14 May 1862, *Life and Letters*, II: 430.

The same topics of cross- and self-fertilization he then analyzed in the second volume of this series, *The Effects of Cross and Self-Fertilisation in the Vegetable Kingdom*, a broader discussion not limited to orchids. One of the book's noteworthy features is that it was based not only on years of careful observation, as in the case of his studies of orchids, but also on a series of experiments that Darwin himself had designed and carried out at Down. Darwin spent more than three decades studying the different ways in which plants are fertilized, which yielded the variegated patterns of adaptation that Darwin loved so much. He expressed his fascination in a letter to his German follower Ernst Haeckel in 1875:

> I am now busy in drawing up an account of ten years' experiments[7] in the growth and fertility of plants raised from crossed and self-fertilised flowers. It is really wonderful what an effect pollen from a distinct seedling plant, which has been exposed to different conditions of life, has on the offspring in comparison with pollen from the same flower or from a distinct individual, but which has been long subjected to the same conditions. The subject bears on the very principle of life, which seems almost to require changes in the conditions.[8]

He then asks why different sexes have developed and concludes that the mating of two distinct individuals yields more vigorous offspring than does reproduction that does not require such pairing. By the same token, the division of species into separate male and female individuals is more advantageous than hermaphroditism, wherein a single plant contains both male and female reproductive organs. Hermaphroditism is adaptive, however, in the case of organisms which lack motility.

The book was important to Darwin. He normally sent proof sheets of his books to selected correspondents so that reviews could be published immediately upon publication. Accordingly, when the proofs were ready, he sent off a set to Asa Gray, the Harvard botanist whose judgment—as we have seen—Darwin greatly valued.

> My Dear Gray:
> I send by this post all the clean sheets as yet printed, and I hope to send the remainder within a fortnight. Please observe that the first six chapters are not readable, and the six last very dull. Still I believe that the results are valuable. If you review the book, I shall be very curious to see what you think of it, for I care more

---

7. The experiments date back to at least 1861, when he studied the intersterility of cowslips and primroses, whose anatomical structures are almost identical; see Desmond and Moore, *Darwin: The Life of a Tormented Evolutionist* (New York, Time Warner, 1991), 510.

8. Darwin to Haeckel, 13 November 1875, *More Letters*, 2: 406.

for your judgement than for that of almost anyone else. I know that you will speak the truth, whether you approve or disapprove. Very few will take the trouble to read the book, and I do not expect you to read the whole, but I hope you will read the latter chapters. . . .

I am so sick of correcting the press and licking my horrid bad style into intelligible English.[9]

The third book of the series considered plant species in which the flowers of different individuals displayed different forms. Darwin is particularly concerned with "heterostyled" plants, plants which have styles of more than one form. (The style is the elongated portion of the pistil, which plays a crucial role in pollination and, as Darwin shows, has also proved to be susceptible to ingenious adaptations produced by natural selection to promote fertilization.) An oddity of heterostyly is that in plant species in which it exists, each individual plant can be fertilized only by an individual of another form. Why, he asks, would the mutual sterility of individuals of the same species be beneficial to that species? The answer is that the arrangement is another "contrivance" whereby natural selection has ensured a maximum of outbreeding in order to enhance the viability and vigor of the offspring produced. When originally hermaphroditic plants evolve into heterostyled forms, the enhanced likelihood of cross-fertilization is the result.

In order to test his hypothesis, Darwin experimented on *Lythrum salicaria*, a trimorphic, heterostyled species—that is, it had three different forms of flowers, one with a long style, another with a mid-sized one, and a third with a short one. He segregated each kind of plant with netting and laboriously, with the help of an assistant, cross-pollinated each type by hand with pollen from the other two types in all possible combinations. He found that the offspring of individuals of the same type ("legitimate crosses") were less vigorous and fertile than those resulting from crossing of different types ("illegitimate crosses").

"I have done nothing which has interested me so much as Lythrum," Darwin wrote to Hooker (26 November 1864, *Life and Letters*, 3: 306), "since making out the complemental males of Cirripedes" (see pp. 118–26 for a discussion of complemental males). Both studies involved ingenious adaptations that enhanced the reproductive vigor of offspring.

Darwin's account of difficulties that arose in the *Lythrum* crosses of 1862 gives a vivid record of his rigorous approach to experimentation:

I endeavoured to prevent pollen dropping from an upper to a lower flower, and I tried to remember to wipe the pincers carefully after each fertilization; but in making eighteen different unions [that is, all the different combinations of this tri-

9. Darwin to Gray, 28 October 1876, *Life and Letters*, 3: 293.

morphic species], sometimes on windy days, and pestered by bees and flies buzzing about, some few errors could hardly be avoided. One day I had to keep a third man by me all the time to prevent the bees visiting the uncovered plants, for in a few seconds' time they might have done irreparable mischief. It was also extremely difficult to exclude minute Diptera from the net. In 1862 I made the great mistake of placing mid-styled and long-styled under the same huge net: in 1863 I avoided this error.[10]

In his *Autobiography*, Darwin states, "no little discovery of mine ever gave me so much pleasure as the making out the meaning of heterostyled flowers."[11]

In all three of these studies, Darwin repeatedly uses the expression "specially adapted," a phrase which recalls the purposeful adaptation that the natural theologian William Paley had stressed in his doctrine of "perfect adaptation"—the notion that God has created each and every species in a form "perfectly adapted" to its environment. But here, Darwin, in effect, presents a parody of Paley's perfect adaptation by describing orchids whose exquisite flowers are molded from apparently useless floral parts, and heterostyly whose adaptations are so seemingly nonsensical that the sober watchmaker of eighteenth-century natural philosophy would never have, and could not have, thought them up.[12]

THE VARIOUS CONTRIVANCES BY WHICH
ORCHIDS ARE FERTILISED BY INSECTS
(SECOND EDITION, 1877)

CONCLUDING REMARKS
ON THE CAUSE OF DIVERSITY AND
OF THE PERFECTION OF CONTRIVANCES

I have now nearly finished this volume, which is perhaps too lengthy. It has, I think, been shown that the Orchideae exhibit an almost endless diversity of beautiful adaptations. When this or that part has been spoken of as adapted for some special purpose, it must not be supposed that it was originally al-

10. *The Different Forms of Flowers in Plants of the Same Species*, 2nd ed. (London, 1884), 156, n. 6.

11. See p. 317, below.

12. See Michael T. Ghiselin's characterization of Darwin's studies of plant adaptation as "A Metaphysical Satire"—a kind of anti-Bridgewater Treatise (the Bridgewater Treatises were an influential series of essays commissioned early in the century to demonstrate God's design in nature); *The Triumph of the Darwinian Method* (Berkeley, University of California Press, 1969), 131–159.

ways formed for this sole purpose. The regular course of events seems to be, that a part which originally served for one purpose, becomes adapted by slow changes for widely different purposes. To give an instance: in all the Ophreae, the long and nearly rigid caudicle manifestly serves for the application of the pollen-grains to the stigma, when the pollinia are transported by insects to another flower; and the anther opens widely in order that the pollinium should be easily withdrawn; but in the Bee Ophrys, the caudicle, by a slight increase in length and decrease in its thickness, and by the anther opening a little more widely, becomes specially adapted for the very different purpose of self-fertilisation, through the combined aid of the weight of the pollen-mass and the vibration of the flower when moved by the wind. Every gradation between these two states is possible — of which we have a partial instance in *O. aranifera*.

Again, the elasticity of the pedicel of the pollinium in some Vandeae is adapted to free the pollen-masses from their anther-cases; but by a further slight modification, the elasticity of the pedicel becomes specially adapted to shoot out the pollinium with considerable force so as to strike the body of the visiting insect. The great cavity in the labellum of many Vandeae is gnawed by insects and thus attracts them; but in *Mormodes ignea* it is greatly reduced in size, and serves in chief part to keep the labellum in its new position on the summit of the column. From the analogy of many plants we may infer that a long spur-like nectary is primarily adapted to secrete and hold a store of nectar; but in many orchids it has so far lost this function, that it contains fluid only in the intercellular spaces. In those orchids in which the nectary contains both free nectar and fluid in the intercellular spaces, we can see how a transition from the one state to the other could be effected, namely, by less and less nectar being secreted from the inner membrane, with more and more retained within the intercellular spaces. Other analogous cases could be given.

Although an organ may not have been originally formed for some special purpose, if it now serves for this end, we are justified in saying that it is specially adapted for it. On the same principle, if a man were to make a machine for some special purpose, but were to use old wheels, springs, and pulleys, only slightly altered, the whole machine, with all its parts, might be said to be specially contrived for its present purpose. Thus throughout nature almost every part of each living being has probably served, in a slightly modified condition, for diverse purposes, and has acted in the living machinery of many ancient and distinct specific forms.

In my examination of orchids, hardly any fact has struck me so much as the endless diversities of structure — the prodigality of resources — for gaining the very same end, namely, the fertilisation of one flower by pollen from another plant. This fact is to a large extent intelligible on the principle of Natural Selection. As all the parts of a flower are co-ordinated, if slight variations in any one part were preserved from being beneficial to the plant, then the other parts would generally have to be modified in some corresponding manner. But these latter parts might not vary at all, or they might not vary in a fitting manner, and these other variations, whatever their nature might be, which tended to bring all the parts into more harmonious action with one another, would be preserved by Natural Selection.

To give a simple illustration: in many orchids the ovarium (but sometimes the foot stalk) becomes for a period twisted, causing the labellum to assume the position of a lower petal, so that insects can easily visit the flower; but from slow changes in the form or position of the petals, or from new sorts of insects visiting the flowers, it might be advantageous to the plant that the labellum should resume its normal position on the upper side of the flower, as is actually the case with *Malaxis paludosa*, and some species of Catasetum, etc. This change, it is obvious, might be simply effected by the continued selection of varieties which had their ovaria less and less twisted; but if the plant only afforded varieties with the ovarium more twisted, the same end could be attained by the selection of such variations, until the flower was turned completely round on its axis. This seems to have actually occurred with *Malaxis paludosa*, for the labellum has acquired its present upward position by the ovarium being twisted twice as much as is usual.

Again, we have seen that in most Vandeae there is a plain relation between the depth of the stigmatic chamber and the length of the pedicel, by which the pollen-masses are inserted; now if the chamber became slightly less deep from any change in the form of the column or other unknown cause, the mere shortening of the pedicel would be the simplest corresponding change; but if the pedicel did not happen to vary in shortness, the slightest tendency to its becoming bowed from elasticity as in Phalaenopsis, or to a backward hygrometric movement as in one of the Maxillarias, would be preserved, and the tendency would be continually augmented by selection; thus the pedicel, as far as its action is concerned, would be modified in the same manner as if it had been shortened. Such processes carried on during many thousand generations in various ways, would create an endless diversity of co-adaped structures in the several parts of the flower for the same general purpose. This view affords, I believe, the key which partly solves the problem of the vast diversity of structure adapted for closely analogous ends in many large groups of organic beings.

The more I study nature, the more I become impressed with ever-increasing force, that the contrivances and beautiful adaptations slowly acquired through each part occasionally varying in a slight degree but in many ways, with the preservation of those variations which were beneficial to the organism under complex and ever-varying conditions of life, transcend in an incomparable manner the contrivances and adaptations which the most fertile imagination of man could invent.

The use of each trifling detail of structure is far from a barren search to those who believe in Natural Selection. When a naturalist casually takes up the study of an organic being, and does not investigate its whole life (imperfect though that study will ever be), he naturally doubts whether each trifling point can be of any use, or indeed whether it be due to any general law. Some naturalists believe that numberless structures have been created for the sake of mere variety and beauty — much as a workman would make different patterns. I, for one, have often and often doubted whether this or that detail of structure in many of the Orchideae and other plants could be of any service; yet, if of no good, these structures could not have been modelled by the natural preservation of useful variations; such details can only be vaguely accounted for by the direct action of the conditions of life, or the mysterious laws of correlated growth.

To give nearly all the instances of trifling details of structure in the flowers of orchids, which are certainly of high importance, would be to recapitulate almost the whole of this volume. But I will recall to the reader's memory a few cases. I do not here refer to the fundamental framework of the plant, such as the remnants of the fifteen primary organs arranged alternately in the five whorls; for almost everyone who believes in the gradual evolution of species will admit that their presence is due to inheritance from a remote parent-form. Innumerable facts with respect to the uses of the variously shaped and placed petals and sepals have been given. So again, the importance of as light difference in the shape of the caudicle of the pollinium of the Bee Ophrys, compared with that of the other species of the same genus, has likewise been referred to; to this might be added the doubly-bent caudicle of the Fly Ophrys. Indeed, the important relation of the length and shape of the caudicle, with reference to the position of the stigma, might be cited throughout many whole tribes. The solid projecting knob of the anther in *Epipactis palustris*, which does not include pollen, liberates the pollen-masses when it is moved by insects. In *Cephalanthera grandiflora*, the upright position of the almost closed flower protects the slightly coherent pillars of pollen from disturbance. The length and elasticity of the filament of the anther in certain species of Dendrobium apparently serves for self-fertilisation, if insects fail to transport the pollen-masses. The slight forward inclination of the crest of the rostellum in Listera prevents the anther-case being caught as soon as the viscid matter is ejected. The elasticity of the lip of the rostellum in Orchis causes it to spring up again when only one of the pollen-masses has been removed, thus keeping the second viscid disc ready for action, which otherwise would be wasted. No one who had not studied orchids would have suspected that these and very many other small details of structure were of the highest importance to each species; and that consequently, if the species were exposed to new conditions of life, and the structure of the several parts varied ever so little, the smallest details of structure might readily be acquired through Natural Selection. These cases afford a good lesson of caution with respect to the importance of apparently trifling particulars of structure in other organic beings.

It may be naturally enquired. Why do the Orchideae exhibit so many perfect contrivances for their fertilisation? From the observations of various botanists and my own, I am sure that many other plants offer analogous adaptations of high perfection; but it seems that they are really more numerous and perfect with the Orchideae than with most other plants. To a certain extent this enquiry can be answered. As each ovule requires at least one, probably several, pollen-grains, and as the seeds produced by orchids are so inordinately numerous, we can see that it is necessary that large masses of pollen should be left on the stigma of each flower. Even in the Neotteae, which have granular pollen, with the grains tied together by weak threads, I have observed that considerable masses of pollen are generally left on the stigmas. This circumstance apparently explains why the grains cohere in packets or large waxy masses, as they do in so many tribes, namely, to prevent waste in the act of transportal. The flowers of most plants produce pollen enough to fertilize several flowers, so as to allow of or to favour cross-fertilization. But with the many orchids which produce only two pollen-masses, and with some of the Malaxeae which produce only one, the pollen from a single flower cannot possibly fertilise more than two flowers or only a single one; and cases of this kind do not occur, as I believe, in any other group of plants. If the Orchideae had elaborated as much pollen as is produced by other plants, relatively to the number of seeds which they yield, they would have had to produce a most extravagant amount, and this would have caused exhaustion. Such exhaustion is avoided by pollen not being produced in any great superfluity owing to the many special contrivances for its safe transportal from plant to plant, and for placing it securely on the stigma. Thus we can understand why the Orchideae are more highly endowed in their mechanism for cross-fertilisation, than are most other plants.

In my work on the *Effects of Cross and Self Fertilisation in the Vegetable Kingdom*, I have shown that when flowers are cross-fertilised they generally receive pollen from a distinct plant and not that from another flower on the same plant; a cross of this latter kind doing little or no good. I have further shown that the benefits derived from a cross between two plants depends altogether on their differing somewhat in constitution; and there is much evidence that each individual seedling possesses its own peculiar constitution. The crossing of distinct plants of the same species is favoured or determined in various ways, as described in the above work, but chiefly by the prepotent action of pollen from another plant over that from the same flower. Now with the Orchideae it is highly probable that such prepotency prevails, for we know from the valuable observations of Mr Scott and Fritz Müller,[1] that with several orchids pollen from their own flower is quite impotent, and is even in some cases poisonous to the stigma. Besides this prepotency, the Orchideae present various special contrivances — such as the pollinia not assuming a proper position for striking the stigma until some time has elapsed after their removal from the anthers — the slow curving forwards and then backwards of the rostellum in Listera and Neottia — the slow movement of the column from the labellum in Sprianthes — the dioecious condition of Catasetum — the fact of some species producing only a single flower, etc. — all render it certain or highly probable that the flowers are habitually fertilized with pollen from a distinct plant.

That cross-fertilisation, to the complete exclusion of self-fertilisation, is the rule with the Orchideae, cannot be doubted from the facts already given in relation to many species in all the tribes throughout the world. I could almost as soon believe the flowers in general were not adapted for the production of seeds, because there are a few plants which have never been known to yield seed, as that the flowers of the Orchideae are not as a general rule adapted so as to ensure cross-fertilisation. Nevertheless, some species are regularly or often self-fertilised; and I will now give a list of all the cases hitherto observed by myself and others. In some of these the flowers appear often to be fertilised by insects, but they are capable of fertilising themselves without aid, though in a more or less incomplete manner; so that they do not remain utterly barren if insects fail to visit them. Under this head may be included three British species, namely, *Cephalanthera grandiflora*, *Neottia nidus-avis*, and perhaps *Listera ovata*. In South Africa *Disa macrantha* often fertilises itself; but Mr Weale believes that it is likewise cross-fertilised by moths. Three species belonging to the Epidendreae rarely open their flowers in the West Indies; nevertheless these flowers fertilise themselves, but it is doubtful whether they are fully fertilised, for a large proportion of the seeds spontaneously produced by some members of this tribe in a hothouse were destitute of an embryo. Some species of Dendrobium, judging from their structure and from their occasionally producing capsules under cultivation, likewise come under this head.

Of species which regularly fertilise themselves without any aid and yield full-sized capsules, hardly any case is more striking than that of *Ophrys apifera*, which was advanced by me in the first edition of this work. To this case may

---

1. A full abstract of these observations is given in my *Variation of Animals and Plants under Domestication*, ch. xvii, 2nd edit., vol. ii, p. 114.

now be added two other European plants, *Orchis* or *Neotinea intacta* and *Epipactis viridiflora*. Two North American species, *Gymnadenia tridentata* and *Platanthera hyperborea* appear to be in the same predicament, but whether when self-fertilised they yield a full complement of capsules containing good seeds has not been ascertained. A curious Epidendrum in South Brazil which bears two additional anthers fertilises itself freely by their aid; and *Dendrobium cretaceum* has been known to produce perfect self-fertilised seeds in a hothouse in England. Lastly, *Spiranthes australis* and two species of Thelymitra, inhabitants of Australia, come under this same head. No doubt other cases will hereafter be added to this short list of about ten species which it appears can fertilise themselves fully, and of about the same number of species which fertilise themselves imperfectly when insects are excluded.

It deserves especial attention that the flowers of all the above-named self-fertile species still retain various structures which it is impossible to doubt are adapted for insuring cross-fertilisation, though they are now rarely or never brought into play. We may therefore conclude that all these plants are descended from species or varieties which were formerly fertilised by insect-aid. Moreover, several of the genera to which these self-fertile species belong, include other species, which are incapable of self-fertilisation. The-lymitra offers indeed the only instance known to me of two species within the same genus which regularly fertilise themselves. Considering such cases as those of Ophrys, Disa, and Epidendrum, in which one species alone in the genus is capable of complete self-fertilisation, whilst the other species are rarely fertilised in any manner owing to the rarity of the visits of the proper insects; bearing also in mind the large number of species in many parts of the world which from this same cause are seldom impregnated, we are led to believe that the above-named self-fertile plants formerly depended on the visits of insects for their fertilisation, and that from such visits failing they did not yield a sufficiency of seed and were verging towards extinction. Under these circumstances it is probable that they were gradually modified, so as to become more or less completely self-fertile; for it would manifestly be more advantageous to a plant to produce self-fertilised seeds rather than none at all or extremely few seeds. Whether any species which is now never cross-fertilised will be able to resist the evil effects of long-continued self-fertilisation, so as to survive for as long an average period as the other species of the same genera which are habitually cross-fertilised, cannot of course be told. But *Ophrys apifera* is still a highly vigorous plant, and *Gymnadenia tridentata* and *Platanthera hyperborea* are said by Asa Gray to be common plants in North America. It is indeed possible that these self-fertile species may revert in the course of time to what was undoubtedly their pristine condition, and in this case their various adaptations for cross-fertilisation would be again brought into action. We may believe that such reversion is possible, when we hear from Mr Moggridge that *Ophrys scolopax* fertilises itself freely in one district of Southern France without the aid of insects, and is completely sterile without such aid in another district.

Finally, if we consider how precious a substance pollen is, and what care has been bestowed on its elaboration and on the accessory parts in the Orchideae — considering how large an amount is necessary for the impregnation of the almost innumerable seeds produced by these plants — considering that the anther stands close behind or above the stigma, self-fertilization would have been an incomparably safer and easier process than the transportal of pollen from flower to flower. Unless we bear in mind the good effects which have been proved to follow in most cases from cross-fertilization, it is an astonishing fact that the flowers of the Orchideae should not have been regularly self-fertilised. It apparently demonstrates that there must be something injurious in this latter process, of which fact I have elsewhere given direct proof. It is hardly an exaggeration to say that Nature tells us, in the most emphatic manner, that she abhors perpetual self-fertilisation.

THE EFFECTS OF CROSS AND SELF-FERTILISATION IN THE VEGETABLE KINGDOM (SECOND EDITION, 1878)

GENERAL REMARKS

Under a theoretical point of view it is some gain to science to know that numberless structures in hermaphrodite plants, and probably in hermaphrodite animals, are special adaptations for securing an occasional cross between two individuals; and that the advantages from such a cross depend altogether on the beings which are united, or their progenitors, having had their sexual elements somewhat differentiated, so that the embryo is benefited in the same manner as is a mature plant or animal by a slight change in its conditions of life, although in a much higher degree.

Another and more important result may be deduced from my observations. Eggs and seeds are highly serviceable as a means of dissemination, but we now know that fertile eggs can be produced without the aid of the male. There are also many other methods by which organisms can be propagated asexually. Why then have the two sexes been developed, and why do males exist which cannot themselves produce offspring? The answer lies, as I can hardly doubt, in the great good which is derived from the fusion of two somewhat differentiated individuals; and with the exception of the lowest organisms this is possible only by means of the sexual elements, these consisting of cells separated from the body, containing the germs of every part, and capable of being fused completely together.

It has been shown in the present volume that the offspring from the union of two distinct individuals, especially if their progenitors have been subjected to very different conditions, have an immense advantage in height, weight, constitutional vigour and fertility over the self-fertilised offspring from one of the same parents. And this fact is amply sufficient to account for the development of the sexual elements, that is, for the genesis of the two sexes.

It is a different question why the two sexes are sometimes combined in the same individual and are sometimes separated. As with many of the lowest plants and animals the conjugation of two individuals which are either quite similar or in some degree different, is a common phenomenon, it seems probable, as remarked in the last chapter, that the sexes were primordially separate. The individual which receives the contents of the other, may be called the female; and the other, which is often smaller and more locomotive, may be called the male; though these sexual names ought hardly to be applied as long as the whole contents of the two forms are blended into one. The object gained by the two sexes becoming united in the same hermaphrodite form probably is to allow of occasional or frequent self-fertilisation, so as to ensure the propagation of the species, more especially in the case of organisms affixed for life to the same spot. There does not seem to be any great difficulty in understanding how an organism, formed by the conjugation of two individuals which represented the two incipient sexes, might give rise by budding first to a monoecious and then to an hermaphrodite form; and in the case of animals even without budding to an hermaphrodite form, for the bilateral structure of animals perhaps indicates that they were aboriginally formed by the fusion of two individuals.

It is a more difficult problem why some plants and apparently all the higher animals, after becoming hermaphrodites, have since had their sexes re-separated. This separation has been attributed by some naturalists to the advantages which follow from a division of physiological labour. The principle is intelligible when the same organ has to perform at the same time diverse functions; but it is not obvious why the male and female glands when placed in different parts of the same compound or simple individual, should not perform their functions equally well as when placed in two distinct individuals. In some instances the sexes may have been re-separated for the sake of preventing too frequent self-fertilization; but this explanation does not seem probable, as the same end might have been gained by other and simpler means, for instance dichogamy. It may be that the production of the male and female reproductive elements and the maturation of the ovules was too great a strain and expenditure of vital force for a single individual to withstand, if endowed with a highly complex organization; and that at the same time there was no need for all the individuals to produce young, and consequently that no injury, on the contrary, good resulted from half of them, or the males, failing to produce offspring.

There is another subject on which some light is thrown by the facts given in this volume, namely, hybridisation. It is notorious that when distinct species of plants are crossed, they produce with the rarest exceptions fewer seeds than the normal number. This unproductiveness varies in different species up to sterility so complete that not even an empty capsule is formed: and all experimentalists have found that it is much influenced by the conditions to which the crossed species are subjected. A plant's own pollen is strongly prepotent over that of any other species, so that if it is placed on the stigma some time after foreign pollen has been applied to it, any effect from the latter is quite obliterated. It is also notorious that not only the parent species, but the hybrids raised from them are more or less sterile; and that their pollen is often in a more or less aborted condition. The degree of sterility of various hybrids does not always strictly correspond with the degree of difficulty in uniting the parent forms. When hybrids are capable of breeding *inter se*, their descendants are more or less sterile, and they often become still more sterile in the later generations; but then close interbreeding has hitherto been practised in all such cases. The more sterile hybrids are sometimes much dwarfed in stature, and have a feeble constitution. Other facts could be given, but these will suffice for us. Naturalists formerly attributed all these results to the difference between species being fundamentally distinct from that between the varieties of the same species; and this is still the verdict of some naturalists.

The results of my experiments in self-fertilising and cross-fertilising the individuals or the varieties of the same species, are strikingly analogous with those just given, though in a reversed manner. With the majority of species flowers fertilised with their own pollen yield fewer sometimes much fewer seeds, than those fertilised with pollen from another individual or variety. Some self-fertilised flowers are absolutely sterile; but the degree of their sterility is largely determined by the conditions to which the parent plants have been exposed, as was well exemplified in the case of Eschscholtzia and Abutilon. The effects of pollen from the same plant are obliterated by the prepotent influence of pollen from another individual or variety, although the latter may have been placed on the stigma some hours afterwards. The offspring from self-fertilised flowers are themselves more or less sterile, sometimes highly sterile, and their pollen is sometimes in an imperfect condition; but I have not met with any case of complete sterility in self-fertilised seedlings, as is so common with hybrids. The degree of their sterility does not correspond with that of the parent-plants when first self-fertilised. The offspring of self-fertilised plants suffer in stature, weight, and constitutional vigour more frequently and in a greater degree than do the hybrid offspring of the greater number of crossed species. Decreased height is transmitted to the next generation, but I did not ascertain whether this applies to decreased fertility.

I have elsewhere shown[1] that by uniting in various ways dimorphic or trimorphic heterostyled plants, which belong to the same undoubted species, we get another series of results exactly parallel with those from crossing distinct species. Plants illegitimately fertilized with pollen from a distinct plant belonging to the same form, yield fewer, often much fewer seeds, than they do when legitimately fertilized with pollen from a plant belonging to a distinct form. They sometimes yield no seed, not even an empty capsule, like a species fertilized with pollen from a distinct genus. The degree of sterility is much affected by the conditions to which the plants have been subjected. The pollen from a distinct form is strongly prepotent over that from the same form, although the former may have been placed on the stigma many hours afterwards. The offspring from a union between plants of the same form are more or less sterile, like hybrids, and have their pollen in a more or less aborted condition; and some of the seedlings are as barren and as dwarfed as the most barren hybrid. They also resemble hybrids in several other respects, which need not here be specified in detail — such as their sterility not corresponding in degree with that of the parent plants — the unequal sterility of the latter, when reciprocally united — and the varying sterility of the seedlings raised from the same seed-capsule.

We thus have two grand classes of cases giving results which correspond in the most striking manner with those which follow from the crossing of so-called true and distinct species. With respect to the difference between seedlings raised from cross and self-fertilised flowers, there is good evidence that this depends altogether on whether the sexual elements of the parents have been sufficiently differentiated, by exposure to different conditions or by spontaneous variation. The manner in which plants have been rendered heterostyled is an obscure subject, but it is probable that the two or three forms first became adapted for mutual fertilisation, that is for cross-fertilisation, through the variation of their stamens and pistils in length, and that afterwards their pollen and ovules became co-adapted; the greater or less sterility of any one form with pollen from the same form being an incidental result.[2] Anyhow, the two or three forms of heterostyled species belong to the same species as certainly as do the two sexes of any one species. We have therefore no right to maintain that the sterility of species when first crossed and of their hybrid offspring, is determined by some cause fundamentally different from that which determines the sterility of the individu-

---

1. *The Different Forms of Flowers on Plants of the same species*, 1877, p. 240.

2. This subject has been discussed in my *Different Forms of Flowers etc.*, pp. 260–68.

als both of ordinary and of heterostyled plants when united in various ways. Nevertheless, I am aware that it will take many years to remove this prejudice.

There is hardly anything more wonderful in nature than the sensitiveness of the sexual elements to external influences, and the delicacy of their affinities. We see this in slight changes in the conditions of life being favourable to the fertility and vigour of the parents, while certain other and not great changes cause them to be quite sterile without any apparent injury to their health. We see how sensitive the sexual elements of those plants must be, which are completely sterile with their own pollen, but are fertile with that of any other individual of the same species. Such plants become either more or less self-sterile if subjected to changed conditions, although the change may be far from great. The ovules of a heterostyled trimorphic plant are affected very differently by pollen from the three sets of stamens belonging to the same species. With ordinary plants the pollen of another variety or merely of another individual of the same variety is often strongly prepotent over its own pollen, when both are placed at the same time on the same stigma. In those great families of plants containing many thousand allied species, the stigma of each distinguishes with unerring certainty its own pollen from that of every other species.

There can be no doubt that the sterility of distinct species when first crossed, and of their hybrid offspring, depends exclusively on the nature or affinities of their sexual elements. We see this in the want of any close correspondence between the degree of sterility and the amount of external difference in the species which are crossed; and still more clearly in the wide difference in the results of crossing reciprocally the same two species; that is, when species A is crossed with pollen from B, and then B is crossed with pollen from A. Bearing in mind what has just been said on the extreme sensitiveness and delicate affinities of the reproductive system, why should we feel any surprise at the sexual elements of those forms, which we call species, having been differentiated in such a manner that they are incapable or only feebly capable of acting on one another? We know that species have generally lived under the same conditions, and have retained their own proper characters, for a much longer period than varieties. Long-continued domestication eliminates, as I have shown in my *Variation under Domestication*, the mutual sterility which distinct species lately taken from a state of nature almost always exhibit when inter-crossed; and we can thus understand the fact that the most different domestic races of animals are not mutually sterile. But whether this holds good with cultivated varieties of plants is not known, though some facts indicate that it does. The elimination of sterility through long-continued domestication may probably be attributed to the varying conditions to which our domestic animals have been subjected; and no doubt it is owing to this same cause that they withstand great and sudden changes in their conditions of life with far less loss of fertility than do natural species. From these several considerations it appears probable that the difference in the affinities of the sexual elements of distinct species, on which their mutual incapacity for breeding together depends, is caused by their having been habituated for a very long period each to its own conditions, and to the sexual elements having thus acquired firmly fixed affinities. However this may be, with the two great classes of cases before us, namely, those relating to the self-fertilisation and cross-fertilisation of the individuals of the same species, and those relating to the illegitimate and legitimate unions of hetero-styled plants, it is quite unjustifiable to assume that the sterility of species when first crossed and of their hybrid offspring, indicates that they differ in some fundamental manner from the varieties or individuals of the same species.

THE DIFFERENT FORMS OF FLOWERS
IN PLANTS OF THE SAME SPECIES,
2ND ED. (1878)

LYTHRUM SALICARIA

Lythrum salicaria. The pistil in each form differs from that in either of the other forms, and in each there are two sets of stamens different in appearance and function. But one set of stamens in each form corresponds with a set in one of the other two forms. Altogether this one species includes three females or female organs and three sets of male organs, all as distinct from one another as if they belonged to different species; and if smaller functional differences are considered, there are five distinct sets of males. Two of the three hermaphrodites must co-exist, and pollen must be carried by insects reciprocally from one to the other, in order that either of the two should be fully fertile; but unless all three forms co-exist, two sets of stamens will be wasted, and the organization of the species, as a whole, will be incomplete. On the other hand, when all three hermaphrodites co-exist, and pollen is carried from one to the other, the scheme is perfect; there is no waste of pollen and no false co-adaptation. In short, nature has ordained a most complex marriage-arrangement, namely a triple union between three hermaphrodites — each hermaphrodite being in its female organ quite distinct from the other two hermaphrodites and partially distinct in its male organs, and each furnished with two sets of males.

The three forms may be conveniently called, from the unequal lengths of their pistils, the *long-styled*, *mid-styled*, and *short-styled*. The stamens also are of unequal length, and these may be called the *longest*, *mid-length*, and *shortest*. Two sets of stamens of different length are found in each form. The existence of the three forms was first observed by Vaucher, and subsequently more carefully by Wirtgen; but these botanists, not being guided by any theory or even suspicion of their functional differences, did not perceive some of the most curious points of difference in their structure. I will first briefly describe the three forms by the aid of the accompanying diagram, which shows the flowers, six times magnified, in their natural position, with their petals and calyx on the near side removed.

Long-styled form. This form can be at once recognized by the length of the pistil, which is (including the ovarium) fully one-third longer than that of the mid-styled, and more than thrice as long as that of the short-styled form. It is so disproportionately long, that it projects in the bud through the folded petals. It stands out considerably beyond the mid-length stamens; its terminal portion depends a little, but the stigma itself is slightly upturned. The globular stigma is considerably larger than that of the other two forms, with the papillae on its surface generally longer. The six mid-length stamens project about two-thirds the length of the pistil, and correspond in length with the pistil of the mid-styled form. Such correspondence in this and the two following forms is generally very close; the difference, where there is any, being usually in a slight excess of length in the stamens. The six shortest stamens lie concealed within the calyx; their ends are turned up, and they are graduated in length, so as to form a double row. The anthers of these stamens are smaller than those of the mid-length ones. The pollen is of the same yellow colour in both sets. H. Müller measured considering the above figures, and from my son telling me that if he had collected in another spot he felt sure that the mid-styled plants would have been in excess. I several times sowed small parcels of seed, and raised all three forms; but I neglected to record the parent form, excepting in one instance, in which I raised from short-styled seed twelve plants, of which only one turned out long-styled, four mid-styled, and seven short-styled.

Two plants of each form were protected from the access of insects during two successive years, and in the autumn they yielded very few capsules and presented a remarkable contrast with the adjoining uncovered plants, which were densely covered with capsules. In 1863 a protected long-styled plant produced only five poor capsules; two mid-styled plants produced together the same number; and two short-styled plants only a single one. These capsules contained very few seeds; yet the plants were fully productive when artificially fertilized under the net. In a state of nature the flowers are incessantly visited for their nectar by hive- and other bees, various Diptera and Lepidoptera. The nectar is secreted all round the base of the ovarium; but a passage is formed along the upper and inner side of the flower by the lateral deflection (not represented in the diagram) of the basal portions of the filaments; so that insects invariably alight on the projecting stamens and pistil, and insert their proboscides along the upper and inner margin of the corolla. We can now see why the ends of the stamens with their anthers, and the ends of the pistils with their stigmas, are a little upturned, so that they may be brushed by the lower hairy surfaces of the insects' bodies. The short-est stamens which lie enclosed within the calyx of the long- and mid-styled forms can be touched only by the proboscis and narrow chin of a bee; hence they have their ends more upturned, and they are graduated in length, so as to fall into a narrow file, sure to be raked by the thin intruding proboscis. The anthers of the longer stamens stand laterally farther apart and are more nearly on the same level, for they have to brush against the whole breadth of the insect's body. In very many other flowers the pistil, or the stamens, or both, are rectangularly bent to one side of the flower. This bending may be permanent, as with Lythrum and many others, or may be effected, as in *Dictamnus fraxinella* and others, by a temporary movement, which occurs in the case of the stamens, when the anthers dehisce, and in the case of the pistil when the stigma is mature; but these two movements do not always take place simultaneously in the same flower. Now I have found no exception to the rule, that when the stamens and pistil are bent, they bend to that side of the flower which secretes nectar, even though there be a rudimentary nectary of large size on the opposite side, as in some species of Corydalis. When nectar is secreted on all sides, they bend to that side where the struc-ture of the flower allows the easiest access to it, as in Lythrum, various Papil-ionaceae, and others. The rule consequently is, that when the pistils and stamens are curved or bent, the stigma and anthers are thus brought into the pathway leading to the nectary. There are a few cases which seem to be ex-ceptions to this rule, but they are not so in truth; for instance, in the Glo-riosa lily, the stigma of the grotesque and rectangularly bent pistil is brought, not into any pathway from the outside towards the nectar-secreting recesses of the flower, but into the circular route which insects follow in proceeding from one nectary to the other. In *Scrophularia aquatica* the pistil is bent down-wards from the mouth of the corolla, but it thus strikes the pollen-dusted breast of the wasps which habitually visit these ill-scented flowers. In all these cases we see the supreme dominating power of insects on the structure of flowers, especially of those which have irregular corollas. Flowers which are fertilized by the wind must of course be excepted; but I do not know of a single instance of an irregular flower which is thus fertilized.

Another point deserves notice. In each of the three forms two sets of stamens correspond in length with the pistils in the other two forms. When bees suck the flowers, the anthers of the longest stamens, bearing the green pollen, are rubbed against the abdomen and the inner sides of the hind legs, as is likewise the stigma of the long-styled form. The anthers of the mid-length stamens and the stigma of the mid-styled form are rubbed against the under side of the thorax and between the front pair of legs. And, lastly, the anthers of the shortest stamens and the stigma of the short-styled form are rubbed against the proboscis and chin; for the bees in sucking the flowers insert only the front part of their heads into the flower. On catching bees, I observed much green pollen on the inner sides of the hind legs and on the abdomen, and much yellow pollen on the under side of the thorax. There was also pollen on the chin, and, it may be presumed, on the proboscis, but this was difficult to observe. I had, however, independent proof that pollen is carried on the proboscis; for a small branch of a protected short-styled plant (which produced spontaneously only two capsules) was accidentally left during several days pressing against the net, and bees were seen inserting their proboscides through the meshes, and in consequence numerous capsules were formed on this one small branch. From these several facts it follows that insects will generally carry the pollen of each form from the stamens to the pistil of corresponding length; and we shall presently see the importance of this adaptation. It must not, however, be supposed that the bees do not get more or less dusted all over with the several kinds of pollen; for this could be seen to occur with the green pollen from the longest stamens. Moreover a case will presently be given of a long-styled plant producing an abundance of capsules, though growing quite by itself, and the flowers must have been fertilised by their own two kinds of pollen; but these capsules contained a very poor average of seed. Hence insects, and chiefly bees, act both as general carriers of pollen, and as special carriers of the right sort.

Wirtgen remarks on the variability of this plant in the branching of the stem, in the length of the bracteae, size of the petals, and in several other characters. The plants which grew in my garden had their leaves, which differed much in shape, arranged oppositely, alternately, or in whorls of three. In this latter case the stems were hexagonal; those of the other plants being quadrangular. But we are concerned chiefly, with the reproductive organs: the upward bending of the pistil is variable, and especially in the short-styled form, in which it is sometimes straight, sometimes slightly curved, but generally bent at right angles. The stigma of the long-styled pistil frequently has longer papillae or is rougher than that of the mid-styled, and the latter than that of the short-styled; but this character, though fixed and uniform in the two forms of *Primula veris*, etc., is here variable, for I have seen mid-styled stigmas rougher than those of the long-styled. The degree to which the longest and mid-length stamens are graduated in length and have their ends upturned is variable; sometimes all are equally long. The colour of the green pollen in the longest stamens is variable, being sometimes pale greenish-yellow; in one short-styled plant it was almost white.

The grains vary a little in size: I examined one short-styled plant with the grains above the average size; and I have seen a long-styled plant with the grains from the mid-length and shortest anthers of the same size. We here see great variability in many important characters; and if any of these variations were of service to the plant, or were correlated with useful functional differences, the species is in that state in which Natural Selection might readily do much for its modification.

Nothing shows more clearly the extraordinary complexity of the reproductive system of this plant, than the necessity of making eighteen distinct unions in order to ascertain the relative fertilising power of the three forms. Thus the long-styled form has to be fertilised with pollen from its own two kinds of anthers, from the two in the mid-styled, and from the two in the short-styled form. The same process has to be repeated with the mid-styled and short-styled forms. It might have been thought sufficient to have tried on each stigma the green pollen, for instance, from either the mid- or short-styled longest stamens, and not from both; but the result proves that this would have been insufficient, and that it was necessary to try all six kinds of pollen on each stigma. As in fertilising flowers there will always be some failures, it would have been advisable to have repeated each of the eighteen unions a score of times; but the labour would have been too great; as it was, I made 223 unions, i.e. on an average I fertilised above a dozen flowers in the eighteen different methods. Each flower was castrated; the adjoining buds had to be removed, so that the flowers might be safely marked with thread, wool, etc.; and after each fertilisation the stigma was examined with a lens to see that there was sufficient pollen on it. Plants of all three forms were protected during two years by large nets on a framework; two plants were used during one or both years, in order to avoid any individual peculiarity in a particular plant. As soon as the flowers had withered, the nets were removed; and in the autumn the capsules were daily inspected and gathered, the ripe seeds being counted under the microscope.

THE MEANS BY WHICH PLANTS MAY HAVE BEEN RENDERED HETEROSTYLED

This is a very obscure subject, on which I can throw little light, but which is worthy of discussion. It has been shown that heterostyled plants occur in fourteen natural families, dispersed throughout the whole vegetable kingdom, and that even within the family of the Rubiaceae they are dispersed in eight of the tribes. We may therefore conclude that this structure has been acquired by various plants independently of inheritance from a common progenitor, and that it can be acquired without any great difficulty — that is, without any very unusual combination of circumstances.

It is probable that the first step towards a species becoming heterostyled is great variability in the length of the pistil and stamens, or of the pistil alone. Such variations are not very rare: with *Amsinckia spectabilis* and *Nolana prostrata* these organs differ so much in length in different individuals that until experimenting on them, I thought both species heterostyled. The stigma of *Gesneria pendulina* sometimes protrudes far beyond, and is sometimes seated beneath the anthers; so it is with *Oxalis acetosella* and various other plants. I have also noticed an extraordinary amount of difference in the length of the pistil in cultivated varieties of *Primula veris* and *vulgaris*.

As most plants are at least occasionally cross-fertilised by the aid of insects, we may assume that this was the case with our supposed varying plant; but that it would have been beneficial to it to have been more regularly cross-fertilized. We should bear in mind how important an advantage it has been proved to be to many plants, though in different degrees and ways, to be cross-fertilised. It might well happen that our supposed species did not vary in function in the right manner, so as to become either dichogamous or completely self-sterile, or in structure so as to ensure cross-fertilisation. If it had thus varied, it would never have been rendered heterostyled, as this state would then have been superfluous. But the parent species of our several existing heterostyled plants may have been, and probably were (judging from their present constitution) in some degree self-sterile; and this would have made regular cross-fertilisation still more desirable.

Now let us take a highly varying species with most or all of the anthers exserted in some individuals and in others seated low down in the corolla; with the stigma also varying in position in like manner. Insects which visited such flowers would have different parts of their bodies dusted with pollen, and it would be a mere chance whether this were left on the stigma of the next flower which was visited. If all the anthers could have been placed on the same level in all the plants, then abundant pollen would have adhered to the same part of the body of the insects which frequented the flowers, and would afterwards have been deposited without loss on the stigma, if it likewise stood on the same unvarying level in all the flowers. But as the stamens and pistils are supposed to have already varied much in length and to be still varying, it might well happen that they could be reduced much more easily through Natural Selection into two sets of different lengths in different individuals, than all to the same length and level in all the individuals. We know from innumerable instances, in which the two sexes and the young of the same species differ, that there is no difficulty in two or more sets of individuals being formed which inherit different characters. In our particular case the law of compensation or balancement (which is admitted by many botanists) would tend to cause the pistil to be reduced in those individuals in which the stamens were greatly developed, and to be increased in length in those which had their stamens but little developed.

Now if in our varying species the longer stamens were to be nearly equalised in length in a considerable body of individuals, with the pistil more or less reduced; and in another body, the shorter stamens to be similarly equalised, with the pistil more or less increased in length, cross-fertilisation would be secured with little loss of pollen; and this change would be so highly beneficial to the species, that there is no difficulty in believing that it could be effected through Natural Selection. Our plant would then make a close approach in structure to a heterostyled dimorphic species; or to a trimorphic species, if the stamens were reduced to two lengths in the same flower in correspondence with that of the pistils in the other two forms. But we have not as yet even touched on the chief difficulty in understanding how heterostyled species could have originated. A completely self-sterile plant or a dichogamous one can fertilise and be fertilised by any other individual of the same species; whereas the essential character of a heterostyled plant is that an individual of one form cannot fully fertilise or be fertilised by an individual of the same form, but only by one belonging to another form.

H. Müller has suggested that ordinary or homostyled plants may have been rendered heterostyled merely through the effects of habit. Whenever pollen from one set of anthers is habitually applied to a pistil of particular length in a varying species, he believes that at last the possibility of fertilisation in any other manner will be nearly or completely lost. He was led to this view by observing that Diptera frequently carried pollen from the long-styled flowers of Hottonia to the stigma of the same form, and that this illegitimate union was not nearly so sterile as the corresponding union in other heterostyled species. But this conclusion is directly opposed by some other cases, for instance by that of *Linum grandiflorum*; for here the long-styled form is utterly barren with its own-form pollen, although from the position of the anthers this pollen is invariably applied to the stigma. It is obvious that with heterostyled dimorphic plants the two female and the two male organs differ in power; for if the same kind of pollen be placed on the stigmas of the two forms, and again if the two kinds of pollen be placed on the stigmas of the same form, the results are in each case widely different. Nor can we see how this differentiation of the two female and two male organs could have been effected merely through each kind of pollen being habitually placed on one of the two stigmas.

Another view seems at first sight probable, namely, that an incapacity to be fertilised in certain ways has been specially acquired by heterostyled plants. We may suppose that our varying species was somewhat sterile (as is often the case) with pollen from its own stamens, whether these were long or short: and that such sterility was transferred to all the individuals with pistils and stamens of the same length, so that these became incapable of intercrossing freely: but that such sterility was eliminated in the case of the individuals which differed in the length of their pistils and stamens. It is, however, incredible that so peculiar a form of mutual infertility should have been specially acquired unless it were highly beneficial to the species; and although it may be beneficial to an individual plant to be sterile with its own pollen, cross-fertilisation being thus ensured, how can it be any advantage to a plant to be sterile with half its brethren, that is, with all the individuals belonging to the same form? Moreover, if the sterility of the unions between plants of the same form had been a special acquirement, we might have expected that the long-styled form fertilised by the long-styled would have been sterile in the same degree as the short-styled fertilised by the short-styled; but this is hardly ever the case. On the contrary, there is sometimes the widest difference in this respect, as between the two illegitimate unions of *Pulmonaria angustifolia* and of *Hottonia palustris*.

It is a more probable view that the male and female organs in two sets of individuals have been by some means specially adapted for reciprocal action: and that the sterility between the individuals of the same set or form is an incidental and purposeless result. The meaning of the term 'incidental' may be illustrated by the greater or less difficulty in grafting or budding together two plants belonging to distinct species: for as this capacity is quite immaterial to the welfare of either, it cannot have been specially acquired, and must be the incidental result of differences in their vegetative systems. But how the sexual elements of heterostyled plants came to differ from what they were whilst the species was homostyled, and how they become co-adapted in two sets of individuals, are very obscure points. We know that in the two forms of our existing heterostyled plants the pistil always differs, and the stamens generally differ in length: so does the stigma in structure, the anthers in size, and the pollen grains in diameter. It appears, therefore, at first sight probable that organs which differ in such important respects could act on one another only in some manner for which they had been specially adapted. The probability of this view is supported by the curious rule that the greater the difference in length between the pistils and stamens of the trimorphic species of Lythrum and Oxalis, the products of which are united for reproduction, by so much the greater is the infertility of the union. The same rule applies to the two illegitimate unions of some dimorphic species, namely, *Primula vulgaris* and *Pulmonaria angustifolia*; but it entirely fails in other cases, as with *Hottonia palustris* and *Linum grandiflorum*. We shall, however, best perceive the difficulty of understanding the nature and origin of the co-adaptation between the reproductive organs of the two forms of heterostyled plants, by considering the case of *Linum grandiflorum*: the two forms of this plant differ exclusively, as far as we can see, in the length of their pistils; in the long-styled form, the stamens equal the pistil in length, but their pollen has no more effect on it than so much inorganic dust; whilst this pollen fully fertilizes the short pistil of the other form. Now, it is scarcely credible that a mere difference in the length of the pistil can make a wide difference in its capacity for being fertilised. We can believe this the less because with some plants, for instance, *Amsinckia spectabilis*, the pistil varies greatly in length without affecting the fertility of the individuals which are intercrossed. So again I observed that the same plants of *Primula veris* and *vulgaris* differed to an extraordinary degree in the length of their pistils during successive seasons; nevertheless they yielded during these seasons exactly the same average number of seeds when left to fertilise themselves spontaneously under a net.

We must therefore look to the appearance of inner or hidden constitutional differences between the individuals of a varying species, of such a nature that the male element of one set is enabled to act efficiently only on the female element of another set. We need not doubt about the possibility of variations in the constitution of the reproductive system of a plant, for we know that some species vary so as to be completely self-sterile or completely self-sterile, either in an apparently spontaneous manner or from slightly changed conditions of life. Gärtner also has shown that the individual plants of the same species vary in their sexual powers in such a manner that one will unite with a distinct species much more readily than another. But what the nature of the inner constitutional differences may be between the sets or forms of the same varying species, or between distinct species, is quite unknown. It seems therefore probable that the species which have become heterostyled at first varied so that two or three sets of individuals were formed differing in the length of their pistils and stamens and in other co-adapted characters, and that almost simultaneously the irreproductive powers became modified in such a manner that the sexual elements in one set were adapted to act on the sexual elements of another set, and consequently that these elements in the same set or form incidentally became ill-adapted for mutual interaction, as in the case of distinct species. I have elsewhere shown[1] that the sterility of species when first crossed and of their hybrid offspring must also be looked at as merely an incidental result, following from the

---

1. *Origin of Species*, 6th edit., p. 247; *Variation of Animals and Plants under Domestication*, 2nd edit., vol. ii, p. 169; *The Effects of Cross and Self-fertilisation*, p. 463. It may be well here to remark that, judging from the remarkable power with which abruptly changed conditions of life act on the reproductive system of most organisms, it is probable that the close adaptation of the male to the female elements in the two forms of the same heterostyled species, or in all the individuals of the same ordinary species, could be acquired only under long-continued nearly uniform conditions of life.

special co-adaptation of the sexual elements of the same species. We can thus understand the striking parallelism, which has been shown to exist between the effects of illegitimately uniting heterostyled plants and of crossing distinct species. The great difference in the degree of sterility between the various heterostyled species when illegitimately fertilised, and between the two forms of the same species when similarly fertilised harmonises well with the view that the result is an incidental one which follows from changes gradually effected in their reproductive systems, in order that the sexual elements of the distinct forms should act perfectly on one another.

FINAL REMARKS

The existence of plants which have been rendered heterostyled is a highly remarkable phenomenon, as the two or three forms of the same undoubted species differ not only in important points of structure, but in the nature of their reproductive powers. As far as structure is concerned, the two sexes of many animals and of some plants differ to an extreme degree; and in both kingdoms the same species may consist of males, females, and hermaphrodites. Certain hermaphrodite cirripedes are aided in their reproduction by a whole cluster of what I have called complemental males, which differ wonderfully from the ordinary hermaphrodite form. With ants we have males and females, and two or three cases of sterile females or workers. With Termites there are, as Fritz Müller has shown, both winged and wingless males and females, besides the workers. But in none of these cases is there any reason to believe that the several males or several females of the same species differ in their sexual powers, except in the atrophied condition of the reproductive organs in the workers of social insects. Many hermaphrodite animals must unite for reproduction, but the necessity of such union apparently depends solely on their structure. On the other hand, with heterostyled dimorphic species there are two females and two sets of males, and with trimorphic species three females and three sets of males, which differ essentially in their sexual powers. We shall, perhaps, best perceive the complex and extraordinary nature of the marriage arrangements of a trimorphic plant by the following illustration. Let us suppose that the individuals of the same species of ant always lived in triple communities; and that in one of these, a large-sized female (differing also in other characters) lived with six middle-sized and six small-sized males; in the second community a middle-sized female lived with six large- and six small-sized males; and in the third, a small-sized female lived with six large- and six middle-sized males. Each of these three females, though enabled to unite with any male, would be nearly sterile with her own two sets of males, and likewise with two other sets of males of the same size with her own which lived in the other two communities; but she would be fully fertile when paired with a male of her own size. Hence the thirty-six males, distributed by half-dozens in the three communities, would be divided into three sets of a dozen each; and these sets, as well as the three females, would differ from one another in their reproductive powers in exactly the same manner as do the distinct species of the same genus. But it is a still more remarkable fact that young ants raised from any one of the three female ants, illegitimately fertilised by a male of a different size, would resemble in a whole series of relations the hybrid off-spring from a cross between two distinct species of ants. They would be dwarfed in stature, and more or less, or even utterly barren. Naturalists are so much accustomed to behold great diversities of structure associated with the two sexes, that they feel no surprise at almost any amount of difference; but differences in sexual nature have been thought to be the very touchstone of specific distinction. We now see that such sexual differences — the greater or less power of fertilising and being fertilised — may characterize the co-existing individuals of the same species, in the same manner as they characterize and have kept separate those groups of individuals, produced during the lapse of ages, which we rank and denominate as distinct species.

CONCLUSIONS

I will now sum up very briefly the chief conclusions which seem to follow from the observations given in this volume. Cleistogamic flowers afford, as just stated, an abundant supply of seeds with little expenditure; and we can hardly doubt that they have had their structure modified and degraded for this special purpose; perfect flowers being still almost always produced so as to allow of occasional cross-fertilisation. Hermaphrodite plants have often been rendered monoecious, dioecious or polygamous; but as the separation of the sexes would have been injurious, had not pollen been already trans-ported habitually by insects, or by the wind from flower to flower, we may assume that the process of separation did not commence and was not com-pleted for the sake of the advantages to be gained from cross-fertilisation. The sole motive for the separation of the sexes which occurs to me, is that the production of a great number of seeds might become superfluous to a plant under changed conditions of life; and it might then be highly benefi-cial to it that the same flower or the same individual should not have its vital powers taxed, under the struggle for life to which all organisms are sub-jected, by producing both pollen and seeds. With respect to the plants be-longing to the gyno-dioecious subclass, or those which co-exist as hermaph-rodites and females, it has been proved that they yield a much larger supply of seeds than they would have done if they had all remained hermaphro-dites; and we may feel sure from the large number of seeds produced by many plants that such production is often necessary or advantageous. It is therefore probable that the two forms in this subclass have been separated or developed for this special end.

Various hermaphrodite plants have become heterostyled, and now exist under two or three forms; and we may confidently believe that this has been effected in order that cross-fertilisation should be assured. For the full and legitimate fertilisation of these plants pollen from the one form must be applied to the stigma of another. If the sexual elements belonging to the same form are united the union is an illegitimate one and more or less sterile. With dimorphic species two illegitimate unions, and with trimorphic species twelve are possible. There is reason to believe that the sterility of these unions has not been specially acquired, but follows as an incidental result from the sexual elements of the two or three forms having been adapted to act on one another in a particular manner, so that any other kind of union is inefficient, like that between distinct species. Another and still more remarkable incidental result is that the seedlings from an illegitimate union are often dwarfed and more or less or completely barren, like hybrids from the union of two widely distinct species.

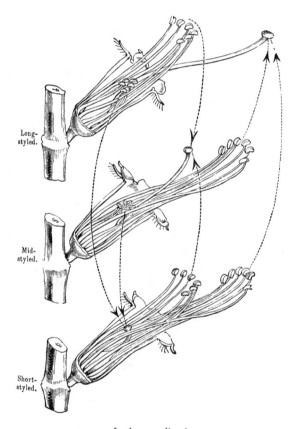

Lythrum salicaria

Chapter 12

AUTOBIOGRAPHY

INTRODUCTION

In order to assess any scholarly autobiography, one should ask the following questions: How clearly is the author able to conceptualize his research program? Does he acknowledge the influences which are brought to bear on all creative thinking or does he portray himself as a kind of isolated genius? Does he see his program as progressing logically or merely as a combination of circumstances and luck?

Judged by these criteria, Darwin's *Autobiography* is a success. Even though Darwin acknowledges that his motive for writing it is the edification and enjoyment of his own family, he is quite explicit regarding the origins and development of his research program. He is particularly explicit regarding his method: He writes that he has collated a "huge pile of notes"; that in keeping notebooks in which to record all relevant facts, he was following Lyell's example; and that, in sum, he "worked on true Baconian principles." Sir Francis Bacon, whose writings greatly influenced all English scientists after the Scientific Revolution, had condemned as speculation hypotheses formed without benefit of observation or experiment; thus, he recommended that scientists proceed purely by induction, from the particular to the general, so that hypotheses would be generated only on the basis of facts gathered with no theoretical end in mind.

Of course, Darwin is romanticizing here. No scientist can be purely inductive in his reasoning: Darwin more likely worked by constantly framing, revising, discarding, and reformulating hypotheses—that is, by means of a mixture of induction and deduction. His insistence here on the impressions gained in South America, on the *Beagle* voyage, which had stuck in his mind regarding the spatial and temporal succession of closely related species, is a case in point.

Expanding on his methodology, he recounts his meticulous interviews with breeders. The information they supplied sensitized him to the role of artificial selection and prepared him for the jolt he received when he picked up a copy of Malthus, which gave him "a theory by which to work." That theory was natural selection. He also describes his later work and makes clear that his botanical studies were closely related to his evolutionary research program, in that they provided case studies of very intricate examples of adaptation, by natural selection at work.

MY SEVERAL PUBLICATIONS

In the early part of 1844, my observations on the Volcanic Islands visited during the voyage of the Beagle were published. In 1845 I took much pains in correcting a new edition of my Journal of Researches, which was originally published in 1839 as part of FitzRoy's work. The success of this my first literary child always tickles my vanity more than that of any of my other books. Even to this day it sells steadily in England and the United States, and has been translated for the second time into German, and into French and other languages. This success of a book of travels, especially of a scientific one, so many years after its first publication, is surprising. Ten thousand copies have now been sold in England of the second edition. In 1846 my geological observations on S. America were published. I record in a little diary which I have always kept, that my three geological books (Coral-Reefs included) consumed four and a half years steady work; 'and now it is ten years since my return to England. How much time have I lost by illness.' I have nothing to say about these three books, except that to my surprise new editions have lately been called for.

In October 1846 I began to work in Cirripedia. When on the coast of Chile I found a most curious form, which burrowed into the shells of Concholepas, and which differed so much from all other cirripedes that I had to form a new sub-order for its sole reception. Lately an allied burrowing genus has been found on the shores of Portugal. To understand the structure of my new cirripede I had to examine and dissect many of the common forms; and this gradually led me on to take up the whole group. I worked steadily on the subject for the next eight years, and ultimately published two thick volumes describing all the known living species and two thin quartos on the extinct species. I do not doubt that Sir E. Lytton Bulwer had me in his mind, when he introduces in one of his novels, a Professor Long who had written two huge volumes on Limpets. Although I was employed during eight years on this work, yet I record in my diary that about two years out of this time was lost by illness. On this account I went in 1848 for some months to Malvern for hydropathic treatment, which did me much good so that on my return home I was able to resume work. So much was I out of health that when my dear Father died on November 13th 1847 I was unable to attend his funeral or to act as one of his executors.

My work on the Cirripedia possesses, I think, considerable value, as besides describing several new and remarkable forms, I made out the homologies of the various parts — I discovered the cementing apparatus, though I blundered dreadfully about the cement-glands — and lastly I proved the existence in certain genera of minute males complemental to and parasitic on the hermaphrodites. This latter discovery has at last been fully confirmed; though at one time a German writer was pleased to attribute the whole account to my fertile imagination. The cirripedes form a highly varying and difficult group of species to class; and my work was of considerable use to me, when I had to discuss in the Origin of Species the principles of a natural classification. Nevertheless I doubt whether the work was worth the consumption of so much time.

From September 1854 onwards I devoted all my time to arranging my huge pile of notes, to observing and experimenting, in relation to the transmutation of species. During the voyage of the Beagle I had been deeply impressed by discovering in the Pampean formation great fossil animals covered with armour like that on the existing armadillos; secondly by the manner in which closely allied animals replace one another in proceeding southwards over the Continent; and thirdly by the South American character of most of the productions of the Galapagos archipelado, and more especially by the manner in which they differ slightly on each island of the group; none of these islands appearing to be very ancient in a geological sense. It was evident that such facts as these as well as many others could be explained on the supposition that species gradually become modified; and the subject haunted me. But it was equally evident that neither the action of the surrounding conditions, nor the will of the organisms (especially in the case of plants), could account for the innumerable cases in which organisms of every kind are beautifully adapted to their habits of life, — for instance a woodpecker or tree-frog to climb trees, or a seed for dispersal by hooks or plumes. I had always been much struck by such adaptations, and until these could be explained it seemed to me almost useless to endeavour to prove by indirect evidence that species have been modified.

After my return to England it appeared to me that by following the example of Lyell in Geology, and by collecting all facts which bore in any way on the variation of animals and plants under domestication and nature, some light might perhaps be thrown on the whole subject. My first note-book was opened in July 1837. I worked on true Baconian principles, and without any theory collected facts on a whole-sale scale, more especially with respect to domesticated productions, by printed enquiries, by conversation with skilful breeders and gardeners, and by extensive reading. When I see the list of books of all kinds which I read and abstracted, including whole series of Journals and Transactions, I am surprised at my industry. I soon perceived that Selection was the key-stone of man's success in making useful races of animals and plants. But how selection could be applied to organisms living in a state of nature remained for some time a mystery to me. In October 1838, that is fifteen months after I had begun my systematic enquiry, I happened to read for amusement 'Malthus on Population', and being well prepared to appreciate the struggle for existence which everywhere goes on from long-continued observation of the habits of animals and plants, it at once struck me that under these circumstances favourable variations would tend to be preserved and unfavourable ones to be destroyed. The results of this would be the formation of new species.

Here then I had at last got a theory by which to work; but I was so anxious to avoid prejudice, that I determined not for some time to write even the briefest sketch of it. In June 1842 I first allowed myself the satisfaction of writing a very brief abstract of my theory in pencil in 35 pages; and this was enlarged during the summer of 1844 into one of 230 pages, which I had fairly copied out and still possess. But at that time I overlooked one problem of great importance; and it is astonishing to me, except on the principle of Columbus and his egg, how I could have overlooked it and its solution. This problem is the tendency in organic beings descended from the same stock to diverge in character as they become modified. That they have diverged greatly is obvious from the manner in which species of all kinds can be classed under genera, genera under families, families under sub-orders, and so forth; and I can remember the very spot in the road whilst in my carriage, when to my joy the solution occurred to me; and this was long after I had come to Down. The solution, as I believe, is that the modified offspring of all dominant and increasing forms tend to become adapted to many and highly diversified places in the economy of nature.

Early in 1856 Lyell advised me to write out my views pretty fully, and I began at once to do so on a scale three or four times as extensive as that which was afterwards followed in my Origin of Species; yet it was only an abstract of the materials which I had collected, and I got through about half the work on this scale. But my plans were overthrown, for early in the summer of 1858 Mr. Wallace, who was then in the Malay Archipelago, sent me an essay 'On the tendency of varieties to depart indefinitely from the original type'; and this essay contained exactly the same theory as mine. Mr. Wallace expressed the wish that if I thought well of his essay, I should send it to Lyell for perusal. The circumstances under which I consented at the request of Lyell and Hooker to allow of an extract from my M.S, together with a letter to Asa Gray dated September 5 1857, to be published at the same time with Wallace's Essay, are given in the Journal of the Proceedings of the Linn. Soc. 1858 p. 45. I was at first very unwilling to consent, as I thought Mr. Wallace might consider my doing so unjustifiable, for I did not then know how generous and noble was his disposition. The extract from my M.S and the letter to Asa Gray had neither been intended for publication and were badly written. Mr. Wallace's essay, on the other hand was admirably expressed and quite clear. Nevertheless our joint productions excited very little attention, and the only published notice of them which I can remember was by Prof. Haughton of Dublin, whose verdict was that all that was new in them was false, and what was true was old. This shows how necessary it is that any new view should be explained at considerable length in order to arouse public attention.

In September 1858 I set to work by the strong advice of Lyell and Hooker to prepare a volume on the transmutation of species, but was often interrupted by ill-health, and short visits to Dr. Lane's delightful hydropathic establishment at Moor Park. I abstracted the M.S begun on a much larger scale in 1856, and completed the volume on the same reduced scale. It cost me 13 months and ten days hard labour. It was published under the title of the 'Origin of Species' in November 1859. Though considerably added to and corrected in the later editions it has remained substantially the same book.

It is no doubt the chief work of my life. It was from the first highly successful. The first small edition of 1250 copies was sold on the day of publication, and a second edition of 3000 copies soon afterwards. Sixteen thousand copies have now (1876) been sold in England, and considering how stiff a book it is this is a large sale. It has been translated into almost every European tongue, even into such languages as Spanish, Bohemian, Polish and Russian. It has also, according to Miss Bird, been translated into Japanese, and is there much studied. Even an essay in Hebrew has appeared on it, showing that the theory is contained in the Old Testament! The Reviews were very numerous; for a time I collected all that appeared on the Origin and on my related books, and these amount (excluding newspaper reviews) to 265; but after a time I gave up the attempt in despair. Many separate essays and books on the subject have appeared; and in Germany a catalogue or Bibliography on 'Darwinismus' has appeared every year or two.

The success of the Origin may, I think, be attributed in large part to my having long before written two condensed sketches, and to my having finally abstracted a much larger manuscript, which was itself an abstract. By this means I was enabled to select the more striking facts and conclusions. I had, also, during many years, followed a golden rule, namely that whenever a published fact, a new observation or thought came across me, which was opposed to my general results, to make a memorandum of it without fail and at once; for I had found by experience that such facts and thoughts were far more apt to escape from the memory, than favourable ones. Owing to this habit, very few objections were raised against my views which I had not at least noticed and attempted to answer. It has sometimes been said that the success of the Origin proved 'that the subject was in the air' or 'that men's minds were prepared for it.' I do not think that this is strictly true, for I occasionally sounded not a few naturalists, and never happened to come across a single one who seemed to doubt about the permanence of species. Even Lyell and Hooker, though they would listen with interest to me never seemed to agree. I tried once or twice to explain to able men what I meant by natural selection, but signally failed. What I believe was strictly true is that innumerable well-observed facts were stored in the minds of naturalists ready to take their proper places, as soon as any theory which would receive them was sufficiently explained. Another element in the success of the book was its moderate size, and this I owe to the appearance of Wallace's essay; had I published on the scale in which I began to write in 1856, the book would have been four or five times as large as the Origin, and very few would have had the patience to read it.

I gained much by my delay in publishing from about 1839, when the theory was clearly conceived, to 1859; and I lost nothing by it, for I cared very little whether men attributed most originality to me or Wallace, and his essay no doubt aided in the reception of the theory. I was forestalled in only one important point, which my vanity has always made me regret, namely the explanation by means of the Glacial period of the presence of the same species of plants and of some few animals on distant mountain-summits and in the Arctic regions. This view pleased me so much that I wrote it out in extenso, and it was read by Hooker, some years before E. Forbes published his celebrated memoir on the subject. In the very few points in which we differed, I still think that I was in the right. I have never of course alluded in print to my having independently worked out this view.

Hardly any point gave me so much satisfaction when I was at work on the Origin, as the explanation of the wide difference in many classes between the embryo and the adult animal, and of the close resemblance of the embryos within the same class. No notice of this point was taken, as far as I remember, in the early reviews of the Origin, and I recollect expressing my surprise on this head in a letter to Asa Gray. Within late years several reviewers have given the whole credit of the idea to Fritz Müller and Häckel, who undoubtedly have worked it out much more fully, and in some respects more correctly than I did. I had materials for a whole chapter on the subject, and I ought to have made the discussion longer; for it is clear that I failed to impress my readers; and he who succeeds in doing so deserves in my opinion all the credit.

This leads me to remark that I have almost always been treated honestly by my reviewers, passing over those without scientific knowledge as not worthy of notice. My views have often been grossly misrepresented, bitterly opposed and ridiculed, but this has been generally done as I believe in good faith. I must however except Mr. Mivart, who as an American expressed it in a letter has acted towards me 'like a pettifogger', or as Huxley has said 'like an Old Bailey lawyer.' On the whole I do not doubt that my work has been over and over again greatly overpraised. I rejoice that I have avoided controversies, and this I owe to Lyell, who many years ago in reference to my geological works strongly advised me never to get entangled in a controversy, as it rarely did any good and caused a miserable loss of time and temper.

Whenever I have found out that I have blundered or that my work has been imperfect, and when I have been contemptuously criticised and even when I have been overpraised, so that I have felt mortified, it has been my greatest comfort to say hundreds of times to myself that 'I have worked as hard and as well as I could, and no man can do more than this.' I remember when in Good Success Bay in Tierra del Fuego, thinking, (and I believe that I wrote home to this effect) that I could not employ my life better than in adding a little to natural science. This I have done to the best of my abilities, and critics may say what they like, but they cannot destroy this conviction.

During the two last months of the year 1859 I was fully occupied in preparing a second edition of the Origin, and by an enormous correspondence. On January 7th 1860 I began arranging my notes for my work on the Variation of Animals and Plants under domestication; but it was not published until the beginning of 1868; the delay having been caused partly by frequent illnesses, one of which lasted seven months, and partly by having been tempted to publish on other subjects which at the time interested me more.

On May 15th 1862, my little book on the Fertilisation of Orchids, which cost me ten months work, was published: most of the facts had been slowly accumulated during several previous years. During the summer of 1839 and I believe during the previous summer, I was led to attend to the cross-fertilisation of flowers by the aid of insects, from having come to the conclusion in my speculations on the origin of species, that crossing played an important part in keeping specific forms constant. I attended to the subject more or less during every subsequent summer; and my interest in it was greatly enhanced by having procured and read in November 1841, through the advice of Robert Brown, a copy of C. K. Sprengel's wonderful book 'Das entdeckte Geheimnis der Natur'. For some years before 1862 I had specially attended to the fertilisation of our British orchids; and it seemed to me the best plan to prepare as complete a treatise on this group of plants as well as I could, rather than to utilise the great mass of matter which I had slowly collected with respect to other plants. My resolve proved a wise one; for since the appearance of my book, a surprising number of papers and separate works on the fertilisation of all kinds of flowers have appeared; and these are far better done than I could possibly have effected. The merits of poor old Sprengel, so long overlooked, are now fully recognised many years after his death.

During this same year I published in the Journal of the Linnean Society a paper 'On the two forms or dimorphic condition of primula', and during the next five years, five other papers on dimorphic and trimorphic plants. I do not think anything in my scientific life has given me so much satisfaction as making out the meaning of the structure of these plants. I had noticed in 1838 or 1839 the dimorphism of *Linum flavum*, and had at first thought that it was merely a case of unmeaning variability. But on examining the common species of Primula I found that the two forms were much too regular and constant to be thus viewed. I therefore became almost convinced that the common cowslip and primrose were on the high road to become diœcious; — that the short pistil in the one form, and the short stamens in the other form were tending towards abortion. The plants were therefore subjected under this point of view to trial; but as soon as the flowers with short pistils fertilised with pollen from the short stamens, were found to yield more seeds than any other of the four possible unions, the abortion-theory was knocked on the head. After some additional experiment, it became evident that the two forms, though both were perfect hermaphrodites, bore almost the same relation to one another as do the two sexes of an ordinary animal. With Lythrum we have the still more wonderful case of three forms standing in a similar relation to one another. I afterwards found that the offspring from the union of two plants belonging to the same forms presented a close and curious analogy with hybrids from the union of two distinct species.

In the autumn of 1864 I finished a long paper on Climbing Plants and sent it to the Linnean Society. The writing of this paper cost me four months: but I was so unwell when I received the proof-sheets that I was forced to leave them very badly and often obscurely expressed. The paper was little noticed, but when in 1875 it was corrected and published as a separate book it sold well. I was led to take up this subject by reading a short paper by Asa Gray, published in 1858, on the movements of the tendrils of a Cucurbitacean plant. He sent me seeds, and on raising some plants I was so much fascinated and perplexed by the revolving movements of the tendrils and stems, which movements are really very simple though appearing at first very complex, that I procured various other kinds of Climbing Plants, and studied the whole subject. I was all the more attracted to it, from not being at all satisfied with the explanation which Henslow gave us in his Lectures, about Twining plants, namely that they had a natural tendency to grow up in a spire. This explanaton proved quite erroneous. Some of the adaptations displayed by Climbing Plants are as beautiful as those by Orchids for ensuring cross-fertilisation.

My Variation of Animals and Plants under domestication was begun, as already stated, in the beginning of 1860, but was not published until the beginning of 1868. It is a big book and cost me four years and two months hard labour. It gives all my observations and an immense number of facts collected from various sources about our domestic productions. In the second volume the causes and laws of variation, inheritance, &c., are discussed, as far as our present state of knowledge permits. Towards the end of the work I give my well-abused hypothesis of Pangenesis. An unverified hypothesis is of little or no value; but if anyone should hereafter be led to make observations by which some such hypothesis could be established, I shall have done good service, as an astonishing number of isolated facts can thus be connected together and rendered intelligible. In 1875 a second and largely corrected edition, which cost me a good deal of labour, was brought out.

My Descent of Man was published in Feb. 1871. As soon as I had become in the year 1837 or 1838 convinced that species were mutable productions, I could not avoid the belief that man must come under the same law. Accordingly I collected notes on the subject for my own satisfaction, and not for a long time with any intention of publishing. Although in the Origin of Species, the derivation of any particular species is never discussed, yet I thought it best, in order that no honourable man should accuse me of concealing my views, to add that by the work in question 'light would be thrown on the origin of man and his history.' It would have been useless and injurious to the success of the book to have paraded without giving any evidence my conviction with respect to his origin. But when I found that many naturalists fully accepted the doctrine of the evolution of species, it seemed to me advisable to work up such notes as I possessed and to publish a special treatise on the origin of man. I was the more glad to do so, as it gave me an opportunity of fully discussing sexual selection, — a subject which had always greatly interested me. This subject, and that of the variation of our domestic productions together with the causes and laws of variation, inheritance, &c., and the intercrossing of Plants are the sole subjects which I have been able to write about in full, so as to use all the materials which I had collected. The Descent of Man took me three years to write, but then as usual some of this time was lost by ill-health, and some was consumed by preparing new editions and other minor works. A second and largely corrected edition of the Descent appeared in 1874.

My book on the Expression of the Emotions in Men and Animals was published in the autumn of 1872. I had intended to give only a chapter on the subject in the Descent of Man, but as soon as I began to put my notes together, I saw that it would require a separate Treatise. My first child was born on December 27th 1839, and I at once commenced to make notes on the first dawn of the various expressions which he exhibited, for I felt convinced, even at this early period, that the most complex and fine shades of expression must all have had a gradual and natural origin. During the summer of the following year, 1840, I read Sir C. Bell's admirable work on Expression, and this greatly increased the interest which I felt in the subject, though I could not at all agree with his belief that various muscles had been specially created for the sake of expression. From this time forward I occasionally attended to the subject, both with respect to man and our domestic animals. My book sold largely; 5267 copies having been disposed of on the day of publication.

In the summer of 1860 I was idling and resting near Hartfield, where two species of Drosera abound; and I noticed that numerous insects had been entrapped by the leaves. I carried home some plants and on giving them insects saw the movements of the tentacles, and this made me think it probable that the insects were caught for some special purpose. Fortunately a crucial test occurred to me, that of placing a large number of leaves in various nitrogenous and non-nitrogenous fluids of equal density; and as soon as I found that the former alone excited energetic movements, it was obvious that here was a fine new field for investigation. During subsequent years, whenever I had leisure I pursued my experiments, and my book on 'Insectivorous Plants' was published July 1875, — that is 16 years after my first observations. The delay in this case, as with all my other books has been a great advantage to me; for a man after a long interval can criticise his own work, almost as well as if it were that of another person. The fact that a plant should secrete when properly excited a fluid containing an acid and ferment, closely analogous to the digestive fluid of an animal, was certainly a remarkable discovery.

During this autumn of 1876 I shall publish on the 'Effects of Cross and Self Fertilisation in the Vegetable Kingdom.' This book will form a complement to that on the Fertilisation of Orchids, in which I showed how perfect were the means for cross-fertilisation, and here I shall show how important are the results. I was led to make during eleven years the numerous experiments recorded in this volume by a mere accidental observation; and indeed it required the accident to be repeated before my attention was thoroughly aroused to the remarkable fact that seedlings of self-fertilised parentage are inferior even in the first generation in height and vigour to seedlings of cross-fertilised parentage. I hope also to republish a revised edition of my book on

Orchids, and hereafter my papers on dimorphic and trimorphic plants, together with some additional observations on allied points which I never have had time to arrange. My strength will then probably be exhausted, and I shall be ready to exclaim 'nunc dimittis.'

'The Effects of Cross and Self Fertilisation' was published in the autumn of 1876; and the results there arrived at explain, as I believe, the endless and wonderful contrivances for the transportal of pollen from one plant to another of the same species. I now believe, however, chiefly from the observations of Hermann Müller, that I ought to have insisted more strongly than I did on the many adaptations for self-fertilisation; though I was well aware of many such adaptations. A much enlarged Edit. of my Fertilisation of Orchids was published in 1877.

In this same year 'The Different Forms of Flowers, &c.' appeared, and in 1880 a 2$^{\underline{d}}$ edition. This book consists chiefly of the several papers on heterostyled flowers, originally published by the Linnean Soc.^{y}, corrected with much new matter added, together with observations on some other cases in which the same plant bears two kinds of flowers. As before remarked no little discovery of mine ever gave me so much pleasure as the making out the meaning of heterostyled flowers. The results of crossing such flowers in an illegitimate manner, I believe to be very important as bearing on the sterility of hybrids; although these results have been noticed by only a few persons.

In 1879, I had a translation of Dr. Ernst Krause's life of Erasmus Darwin published, and I added a sketch of his character and habits from materials in my possession. Many persons have been much interested by this little life, and I am surprised that only 800 or 900 copies were sold. Owing to my having accidentally omitted to mention that Dr. Krause had enlarged and corrected his article in German before it was translated, Mr. Samuel Butler abused me with almost insane virulence. How I offended him so bitterly, I have never been able to understand. The subject gave rise to some controversy in the Athenaeum newspaper and Nature. I laid all the documents before some good judges, viz. Huxley, Leslie Stephen, Litchfield, &c., and they were all unanimous that the attack was so baseless that it did not deserve any public answer; for I had already expressed privately my regret to Mr. Butler for my accidental omission. Huxley consoled me by quoting some German lines from Goethe, who had been attacked by someone, to the effect 'that every Whale has its Louse.'

In 1880 I published, with Frank's assistance, our 'Power of Movement in Plants'. This was a tough piece of work. The book bears somewhat the same relation to my little book on Climbing Plants, which 'Cross-Fertilisation' did to the 'Fertilisation of Orchids'; for in accordance with the principles of evolution it was impossible to account for climbing plants having been developed in so many widely different groups, unless all kinds of plants possess some slight power of movement of an analogous kind. This I proved to be the case, and I was further led to a rather wide generalisation, viz., that the great and important classes of movement, excited by light, the attraction of gravity, &c., are all modified forms of the fundamental movement of circumnutation. It has always pleased me to exalt plants in the scale of organised beings; and I therefore felt an especial pleasure in showing how many and what admirably well adapted movements the tip of a root possesses.

I have now (May 1st 1881) sent to the printers the M.S of a little book on 'The Formation of Vegetable Mould, through the action of worms'. This is a subject of but small importance; and I know not whether it will interest any readers, but it has interested me. It is the completion of a short paper read before the Geological Society more than 40 years ago, and has revived old geological thoughts.

I have now mentioned all the books which I have published, and these have been the mile-stones in my life, so that little remains to be said. I am not conscious of any change in my mind during the last 30 years, excepting in one point presently to be mentioned; nor indeed could any change have been expected unless one of general deterioration. But my Father lived to his 83^d years with his mind as lively as ever it was, and all his faculties undimmed; and I hope that I may die before mine fails to a sensible extent. I think that I have become a little more skilful in guessing right explanations and in devising experimental tests; but this may probably be the result of more practice and of a larger store of knowledge. I have as much difficulty as ever in expressing myself clearly and concisely; and this difficulty has caused me a very great loss of time; but it has had the compensating advantage of forcing me to think long and intently about every sentence, and thus I have been often led to see errors in reasoning and in my own observations or those of others. There seems to be a sort of fatality in my mind leading me to put at first my statement and proposition in a wrong or awkward form. Formerly I used to think about my sentences before writing them down; but for several years I have found that it saves time to scribble in a vile hand whole pages as quickly as I possibly can, contracting half the words; and then correct deliberately. Sentences thus scribbled down are often better ones than I could have written deliberately.

Having said this much about my manner of writing, I will add that with my larger books I spend a good deal of time over the general arrangement of the matter. I first make the rudest outline in two or three pages, and then a larger one in several pages, a few words or one word standing for a whole discussion or series of facts. Each of these headings is again enlarged and often transformed before I begin to write in extenso. As in several of my books facts observed by others have been very extensively used, and as I have always had several quite distinct subjects in hand at the same time, I may mention that I keep from 30 to 40 large portfolios, in cabinets with labelled shelves, into which I can at once put a detached reference or memorandum. I have bought many books and at their ends I make an index of all the facts that concern my work; or if the book is not my own, write out a separate abstract, and of such abstracts I have a large drawer full. Before beginning on any subject I look to all the short indexes and make a general and classified index, and by taking the one or more proper portfolios I have all the information collected during my life ready for use.

I have said that in one respect my mind has changed during the last 20 or 30 years. Up to the age of thirty, or beyond it, poetry of many kinds, such as the works of Milton, Gray, Byron, Wordsworth, Coleridge and Shelley, gave me great pleasure, and even as a schoolboy I took intense delight in Shakespeare especially in the historical plays. I have also said that formerly Pictures gave me considerable, and music very great delight. But now for many years I cannot endure to read a line of poetry: I have tried lately to read Shakespeare and found it so intolerably dull that it nauseated me. I have also almost lost any taste for pictures or music. — Music generally sets me thinking too energetically on what I have been at work on, instead of giving me pleasure. I retain some taste for fine scenery, but it does not cause me the exquisite delight which it formerly did. On the other hand, novels which are works of the imagination, though not of a very high order, have been for years a wonderful relief and pleasure to me, and I often bless all novelists. A surprising number have been read aloud to me, and I like all if moderately good, and if they do not end unhappily — against which a law ought to be passed. A novel, according to my taste, does not come into the first class, unless it contains some person whom one can thoroughly love, and if it be a pretty woman all the better.

This curious and lamentable loss of the higher aesthetic tastes is all the odder, as books on history, biographies and travels (independently of any scientific facts which they may contain), and essays on all sorts of subjects interest me as much as ever they did. My mind seems to have become a kind of machine for grinding general laws out of large collections of facts, but why this should have caused the atrophy of that part of the brain alone, on which the higher tastes depend, I cannot conceive. A man with a mind more highly organised or better constituted than mine, would not I suppose have thus suffered; and if I had to live my life again I would have made a rule to read some poetry and listen to some music at least once every week; for perhaps the parts of my brain now atrophied could thus have been kept active through use. The loss of these tastes is a loss of happiness, and may possibly be injurious to the intellect, and more probably to the moral character by enfeebling the emotional part of our nature.

My books have sold largely in England, have been translated into many languages and passed through several editions in foreign countries. I have heard it said that the success of a work abroad is the best test of its enduring value. I doubt whether this is at all trustworthy, but judged by this standard my name ought to last for a few years. Therefore it may be worth while for me to try to analyse the mental qualities and the conditions on which my success has depended; though I am aware that no man can do this correctly. I have no great quickness of apprehension or wit which is so remarkable in some clever men, for instance Huxley. I am therefore a poor critic: a paper or book, when first read, generally excites my admiration, and it is only after considerable reflection that I perceive the weak points. My power to follow a long and purely abstract train of thought is very limited; I should, therefore, never have succeeded with metaphysics or mathematics. My memory is extensive, yet hazy: it suffices to make me cautious by vaguely telling me that I have observed or read something opposed to the conclusion which I am drawing, or on the other hand in favour of it; and after a time I can generally recollect where to search for my authority. So poor in one sense is my memory, that I have never been able to remember for more than a few days a single date or a line of poetry. Some of my critics have said, 'Oh, he is a good observer but has no power of reasoning.' I do not think that this can be true, for the Origin of Species is one long argument from the beginning to the end, and it has convinced not a few able men. No one could have written it without having some power of reasoning. I have a fair share of invention and of common sense or judgement, such as every fairly successful lawyer or doctor must have, but not I believe in any higher degree.

On the favourable side of the balance I think that I am superior to the common run of men in noticing things which easily escape attention, and in observing them carefully. My industry has been nearly as great as it could have been in the observation and collection of facts. What is far more important my love of natural science has been steady and ardent. This pure love has however been much aided by the ambition to be esteemed by my fellow naturalists. From my early youth I have had the strongest desire to understand or explain whatever I observed, — that is to group all facts under some general laws. These causes combined have given me the practice to reflect or ponder for any number of years over any unexplained problem. As far as I can judge, I am not apt to follow blindly the lead of other men. I have steadily endeavoured to keep my mind free, so as to give up any hypothesis, however much beloved (and I cannot resist forming one on every subject) as soon as facts are shown to be opposed to it. Indeed I have had no choice but to act in this manner, for with the exception of the Coral Reefs I cannot remember a single first-formed hypothesis which had not after a time to be given up or greatly modified. This has naturally led me to distrust greatly deductive reasoning in the mixed sciences. On the other hand I am not very sceptical, — a frame of mind which I believe to be injurious to the progress of science; a good deal of scepticism in a scientific man is advisable to avoid much loss of time; for I have met with not a few men, who I feel sure have often thus been deterred from experiment or observations, which would have proved directly or indirectly serviceable.

In illustration, I will give the oddest case which I have known. A gentleman (who as I afterwards heard was a good local botanist) wrote to me from the Eastern countries that the seeds or beans in the common field-bean had this year everywhere grown on the wrong side of the pod. I wrote back, asking for further information, as I did not understand what was meant; but I did not receive any answer for a long time. I then saw in two newspapers, one published in Kent and the other in Yorkshire, paragraphs stating that it was a most remarkable fact that the beans this year had all grown on the wrong side. So I thought that there must be some foundation for so general a statement. Accordingly I went to my gardener, an old Kentish man, and asked him whether he had heard anything about it; and he answered 'Oh no Sir, it must be a mistake, for the beans grow on the wrong side only on Leap-year, and this is not Leap-year.' I then asked him how they grew on common years and how on leap-years, but soon found out that he knew absolutely nothing of how they grew at any time; but he stuck to his belief. After a time I heard from my first informant, who with many apologies said that he should not have written to me had he not heard the statement from several intelligent farmers; but that he had since spoken again to every one of them, and not one knew in the least what he had himself meant. So that here a belief — if indeed a statement with no definite idea attached to it can be called a belief — had spread over almost the whole of England without any vestige of evidence. I have known in the course of my life only three intentionally falsified statements, and one of these may have been a hoax (and there have been several scientific hoaxes) which, however, took in an American Agricultural Journal. It related to the formation in Holland of a new breed of oxen by the crossing of distinct species of Bos (some of which I happen to know are sterile together), and the author had the impudence to state that he had corresponded with me and that I had been deeply impressed with the importance of his results. The article was sent to me by the editor of an English Agricult. Journal, asking for my opinion before republishing it.

A second case was an account of several varieties raised by the author from several species of Primula, which had spontaneously yielded a full complement of seed, although the parent-plants had been carefully protected from the access of insects. This account was published before I had discovered the meaning of heterostylism, and the whole statement must have been fraudulent, or there was neglect in excluding insects so gross as to be scarcely credible.

The third case was more curious: Mr. Huth published in his book on consanguineous marriage some long extracts from a Belgian author, who stated that he had interbred rabbits in the closest manner for very many generations without the least injurious effects. The account was published in a most respectable Journal, that of the R. Medical Soc. of Belgium; but I could not avoid feeling doubts, — I hardly know why, except that there were no accidents of any kind, and my experience in breeding animals made me think this improbable.

So with much hesitation I wrote to Prof. Van Beneden asking him whether the author was a trustworthy man. I soon heard in answer that the Society had been greatly shocked by discovering that the whole account was a fraud. The writer had been publicly challenged in the Journal to say where he had resided and kept his large stock of rabbits while carrying on his experiments, which must have consumed several years, and no answer could be extracted from him. I informed poor Mr. Huth, that the account which formed the corner-stone of his argument was fraudulent; and he in the most honourable manner immediately had a slip printed to this effect to be inserted in all future copies of his book which might be sold.

My habits are methodical, and this has been of not a little use for my particular line of work. Lastly, I have had ample leisure from not having to earn my own bread. Even ill-health, though it has annihilated several years of my life, has saved me from the distractions of society and amusement.

Therefore, my success as a man of science, whatever this may have amounted to, has been determined, as far as I can judge, by complex and diversified mental qualities and conditions. Of these the most important have been — the love of science — unbounded patience in long reflecting over any subject — industry in observing and collecting facts — and a fair share of invention as well as of common sense. With such moderate abilities as I possess, it is truly surprising that thus I should have influenced to a considerable extent the beliefs of scientific men on some important points.

Appendix 1

SELECTIONS FROM

MALTHUS'S *AN ESSAY*

ON THE PRINCIPLE

OF POPULATION

INTRODUCTION

Thomas Robert Malthus (1766–1834), a clergyman by training, was the first professor of political economy in England, teaching that subject at the East India Company's college in Hertford. He was not the first writer to suggest that populations tend to increase proportionally faster than their food supply, but he was the first demographic thinker who drew unremittingly pessimistic conclusions from that observation: Poverty was inevitable, and the poor were irresponsible because they failed to control their passions and spare their children extreme suffering. In *An Essay on the Principle of Population* (1798), Malthus described the positive and negative "checks" to population growth that Darwin would later put to use in support of his own theory. A "positive" check is anything that causes death, such as famine or disease. A "preventive" check is one that affects the birth rate, such as—in human populations—age of marriage or contraception. Malthus believed that such checks acted constantly, but Darwin's reading of Malthus—he read the sixth edition of 1826, which is the text from which we have drawn our selections—was conditioned by later thought and by his own observations, which led him to believe that the application of those checks was intermittent.

Like virtually all English social thinkers of the period, Malthus thought of himself as a Newtonian. He believed that his laws of population were "laws of nature" and were—like Newton's law of universal gravitation—self-regulating. In Malthus's formula, the proportion of births to deaths rises until checks cut in to bring the population back into balance with the food supply. He also presents a self-regulating mechanism (much like a similar formulation by Adam Smith) that explains the relationship of wages, subsistence, and the demand for labor.

Malthus's ideas were widely discussed in the first half of the nineteenth century; everyone was familiar with the fundamentals of his theory. As Malthus stressed, the geometrical increase in the population of human beings—not a very fecund species—allowed for a doubling of population every twenty-five years. In Notebook D Darwin wrote, "yet until the one sentence of Malthus no one clearly perceived the great check amongst men. Even a *few* years plenty, makes population in Men increase. . . ." There has been speculation as to which "one sentence" Darwin had singled out. In the text of Malthus presented here, we have italicized the two most likely candidates.

Some authors have argued that Darwin had been prepared for his reading of Malthus because he had already read an article by Christian Ehrenberg on the amazing reproductive powers of microscopic animals (for Darwin's rumination on Ehrenberg—Notebook D:167—see p. 74). "One invisible animacule," Darwin writes, "in four days could for 2. cubic stone. like that of Billin."—Darwin's shorthand for Ehrenberg's statement that "an imperceptible corpuscle can become in four days 170 billions, or as many single animacules as contained in 2 cubic feet of the stone for the polishing slate of Bilin."[1] But in fact, Ehrenberg's article was published in October 1838, and there is no way that Darwin could have read it before 28 September. Therefore, Darwin's account can safely be assessed on its own merits.

CHAPTER I

STATEMENT OF THE SUBJECT
RATIOS OF THE INCREASE
OF POPULATION AND FOOD

In an inquiry concerning the improvement of society, the mode of conducting the subject which naturally presents itself, is:

1. To investigate the causes that have hitherto impeded the progress of mankind towards happiness; and,

---

1. Christian Ehrenberg, "Communication Respecting Fossil and Recent Infusoria Made to the British Association at Newcastle," *Annals of Natural History* 2 (1839): 121–124. Howard E. Gruber, *Darwin on Man*, 2nd ed. (Chicago, University of Chicago Press, 1981), 162, assumes that Darwin read Ehrenberg's article some days before reading Malthus. Ernst Mayr, *One Long Argument: Charles Darwin and the Genesis of Modern Evolutionary Thought* (Cambridge, Mass., Harvard University Press, 1991), 76, accepts Gruber's conclusion. Ehrenberg's article was published in the 8 October number of the *Annals*.

2. To examine the probability of the total or partial removal of these causes in future.

To enter fully into this question, and to enumerate all the causes that have hitherto influenced human improvement, would be much beyond the power of an individual. The principal object of the present essay is to examine the effects of one great cause intimately united with the very nature of man; which, though it has been constantly and powerfully operating since the commencement of society, has been little noticed by the writers who have treated this subject. The facts which establish the existence of this cause have, indeed, been repeatedly stated and acknowledged; but its natural and necessary effects have been almost totally overlooked; though probably among these effects may be reckoned a very considerable portion of that vice and misery, and of that unequal distribution of the bounties of nature, which it has been the unceasing object of the enlightened philanthropist in all ages to correct.

The cause to which I allude, is the constant tendency in all animated life to increase beyond the nourishment prepared for it.

It is observed by Dr Franklin, that there is no bound to the prolific nature of plants or animals, but what is made by their crowding and interfering with each other's means of subsistence. Were the face of the earth, he says, vacant of other plants, it might be gradually sowed and overspread with one kind only, as for instance with fennel; and were it empty of other inhabitants, it might in a few ages be replenished from one nation only, as for instance with Englishmen.

This is incontrovertibly true. Through the animal and vegetable kingdoms nature has scattered the seeds of life abroad with the most profuse and liberal hand; but has been comparatively sparing in the room and the nourishment necessary to rear them. The germs of existence contained in this earth, if they could freely develop themselves, would fill millions of worlds in the course of a few thousand years. Necessity, that imperious, all-pervading law of nature, restrains them within the prescribed bounds. The race of plants and the race of animals shrink under this great restrictive law; and man cannot by any efforts of reason escape from it.

In plants and irrational animals, the view of the subject is simple. They are all impelled by a powerful instinct to the increase of their species; and this instinct is interrupted by no doubts about providing for their offspring. Wherever therefore there is liberty, the power of increase is exerted; and the superabundant effects are repressed afterwards by want of room and nourishment.

The effects of this check on man are more complicated. Impelled to the increase of his species by an equally powerful instinct, reason interrupts his career, and asks him whether he may not bring beings into the world, for whom he cannot provide the means of support. If he attend to this natural suggestion, the restriction too frequently produces vice. If he hear it not, the human race will be constantly endeavouring to increase beyond the means of subsistence. But as, by that law of our nature which makes food necessary to the life of man, population can never actually increase beyond the lowest nourishment capable of supporting it, a strong check on population, from the difficulty of acquiring food, must be constantly in operation. This difficulty must fall somewhere, and must necessarily be severely felt in some or other of the various forms of misery, or the fear of misery, by a large portion of mankind.

That population has this constant tendency to increase beyond the means of subsistence, and that it is kept to its necessary level by these causes, will sufficiently appear from a review of the different states of society in which man has existed. But, before we proceed to this review, the subject will, perhaps, be seen in a clearer light, if we endeavour to ascertain what would be the natural increase of population, if left to exert itself with perfect freedom; and what might be expected to be the rate of increase in the productions of the earth, under the most favourable circumstances of human industry.

It will be allowed that no country has hitherto been known, where the manners were so pure and simple, and the means of subsistence so abundant, that no check whatever has existed to early marriages from the difficulty of providing for a family, and that no waste of the human species has been occasioned by vicious customs, by towns, by unhealthy occupations, or too severe labour. Consequently in no state that we have yet known, has the power of population been left to exert itself with perfect freedom.

Whether the law of marriage be instituted, or not, the dictate of nature and virtue seems to be an early attachment to one woman; and where there were no impediments of any kind in the way of an union to which such an attachment would lead, and no causes of depopulation afterwards, the increase of the human species would be evidently much greater than any increase which has been hitherto known.

In the northern states of America, where the means of subsistence have been more ample, the manners of the people more pure, and the checks to early marriages fewer, than in any of the modern states of Europe, the population has been found to double itself, for above a century and a half successively, in less than twenty five years. Yet, even during these periods, in some of the towns, the deaths exceeded the births, a circumstance which clearly proves that, in those parts of the country which supplied this deficiency, the increase must have been much more rapid than the general average.

In the back settlements, where the sole employment is agriculture, and vicious customs and unwholesome occupations are little known, the population has been found to double itself in fifteen years. Even this extraordinary rate of increase is probably short of the utmost power of population. Very severe labour is requisite to clear a fresh country; such situations are not in general considered as particularly healthy; and the inhabitants, probably, are occasionally subject to the incursions of the Indians, which may destroy some lives, or at any rate diminish the fruits of industry.

According to a table of Euler, calculated on a mortality of 1 in 36, if the births be to the deaths in the proportion of 3 to 1, the period of doubling will be only 12⅘ years. And this proportion is not only a possible supposition, but has actually occurred for short periods in more countries than one.

Sir William Perry supposes a doubling possible in so short a time as ten years.

But, to be perfectly sure that we are far within the truth, we will take the slowest of these rates of increase, a rate in which all concurring testimonies agree, and which has been repeatedly ascertained to be from procreation only.

It may safely be pronounced, therefore, that population, when unchecked, goes on doubling itself every twenty five years, or increases in a geometrical ratio.

The rate according to which the productions of the earth may be supposed to increase, it will not be so easy to determine. Of this, however, we may be perfectly certain, that the ratio of their increase in a limited territory must be of a totally different nature from the ratio of the increase of population. *A thousand millions are just as easily doubled every twenty five years by the power of population as a thousand. But the food to support the increase from the greater number will by no means be obtained with the same facility. Man is necessarily confined in room.* When acre has been added to acre till all the fertile land is occupied, the yearly increase of food must depend upon the melioration of the land already in possession. This is a fund, which, from the nature of all soils, instead of increasing, must be gradually diminishing. But population, could it be supplied with food, would go on with unexhausted vigour; and the increase of one period would furnish the power of a greater increase the next, and this without any limit.

From the accounts we have of China and Japan, it may be fairly doubted, whether the best directed efforts of human industry could double the produce of these countries even once in any number of years. There are many parts of the globe, indeed, hitherto uncultivated, and almost unoccupied; but the right of exterminating, or driving into a corner where they must starve, even the inhabitants of these thinly peopled regions, will be questioned in a moral view. The process of improving their minds and directing their industry would necessarily be slow; and during this time, as population would regularly keep pace with the increasing produce, it would rarely happen that a great degree of knowledge and industry would have to operate at once upon rich unappropriated soil. Even where this might take place, as it does sometimes in new colonies, a geometrical ratio increases with such extraordinary rapidity, that the advantage could not last long. If the United States of America continue increasing, which they certainly will do, though not with the same rapidity as formerly, the Indians will be driven further and further back into the country, till the whole race is ultimately exterminated, and the territory is incapable of further extension.

These observations are, in a degree, applicable to all the parts of the earth, where the soil is imperfectly cultivated. To exterminate the inhabitants of the greatest part of Asia and Africa, is a thought that could not be admitted for a moment. To civilize and direct the industry of the various tribes of Tartars and negroes, would certainly be a work of considerable time, and of variable and uncertain success.

Europe is by no means so fully peopled as it might be. In Europe there is the fairest chance that human industry may receive its best direction. The science of agriculture has been much studied in England and Scotland; and there is still a great portion of uncultivated land in these countries. Let us consider at what rate the produce of this island might be supposed to increase under circumstances the most favourable to improvement.

If it be allowed that by the best possible policy, and great encouragements to agriculture, the average produce of the island could be doubled in the first twenty five years, it will be allowing, probably, a greater increase than could with reason be expected.

In the next twenty five years, it is impossible to suppose that the produce could be quadrupled. It would be contrary to all our knowledge of the properties of land. The improvement of the barren parts would be a work of time and labours and it must be evident to those who have the slightest acquaintance with agricultural subjects, that in proportion as cultivation extended, the additions that could yearly be made to the former average produce must be gradually and regularly diminishing. That we may be the better able to compare the increase of population and food, let us make a supposition, which, without pretending to accuracy, is clearly more favourable to the power of production in the earth, than any experience we have had of its qualities will warrant.

Let us suppose that the yearly additions which might be made to the former average produce, instead of decreasing, which they certainly would do, were to remain the same; and that the produce of this island might be increased every twenty five years, by a quantity equal to what it at present produces. The most enthusiastic speculator cannot suppose a greater increase than this. In a few centuries it would make every acre of land in the island like a garden.

If this supposition be applied to the whole earth, and if it be allowed that the subsistence for man which the earth affords might be increased every twenty five years by a quantity equal to what it at present produces, this will be supposing a rate of increase much greater than we can imagine that any possible exertions of mankind could make it.

It may be fairly pronounced, therefore, that, considering the present average state of the earth, the means of subsistence, under circumstances the most favourable to human industry, could not possibly be made to increase faster than in an arithmetical ratio.

The necessary effects of these two different rates of increase, when brought together, will be very striking. Let us call the population of this island 11 millions; and suppose the present produce equal to the easy support of such a number. In the first twenty five years the population would be 22 millions, and the food being also doubled, the means of subsistence would be equal to this increase. In the next twenty five years, the population would be 44 millions, and the means of subsistence only equal to the support of 33 millions. In the next period the population would be 88 millions, and the means of subsistence just equal to the support of half that number. And, at the conclusion of the first century, the population would be 176 millions, and the means of subsistence only equal to the support of 55 millions, leaving a population of 121 millions totally unprovided for.

Taking the whole earth, instead of this island, emigration would of course be excluded; and, supposing the present population equal to a thousand millions, the human species would increase as the numbers, 1, 2, 4, 8, 16, 32, 64, 128, 256, and subsistence as 1, 2, 3, 4, 5, 6, 7, 8, 9. In two centuries the population would be to the mean of subsistence as 256 to 9; in three centuries as 4096 to 13, and in two thousand years the difference would be almost incalculable.

In this supposition no limits whatever are placed to the produce of the earth. It may increase for ever and be greater than any assignable quantity; yet still the power of population being in every period so much superior, the increase of the human species can only be kept down to the level of the means of subsistence by the constant operation of the strong law of necessity, acting as a check upon the greater power.

CHAPTER II

OF THE GENERAL CHECKS TO POPULATION, AND THE MODE OF THEIR OPERATION

The ultimate check to population appears then to be a want of food, arising necessarily from the different ratios according to which population and food increase. But this ultimate check is never the immediate check, except in cases of actual famine.

The immediate check may be stated to consist in all those customs, and all those diseases, which seem to be generated by a scarcity of the means of subsistence, and all those causes, independent of this scarcity, whether of a moral or physical nature, which tend prematurely to weaken and destroy the human frame.

These checks to population, which are constantly operating with more or less force in every society, and keep down the number to the level of the means of subsistence, may be classed under two general heads — the preventive, and the positive checks.

The preventive check, as far as it is voluntary, is peculiar to man, and arises from that distinctive superiority in his reasoning faculties, which enables him to calculate distant consequences. The checks to the indefinite increase of plants and irrational animals are all either positive, or, if preventive, involuntary. But man cannot look around him, and see the distress which frequently presses upon those who have large families; he cannot contemplate his present possessions or earnings, which he now nearly consumes himself, and calculate the amount of each share, when with very little addition they must be divided, perhaps, among seven or eight, without feeling a doubt whether, if he follow the bent of his inclinations, he may be able to support the offspring which he will probably bring into the world. In a state of equality, if such can exist, this would be the simple question. In the present state of society other considerations occur. Will he not lower his rank in life, and be obliged to give up in great measure his former habits? Does any mode of employment present itself by which he may reasonably hope to maintain a family? Will he not at any rate subject himself to greater difficulties, and more severe labour, than in his single state? Will he not be unable to transmit to his children the same advantages of education and improvement that he had himself possessed? Does he even feel secure that, should he have a large family, his utmost exertions can save them from rags and squalid poverty, and their consequent degradation in the community? And may he not be reduced to the grating necessity of forfeiting his independence, and of being obliged to the sparing hand of charity for support?

These considerations are calculated to prevent, and certainly do prevent, a great number of persons in all civilized nations from pursuing the dictate of nature in an early attachment to one woman.

If this restraint do not produce vice, it is undoubtedly the least evil that can arise from the principle of population. Considered as a restraint on a strong natural inclination, it must be allowed to produce a certain degree of temporary unhappiness; but evidently slight, compared with the evils which result from any of the other checks to population; and merely of the same nature as many other sacrifices of temporary to permanent gratification, which it is the business of a moral agent continually to make.

When this restraint produces vice, the evils which follow are but too conspicuous. A promiscuous intercourse to such a degree as to prevent the birth of children, seems to lower, in the most marked manner, the dignity of human nature. It cannot be without its effect on men, and nothing can be more obvious than its tendency to degrade the female character, and to destroy all its most amiable and distinguishing characteristics. Add to which, that among those unfortunate females, with which all great towns abound, more real distress and aggravated misery are, perhaps, to be found, than in any other department of human life.

When a general corruption of morals, with regard to the sex, pervades all the classes of society, its effects must necessarily be, to poison the springs of domestic happiness, to weaken conjugal and parental affection, and to lessen the united exertions and ardour of parents in the care and education of their children; effects which cannot take place without a decided diminution of the general happiness and virtue of the society; particularly as the necessity of art in the accomplishment and conduct of intrigues, and in the concealment of their consequences necessarily leads to many other vices.

The positive checks to population are extremely various, and include every cause, whether arising from vice or misery, which in any degree contributes to shorten the natural duration of human life. Under this head, therefore, may be enumerated all unwholesome occupations, severe labour and exposure to the seasons, extreme poverty, bad nursing of children, great towns, excesses of all kinds, the whole train of common diseases and epidemics, wars, plague, and famine.

On examining these obstacles to the increase of population which I have classed under the heads of preventive and positive checks, it will appear that they are all resolvable into moral restraint, vice, and misery.

Of the preventive checks, the restraint from marriage which is not followed by irregular gratifications may properly be termed moral restraint.

Promiscuous intercourse, unnatural passions, violations of the marriage bed, and improper arts to conceal the consequences of irregular connections, are preventive checks that clearly come under the head of vice.

Of the positive checks, those which appear to arise unavoidably from the laws of nature, may be called exclusively misery; and those which we obviously bring upon ourselves, such as wars, excesses, and many others which it would be in our power to avoid, are of a mixed nature. They are brought upon us by vice, and their consequences are misery.

The sum of all these preventive and positive checks, taken together, forms the immediate check to population; and it is evident that, in every country where the whole of the procreative power cannot be called into action, the preventive and the positive checks must vary inversely as each other; that is, in countries either naturally unhealthy, or subject to a great mortality, from whatever cause it may arise, the preventive check will prevail very little. In those countries, on the contrary, which are naturally healthy, and where the preventive check is found to prevail with considerable force, the positive check will prevail very little, or the mortality be very small.

In every country some of these checks are, with more or less force, in constant operation; yet, notwithstanding their general prevalence, there are few states in which there is not a constant effort in the population to increase beyond the means of subsistence. This constant effort as constantly tends to subject the lower classes of society to distress, and to prevent any great permanent [melioration] of their condition.

These effects, in the present state of society, seem to be produced in the following manner. We will suppose the means of subsistence in any country just equal to the easy support of its inhabitants. The constant effort towards population, which is found to act even in the most vicious societies, increases the number of people before the means of subsistence are increased. The food, therefore, which before supported eleven millions, must now be divided among eleven millions and a half. The poor consequently must live much worse, and many of them be reduced to severe distress. The number of labourers also being above the proportion of work in the market, the price of labour must tend to fall, while the price of provisions would at the same time tend to rise. The labourer therefore must do more work, to earn the same as he did before. During this season of distress, the discouragements to marriage and the difficulty of rearing a family are so great, that the progress of population is retarded. In the meantime, the cheapness of labour, the plenty of labourers, and the necessity of an increased industry among them, encourage cultivators to employ more labour upon their land, to turn up fresh soil, and to manure and improve more completely what is already in tillage, till ultimately the means of subsistence may become in the same proportion to the population, as at the period from which we set out. The situation of the labourers being then again tolerably comfortable, the restraints to population are in some degree loosened; and, after a short period, the same retrograde and progressive movements, with respect to happiness, are repeated.

This sort of oscillation will not probably be obvious to common view; and it may be difficult even for the most attentive observer to calculate its periods. Yet that, in the generality of old states, some alternation of this kind does exist though in a much less marked, and in a much more irregular manner, than I have described it, no reflecting man, who considers the subject deeply, can well doubt.

One principal reason why this oscillation has been less remarked, and less decidedly confirmed by experience than might naturally be expected, is, that the histories of mankind which we possess are, in general, histories only of the highest classes. We have not many accounts that can be depended upon, of the manners and customs of that part of mankind, where these retrograde and progressive movements chiefly take place. A satisfactory history of this kind, of one people and of one period, would require the constant and minute attention of many observing minds in local and general remarks on the state of the lower classes of society, and the causes that influenced it; and, to draw accurate inferences upon this subject, a succession of such historians for some centuries would be necessary. This branch of statistical knowledge has, of late years, been attended to in some countries, and we may promise ourselves a clearer insight into the internal structure of human society from the progress of these inquiries. But the science may be said yet to be in its infancy, and many of the objects, on which it would be desirable to have information, have been either omitted or not stated with sufficient accuracy. Among these, perhaps, may be reckoned the proportion of the number of adults to the number of marriages; the extent to which vicious customs have prevailed in consequence of the restraints upon matrimony; the comparative mortality among the children of the most distressed part of the community, and of those who live rather more at their ease; the variations in the real price of labour; the observable differences in the state of the lower classes of society, with respect to ease and happiness, at different times during a certain period; and very accurate registers of births, deaths, and marriages, which are of the utmost importance in this subject.

A faithful history, including such particulars, would tend greatly to elucidate the manner in which the constant check upon population acts; and would probably prove the existence of the retrograde and progressive movements that have been mentioned; though the times of their vibration must necessarily be rendered irregular from the operation of many interrupting causes; such as, the introduction or failure of certain manufactures; a greater or less prevalent spirit of agricultural enterprise; years of plenty, or years of scarcity; wars, sickly seasons, poor laws, emigrations and other causes of a similar nature.

A circumstance which has, perhaps, more than any other, contributed to conceal this oscillation from common view, is the difference between the nominal and real price of labour. It very rarely happens that the nominal price of labour universally falls; but we well know that it frequently remains the same, while the nominal price of provisions has been gradually rising. This, indeed, will generally be the case, if the increase of manufactures and commerce be sufficient to employ the new labourers that are thrown into the market, and to prevent the increased supply from lowering the money price.

But an increased number of labourers receiving the same money wages will necessarily, by their competition, increase the money price of corn. This is, in fact, a real fall in the price of labour; and, during this period, the condition of the lower classes of the community must be gradually growing worse. But the farmers and capitalists are growing rich from the real cheapness of labour. Their increasing capitals enable them to employ a greater number of men; and, as the population had probably suffered some check from the greater difficulty of supporting a family, the demand for labour, after a certain period, would be great in proportion to the supply, and its price would of course rise, if left to find its natural level; and thus the wages of labour, and consequently the condition of the lower classes of society, might have progressive and retrograde movements, though the price of labour might never nominally fall.

In savage life, where there is no regular price of labour, it is little to be doubted that similar oscillations took place. When population has increased nearly to the utmost limits of the food, all the preventive and the positive checks will naturally operate with increased force. Vicious habits with respect to the sex will be more general, the exposing of children more frequent, and both the probability and fatality of wars and epidemics will be considerably greater; and these causes will probably continue their operation till the population is sunk below the level of the food; and then the return to comparative plenty will again produce an increase, and, after a certain period, its further progress will again be checked by the same causes.

But without attempting to establish these progressive and retrograde movements in different countries, which would evidently require more minute histories than we possess, and which the progress of civilization naturally tends to counteract, the following propositions are intended to be proved:

1. Population is necessarily limited by the means of subsistence.

2. Population invariably increases where the means of subsistence increase, unless prevented by some very powerful and obvious checks.

3. These checks, and the checks which repress the superior power of population, and keep its effects on a level with the means of subsistence, are all resolvable into moral restraint, vice and misery.

The first of these propositions scarcely needs illustration. The second and third will be sufficiently established by a review of the immediate checks to population in the past and present state of society.

This review will be the subjects of the following chapters.

Appendix 2

ALFRED RUSSEL WALLACE "ON THE TENDENCY OF VARIETIES TO DEPART INDEFINITELY FROM THE ORIGINAL TYPE" (1858)

INTRODUCTION

Alfred Russel Wallace (1823–1913) was a self-trained naturalist who worked as a surveyor and schoolmaster. In the mid-1840s, he began to correspond with the naturalist Henry Bates, with whom he subsequently made an expedition to the Amazon basin from 1848 to 1852. After losing most of his collections in a shipwreck, he set out again for the Malay Archipelago, where he explored from 1854 to 1862. Wallace had been an evolutionist since 1845, when he had read Robert Chambers's *Vestiges of the Natural History of Creation*, but he could determine no mechanism for the rise of new species until one day in February 1858, when, while fighting off a malarial fever, he made the same connection between Malthus's notions of population pressure and the survival of the fittest that Darwin himself had made twenty years earlier. As Wallace later recalled the moment:

> One day something brought to my recollection Malthus's *Principles of Population*, which I had read about twelve years before. I thought of his clear exposition of "the positive checks to increase"—disease, accidents, wars, and famine— which kept down the population of savage races to so much lower an average than that of more civilized peoples. It then occurred to me that these causes or their equivalents are continually acting in the case of animals also; and as animals usually breed much more rapidly than does mankind, the destruction every year from these causes must be enormous in order to keep down the numbers of each species, since they evidently do not increase regularly from year to year, as otherwise the world would

335

long ago have been densely crowded with those that breed most quickly. Vaguely thinking over the enormous and constant destruction which this implied, it occurred to me to ask the question, why do some die and some live? And the answer was clearly, that on the whole the best fitted lived. From the effects of disease the most healthy escaped; from enemies, the strongest, the swiftest, or the most cunning; from famine, the best hunters or those with best digestion; and so on. Then it suddenly flashed upon me that this self-acting process would necessarily improve the race, because in every generation the inferior would inevitably be killed off and the superior would remain—that is, *the fittest would survive*. Then at once I seemed to see the whole effect of this, that when changes of land and sea, or of climate, or of food-supply, or of enemies occurred—and we know that such changes have always been taking place—and considering the amount of individual variation that my experience as a collector had shown me to exist,[1] then it followed that all the changes necessary for the adaptation of the species to the changing conditions would be brought about; and as great changes in the environment are always slow, there would be ample time for the change to be effected by the survival of the best fitted in every generation. In this way every part of an animal's organization could be modified exactly as required, and in the very process of this modification the unmodified would die out, and thus the *definite* characters and the clear *isolation* of new species would be explained. The more I thought it over the more I became convinced that I had at length solved the problem of the origin of species. For the next hour I thought over the deficiencies of the theories of Lamarck and of the author of the 'Vestiges,' and I saw that my new theory obviated every important difficulty.[2]

Wallace accordingly wrote up a sketch of his ideas—the essay reproduced here—and mailed it to Darwin, who received it in early June 1858. The latter's reaction was that "if Wallace had my M.S. sketch [of 1844] . . . he could not have made a better short abstract!"[3] Yet there are telling differences in the way the two men portrayed the action of natural selection. Darwin, ever the moderate gradualist, emphasized the delicate balance of forces in the struggle for existence. As he says in Notebook E, in an image that recurs in the *Origin*, "It is difficult to believe in the dreadful but quiet war of organic

1. Compare the similar phrasing of Darwin's account of his preparation to appreciate Malthus's point: "and being prepared to appreciate the struggle for existence which everywhere goes on from long-continued observation of the habits of animals and plants. . . . (*Autobiography*, quoted on p. 310 of this volume). Wallace must have read Darwin's *Autobiography*, which was published in 1887, and it is not illogical to suppose that Wallace recognized the parallelism between his experience and Darwin's and mimicked the latter's style, whether consciously or not.

2. A. R. Wallace, *My Life*, 2 vols. (London, 1905), 1: 362–363.

3. Darwin, *Correspondence*, 7: 107.

beings, . . . there is a contest & a grain of sand turns the balance."[4] Wallace, in contrast, sets his most compelling depiction of struggle not in terms of grains of sand, but in terms of drastic environmental shifts: "Now, let some alteration of physical conditions occur in the district—a long period of drought, a destruction of vegetation by locusts, the irruption of some new carnivorous animal. . . ." And such striking changes have drastic outcomes: "it is evident that . . . those forming the least numerous and most feebly organized variety would suffer first, and were the pressure severe, must soon become extinct. The same causes continuing in action, the parent species . . . might also become extinct. The superior variety would then alone remain . . . and occupy the place of the extinct species and variety" (p. 342).

Thus, as Darwin conceives the action of natural selection, change occurs as small accommodations to altering conditions, while for Wallace the alterations are large and the consequences are explicit extinction. Perhaps these alternative portraits of competition in nature reflect different experiences of economic competition. For the Whig reformer Darwin, born to wealth and himself an adroit investor, the ruthlessness of struggle was something that was "difficult to believe in" and needed constantly to be recalled. For Wallace, the artisan turned naturalist, who lost all his savings more than once in the rough-and-tumble of the market economy, the struggle for existence was an untempered process that depended on major—boom and bust—environmental shifts and led to clear winners and broken losers.

In order to give Wallace his due and preserve Darwin's priority of discovery, Charles Lyell and Joseph Hooker arranged a special session of the Linnean Society on July 30, 1858, at which Wallace's paper was read, along with part of the abstract Darwin had sent to Asa Gray in 1857 (see chapter 7), and a selection on natural selection from the 1844 Essay.

ON THE TENDENCY OF VARIETIES TO DEPART INDEFINITELY FROM THE ORIGINAL TYPE
BY ALFRED RUSSEL WALLACE

One of the strongest arguments which have been adduced to prove the original and permanent distinctness of species is, that *varieties* produced in a state of domesticity are more or less unstable, and often have a tendency, if left to themselves, to return to the normal form of the parent species; and this instability is considered to be a distinctive peculiarity of all varieties, even of those occurring among wild animals in a state of nature, and to constitute a provision for preserving unchanged the originally created distinct species.

---

4. Notebook E:114; (Chapter 2, 75).

In the absence or scarcity of facts and observations as to *varieties* occurring among wild animals, this argument has had great weight with naturalists, and has led to a very general and somewhat prejudiced belief in the stability of species. Equally general, however, is the belief in what are called "permanent or true varieties," — races of animals which continually propagate their like, but which differ so slightly (although constantly) from some other race, that the one is considered to be a *variety* of the other. Which is the *variety* and which the original *species*, there is generally no means of determining, except in those rare cases in which the one race has been known to produce an offspring unlike itself and resembling the other. This, however, would seem quite incompatible with the "permanent invariability of species," but the difficulty is overcome by assuming that such varieties have strict limits, and can never again vary further from the original type, although they may return to it, which, from the analogy of the domesticated animals, is considered to be highly probable, if not certainly proved.

It will be observed that this argument rests entirely on the assumption, that *varieties* occurring in a state of nature are in all respects analogous to or even identical with those of domestic animals, and are governed by the same laws as regards their permanence or further variation. But it is the object of the present paper to show that this assumption is altogether false, that there is a general principle in nature which will cause many *varieties* to survive the parent species, and to give rise to successive variations departing further and further from the original type, and which also produces, in domesticated animals, the tendency of varieties to return to the parent form.

The life of wild animals is a struggle for existence. The full exertion of all their faculties and all their energies is required to preserve their own existence and provide for that of their infant offspring. The possibility of procuring food during the least favourable seasons, and of escaping the attacks of their most dangerous enemies, are the primary conditions which determine the existence both of individuals and of entire species. These conditions will also determine the population of a species; and by a careful consideration of all the circumstances we may be enabled to comprehend, and in some degree to explain, what at first sight appears so inexplicable — the excessive abundance of some species, while others closely allied to them are very rare.

The general proportion that must obtain between certain groups of animals is readily seen. Large animals cannot be so abundant as small ones; the carnivora must be less numerous than the herbivora; eagles and lions can never be so plentiful as pigeons and antelopes; the wild asses of the Tartarian deserts cannot equal in numbers the horses of the more luxuriant prairies and pampas of America. The greater or less fecundity of an animal is often considered to be one of the chief causes of its abundance or scarcity;

but a consideration of the facts will show us that it really has little or nothing to do with the matter. Even the least prolific of animals would increase rapidly if unchecked, whereas it is evident that the animal population of the globe must be stationary, or perhaps, through the influence of man, decreasing. Fluctuations there may be; but permanent increase, except in restricted localities, is almost impossible. For example, our own observation must convince us that birds do not go on increasing every year in a geometrical ratio, as they would do, were there not some powerful check to their natural increase. Very few birds produce less than two young ones each year, while many have six, eight, or ten; four will certainly be below the average; and if we suppose that each pair produce young only four times in their life, that will also be below the average, supposing them not to die either by violence or want of food. Yet at this rate how tremendous would be the increase in a few years from a single pair! A simple calculation will show that in fifteen years each pair of birds would have increased to nearly ten millions! whereas we have no reason to believe that the number of the birds of any country increases at all in fifteen or in one hundred and fifty years. With such powers of increase the population must have reached its limits, and have become stationary, in a very few years after the origin of each species. It is evident, therefore, that each year an immense number of birds must perish — as many in fact as are born; and as on the lowest calculation the progeny are each year twice as numerous as their parents, it follows that, whatever be the average number of individuals existing in any given country, *twice that number must perish annually,* — a striking result, but one which seems at least highly probable, and is perhaps under rather than over the truth. It would therefore appear that, as far as the continuance of the species and the keeping up the average number of individuals are concerned, large broods are superfluous. On the average all above *one* become food for hawks and kites, wild cats and weasels, or perish of cold and hunger as winter comes on. This is strikingly proved by the case of particular species; for we find that their abundance in individuals bears no relation whatever to their fertility in producing offspring. Perhaps the most remarkable instance of an immerse bird population is that of the passenger pigeon of the United States, which lays only one, or at most two eggs, and is said to rear generally but one young one. Why is this bird so extraordinarily abundant, while others producing two or three times as many young are much less plentiful? The explanation is not difficult. The food most congenial to this species, and on which it thrives best, is abundantly distributed over a very extensive region, offering such differences of soil and climate, that in one part or another of the area the supply never fails. The bird is capable of a very rapid and long-continued flight, so that it can pass without fatigue over the whole of the district it inhabits, and as soon as the supply of food begins to fail in one place is able to discover a fresh feeding-ground. This example strikingly shows us that the procuring a constant supply of wholesome food is almost the sole condition requisite for ensuring the rapid increase of a given species, since neither the limited fecundity, nor the unrestrained attacks of birds of prey and of man are here sufficient to check it. In no other birds are these peculiar circumstances so strikingly combined. Either their food is more liable to failure, or they have not sufficient power of wing to search for it over an extensive area, or during some season of the year it becomes very scarce, and less wholesome substitutes have to be found; and thus, though more fertile in offspring, they can never increase beyond the supply of food in the least favourable seasons. Many birds can only exist by migrating, when their food becomes scarce, to regions possessing a milder, or at least a different climate, though, as these migrating birds are seldom excessively abundant, it is evident that the countries they visit are still deficient in a constant and abundant supply of wholesome food. Those whose organization does not permit them to migrate when their food becomes periodically scarce, can never attain a large population. This is probably the reason why woodpeckers are scarce with us, while in the tropics they are among the most abundant of solitary birds. Thus the house sparrow is more abundant than the redbreast, because its food is more constant and plentiful, — seeds of grasses being preserved during the winter, and our farmyards and stubble-fields furnishing an almost inexhaustible supply. Why, as a general rule, are aquatic, and especially sea birds, very numerous in individuals? Not because they are more prolific than others, generally the contrary; but because their food never fails, the sea-shores and river-banks daily swarming with a fresh supply of small mollusca and crustacea. Exactly the same laws will apply to mammals. Wild cats are prolific and have few enemies; why then are they never as abundant as rabbits? The only intelligible answer is, that their supply of food is more precarious. It appears evident, therefore, that so long as a country remains physically unchanged, the numbers of its animal population cannot materially increase. If one species does so, some others requiring the same kind of food must diminish in proportion. The numbers that die annually must be immense; and as the individual existence of each animal depends upon itself, those that die must be the weakest — the very young, the aged, and the diseased, — while those that prolong their existence can only be the most perfect in health and vigour — those who are best able to obtain food regularly, and avoid their numerous enemies. It is, as we commenced by remarking, "a struggle for existence," in which the weakest and least perfectly organized must always succumb.

Now it is clear that what takes place among the individuals of a species must also occur among the several allied species of a group, — viz. that those which are best adapted to obtain a regular supply of food, and to defend themselves against the attacks of their enemies and the vicissitudes of the seasons, must necessarily obtain and preserve a superiority in population; while those species which from some defect of power or organization are the least capable of counteracting the vicissitudes of food, supply, &c., must diminish in numbers, and, in extreme cases, become altogether extinct. Between these extremes the species will present various degrees of capacity for ensuring the means of preserving life; and it is thus we account for the abundance or rarity of species. Our ignorance will generally prevent us from accurately tracing the effects to their causes; but could we become perfectly acquainted with the organization and habits of the various species of animals, and could we measure the capacity of each for performing the different acts necessary to its safety and existence under all the varying circumstances by which it is surrounded, we might be able even to calculate the proportionate abundance of individuals which is the necessary result.

If now we have succeeded in establishing these two points — 1st, *that the animal population of a country is generally stationary, being kept down by a periodical deficiency of food, and other checks;* and, 2nd, *that the comparative abundance or scarcity of the individuals of the several species is entirely due to their organization and resulting habits, which, rendering it more difficult to procure a regular supply of food and to provide for their personal safety in some cases than in others, can only be balanced by a difference in the population which have to exist in a given area* — we shall be in a condition to proceed to the consideration of *varieties*, to which the preceding remarks have a direct and very important application.

Most or perhaps all the variations from the typical form of a species must have some definite effect, however slight, on the habits or capacities of the individuals. Even a change of colour might, by rendering them more or less distinguishable, affect their safety; a greater or less development of hair might modify their habits. More important changes, such as an increase in the power or dimensions of the limbs or any of the eternal organs, would more or less affect their mode of procuring food or the range of country which they inhabit. It is also evident that most changes would affect, either favourable or adversely, the powers of prolonging existence. An antelope with shorter or weaker legs must necessarily suffer more from the attacks of the feline carnivora; the passenger pigeon with less powerful wings would sooner or later be affected in its powers of procuring a regular supply of food; and in both cases the result must necessarily be a diminution of the population of the modified species. If, on the other hand, any species should produce a variety having slightly increased powers of preserving existence, that variety must inevitably in time acquire a superiority in numbers. These results must follow as surely as old age, intemperance, or scarcity of food produce an increased mortality. In both cases there may be many individual exceptions;

but on the average the rule will invariably be found to hold good. All varieties will therefore fall into two classes — those which under the same conditions would never reach the population of the parent species, and those which would in time obtain and keep a numerical superiority. Now, let some alteration of physical conditions occur in the district — a long period of drought, a destruction of vegetation by locusts, the irruption of some new carnivorous animal seeking "pastures new" — any change in fact tending to render existence more difficult to the species in question, and tasking its utmost powers to avoid complete extermination; it is evident that, of all the individuals composing the species, those forming the least numerous and most feebly organized variety would suffer first, and, were the pressure severe, must soon become extinct. The same causes continuing in action, the parent species would next suffer, would gradually diminish in numbers, and with a recurrence of similar unfavourable conditions might also become extinct. The superior variety would then alone remain, and on a return to favourable circumstances would rapidly increase in numbers and occupy the place of the extinct species and variety.

The *variety* would now have replaced the *species*, of which it would be a more perfectly developed and more highly organized form. It would be in all respects better adapted to secure its safety, and to prolong its individual existence and that of the race. Such a variety *could not* return to the original form; for that form is an inferior one, and could never compete with it for existence. Granted, therefore, a "tendency" to reproduce the original type of the species, still the variety must ever remain preponderant in numbers, and under adverse physical conditions *again alone survive*. But this new, improved, and populous race might itself, in course of time, give rise to new varieties, exhibiting several diverging modifictions of form, any of which, tending to increase the facilities for preserving existence, must, by the same general law, in their turn become predominant. Here, then, we have *progression and continued divergence* deduced from the general laws which regulate the existence of animals in a state of nature, and from the undisputed fact that varieties do frequently occur. It is not, however, contended that this result would be invariable; a change of physical conditions in the district might at times materially modify it, rendering the race which had been the most capable of supporting existence under the former conditions now the least so, and even causing the extinction of the newer and, for a time, superior race, while the old or parent species and its first inferior varieties continued to flourish. Variations in unimportant parts might also occur, having no perceptible effect on the life-preserving powers; and the varieties so furnished might run a course parallel with the parent species, either giving rise to further variations or returning to the former type. All we argue for is, that certain varieties have a tendency to maintain their existence longer than the original species, and this tendency must make itself felt; for though the doctrine of chances or averages can never be trusted to on a limited scale, yet, if applied to higher numbers, the results come nearer to what theory demands, and, as we approach to an infinity of examples, become strictly accurate. Now the scale on which nature works is so vast — the numbers of individuals and periods of time with which she deals approach so near to infinity, that any cause, however slight, and however liable to be veiled and counteracted by accidental circumstances, must in the end produce its full legitimate results.

Let us now turn to domesticated animals, and inquire how varieties produced among them are affected by the principles here enunciated. The essential difference in the condition of wild and domestic animals is this, — that among the former, their well-being and very existence depend upon the full exercise and healthy condition of all their senses and physical powers, whereas, among the latter, these are only partially exercised, and in some cases are absolutely unused. A wild animal has to search, and often to labour, for every mouthful of food — to exercise sight, hearing, and smell in seeking it, and in avoiding dangers, in procuring shelter from the inclemency of the seasons, and in providing for the subsistence and safety of its offspring. There is no muscle of its body that is not called into daily and hourly activity; there is no sense or faculty that is not strengthened by continual exercise. The domestic animal, on the other hand, has food provided for it, is sheltered, and often confined, to guard it against the vicissitudes of the seasons, is carefully secured from the attacks of its natural enemies, and seldom even rears its young without human assistance. Half of its senses and faculties are quite useless; and the other half are but occasionally called into feeble exercise, while even its muscular system is only irregularly called into action.

Now when a variety of such an animal occurs, having increased power or capacity in any organ or sense, such increase is totally useless, is never called into action, and may even exist without the animal ever becoming aware of it. In the wild animal, on the contrary, all its faculties and powers being brought into full action for the necessities of existence, any increase becomes immediately available, is strengthened by exercise, and must even slightly modify the food, the habits, and the whole economy of the race. It creates as it were a new animal, one of superior powers, and which will necessarily increase in numbers and outlive those inferior to it.

Again, in the domesticated animal all variations have an equal chance of continuance; and those which would decidedly render a wild animal unable to compete with its fellows and continue its existence are no disadvantage whatever in a state of domesticity. Our quickly fattening pigs, short-legged sheep, pouter pigeons, and poodle dogs could never have come into existence in a state of nature, because the very first step towards such inferior forms would have led to the rapid extinction of the race; still less could they now exist in competition with their wild allies. The great speed but slight endurance of the race horse, the unwieldy strength of the ploughman's team, would both be useless in a state of nature. If turned wild on the pampas, such animals would probably soon become extinct, or under favourable circumstances might each lose those extreme qualities which would never be called into action, and in a few generations would revert to a common type, which must be that in which the various powers and faculties are so proportioned to each other as to be best adapted to procure food and secure safety, — that in which by the full exercise of every part of his organization the animal can alone continue to live. Domestic varieties, when turned wild, *must* return to something near the type of the original wild stock, *or become altogether extinct.*

We see, then, that no inferences as to varieties in a state of nature can be deduced from the observation of those occurring among domestic animals. The two are so much opposed to each other in every circumstance of their existence, that what applies to the one is almost sure not to apply to the other. Domestic animals are abnormal, irregular, artificial; they are subject to varieties which never occur and never can occur in a state of nature; their very existence depends altogether on human care; so far are many of them removed from that just proportion of faculties, that true balance of organization, by means of which alone an animal left to its own resources can preserve its existence and continue its race.

The hypothesis of Lamarck — that progressive changes in species have been produced by the attempts of animals to increase the development of their own organs, and thus modify their structure and habits — has been repeatedly and easily refuted by all writers on the subject of varieties and species, and it seems to have been considered that when this was done the whole question has been finally settled; but the view here developed renders such an hypothesis quite unnecessary, by showing that similar results must be produced by the action of principles constantly at work in nature. The powerful retractile talons of the falcon- and the cat-tribes have not been producd or increased by the volition of those animals; but among the different varieties which occurred in the earlier and less highly organized forms of these groups, *those always survived longest which had the greatest facilities for seizing their prey.* Neither did the giraffe acquire its long neck by desiring to reach the foliage of the more lofty shrubs, and constantly stretching its neck for the purpose, but because any varieties which occurred among its antitypes with a longer neck than usual *at once secured a fresh range of pasture over the same ground as their shorter-necked companions, and on the first scarcity of food were thereby enabled to outlive them.* Even the peculiar colours of many animals, especially insects, so closely resembling the soil or the leaves or the trunks on which they habitually reside, are explained on the same principle; for though in the course of ages varieties of many tints may have occurred, *yet those races having colours best adapted to concealment from their enemies would inevitably survive the longest*. We have also here an acting cause to account for that balance so often observed in nature, — a deficiency in one set of organs always being compensated by an increased development of some others — powerful wings accompanying weak feet, or great velocity making up for the absence of defensive weapons; for it has been shown that all varieties in which an unbalanced deficiency occurred could not long continue their existence. The action of this principle is exactly like that of the centrifugal governor of the steam engine, which checks and corrects any irregularities almost before they become evident; and in like manner no unbalanced deficiency in the animal kingdom can ever reach any conspicuous magnitude, because it would make itself felt at the very first step, by rendering existence difficult and extinction almost sure soon to follow. An origin such as is here advocated will also agree with the peculiar character of the modifications of form and structure which obtain in organized beings — the many lines of divergence from a central type, the increasing efficiency and power of a particular organ through a succession of allied species, and the remarkable persistence of unimportant parts such as colour, texture of plumage and hair, form of horns or crests, through a series of species differing considerably in more essential characters. It also furnishes us with a reason for that "more specialized structure" which Professor Owen states to be a characteristic of recent compared with extinct forms, and which would evidently be the result of the progressive modification of any organ applied to a special purpose in the animal economy.

We believe we have now shown that there is a tendency in nature to the continued progression of certain classes of *varieties* further and further from the original type — a progression to which there appears no reason to assign any definite limits — and that the same principle which produces this result in a state of nature will also explain why domestic varieties have a tendency to revert to the original type. This progression, by minute steps, in various directions, but always checked and balanced by the necessary conditions, subject to which alone existence can be preserved, may, it is believed, be followed out so as to agree with all the phenomena presented by organized beings, their extinction and succession in past ages, and all the extraordinary modifications of form, instinct, and habits which they exhibit.

Ternate, February, 1858.

DARWIN'S WRITINGS AND GENERAL BIBLIOGRAPHY

Bowlby, John. *Charles Darwin: A New Life.* New York: Norton, 1990.

Brent, Peter. *Charles Darwin: A Man of Enlarged Curiosity.* New York: Norton, 1981.

Brown, Frank Burch. "The Evolution of Darwin's Theism." *Journal of the History of Biology* 19 (1986): 1–45.

Browne, Janet. *Charles Darwin. Voyaging.* New York: Knopf, 1995.

Darwin, Charles. *The Autobiography of Charles Darwin.* New York: Norton, 1959. Edited by Nora Barlow.

———. *Charles Darwin's Natural Selection. Being the Second Part of His Big Species Book, 1856-1858.* Cambridge: Cambridge University Press, 1975. Edited by Robert C. Stauffer.

———. *Charles Darwin's Notebooks, 1836-1844. Geology, Transmutation of Species, Metaphysical Enquiries.* Ithaca: British Museum (Natural History) and Cornell University Press, 1987. Edited by Paul H. Barrett, Sandra Herbert, David Kohn, and Sydney Smith.

———. *The Correspondence of Charles Darwin.* 9 vols. to date. Cambridge: Cambridge University Press, 1985–. Edited by Frederick Burkhardt and Sydney Smith.

———. *The Descent of Man, and Selection in Relation to Sex.* 2 vols. London: John Murray, 1874 second edition. [First edition 1870, 1871.]

———. *The Different Forms of Flowers on Plants of the Same Species.* London: John Murray, 1878 second edition. [First edition 1877.]

———. *The Effects of Cross and Self Fertilisation in the Vegetable Kingdom.* London: John Murray, 1878 second edition. [First edition 1876.]

———. *The Expression of the Emotions in Man and Animals.* London: John Murray, 1872.

———. *The Foundations of the Origin of Species: Two Essays Written in 1842 and 1844.* Cambridge: Cambridge University Press, 1909. Edited by Francis Darwin.

————. *Journal of Researches into the Geology and Natural History of the Various Countries Visited by H.M.S. Beagle* etc. London: Henry Colburn, 1839. London: John Murray, 1845 second edition. [Widely known as *The Voyage of the Beagle.*]

————. *The Life and Letters of Charles Darwin.* 3 vols. London: John Murray, 1887. Edited by Francis Darwin.

————. *A Monograph of the Fossil Lepadidae* (London: Ray Society, 1851).

————. *A Monograph of the Fossil Balanidae and Verrucidae.* (London: Ray Society, 1854).

————. *A Monograph of the Sub-Class Cirripedia.* Vol. 1, *The Lepadidae* (London: Ray Society, 1851). Vol. 2, *The Balanidae* (London: Ray Society, 1854).

————. *More Letters of Charles Darwin.* 2 vols. London: John Murray, 1903. Edited by Francis Darwin and A. C. Seward.

————. *On the Origin of Species by Means of Natural Selection, or the Preservation of Favoured Races in the Struggle for Life.* London: John Murray, 1859. [Third edition 1861, with Historical Sketch. Sixth edition 1876.]

————. *The Variation of Animals and Plants Under Domestication.* London: John Murray, 1868. 2 vols. [Second edition 1875.]

————. *The Various Contrivances by which Orchids are Fertilised by Insects.* London: John Murray, 1877 second edition. [First edition 1862.]

Desmond, Adrian. "Robert E. Grant: The Social Predicament of a Pre-Darwinian Transmutationist." *Journal of the History of Biology* 17 (1984): 189–223.

————. *Archetypes and Ancestors: Paleontology in Victorian England, 1850–1875.* Chicago: University of Chicago Press, 1985.

————. *The Politics of Evolution: Morphology, Medicine, and Reform in Radical London.* Chicago: University of Chicago Press, 1989.

————. "Lamarckism and Democracy." In *History, Humanity and Evolution,* ed. James R. Moore, 99–130. Cambridge: Cambridge University Press, 1989.

————, and James Moore. *Darwin: The Life of a Tormented Evolutionist.* New York: Time Warner, 1991.

Di Gregorio, Mario A. *Charles Darwin's Marginalia.* New York: Garland, 1990.

Gale, Barry G. *Evolution Without Evidence: Charles Darwin and the Origin of Species.* Albuquerque: University of New Mexico Press, 1982.

Ghiselin, Michael T. *The Triumph of the Darwinian Method.* Berkeley: University of California Press, 1969.

Glick, Thomas F. *The Comparative Reception of Darwinism.* 2nd ed. Chicago: University of Chicago Press, 1988.

Gordon, Scott. "Darwin and Political Economy: The Connection Reconsidered." *Journal of the History of Biology* 22 (1989): 437–459.

Gould, Stephen J., and Elisabeth A. Vrba. "Exaptation—A Missing Term in the Science of Form." *Paleobiology* 8 (1982): 4–15.

Gruber, Howard E. *Darwin on Man.* 2nd ed. Chicago: University of Chicago Press, 1981.

Herbert, Sandra. "The Place of Man in the Development of Darwin's Theory of Transmutation," *Journal of the History of Biology* 7 (1974), 217–58; 10 (1977), 155–227.

Himmelfarb, Gertrude. *Darwin and the Darwinian Revolution.* New York: Norton, 1968.

Hodge, M.J.S. "Darwin as a Lifelong Generation Theorist." In *The Darwinian Heritage,* ed. D. Kohn, 207–243. Princeton: Princeton University Press, 1985.

Hull, David. *Darwin and His Critics.* Cambridge, Mass.: Harvard University Press, 1974.

———. "Lamarck Among the Anglos." In *Zoological Philosophy,* ed. J. B. Lamarck, xl–lxvi. Chicago: University of Chicago Press, 1984.

Kohn, David. "Theories to Work By: Rejected Theories, Reproduction, and Darwin's Path to Natural Selection." *Studies in the History of Biology* 4 (1980): 67–170.

———, ed. *The Darwinian Heritage.* Princeton: Princeton University Press, 1985.

Malthus, Thomas Robert. *An Essay on the Principle of Population.* 6th ed. London: John Murray, 1826.

Mayr, Ernst. *One Long Argument: Charles Darwin and the Genesis of Modern Evolutionary Thought.* Cambridge, Mass.: Harvard University Press, 1991.

Moore, James R. "Of Love and Death: Why Darwin Gave Up Christianity." In *History, Humanity, and Evolution,* ed. James R. Moore, 195–229. Cambridge: Cambridge University Press, 1989.

Olby, Robert. *Origins of Mendelism.* 2nd ed. Chicago: University of Chicago Press, 1985.

Oldroyd, David. R. "How Did Darwin Arrive at His Theory? The Secondary Literature to 1982." *History of Science* 22 (1984): 325–374.

Ospovat, Dov. *The Development of Darwin's Theory: Natural History, Natural Theology, and Natural Selection, 1838–1859.* Cambridge: Cambridge University Press, 1981.

Richards, Evelleen. "Darwin and the Descent of Woman." In *The Wider Domain of Evolutionary Thought*, ed. Oldroyd and Langham, 57–111. Dordrecht: D. Reidel, 1983.

Richards, Robert J. *Darwin and the Emergence of Evolutionary Theories of Mind and Behavior.* Chicago: University of Chicago Press, 1987.

Schweber, Sylvan S. "Darwin and the Political Economists: Divergence of Character." *Journal of the History of Biology* 13 (1980): 195–289.

———. "The Wider British Context in Darwin's Theorizing." In *The Darwinian Heritage*, ed. D. Kohn, 35–69. Princeton, Princeton University Press, 1985.

Sulloway, Frank J. "Darwin and His Finches: The Evolution of a Legend." *Journal of the History of Biology* 15 (1982): 1–53.

———. "Darwin's Conversion: The *Beagle* Voyage and Its Aftermath." *Journal of the History of Biology* 15 (1982): 325–396.

———. "Darwin and the Galapagos." *Biological Journal of the Linnean Society* 21 (1984): 29–59.

Vorzimmer, Peter J. *Charles Darwin: The Years of Controversy. The Origin of Species and Its Critics, 1859–1882.* Philadelphia: Temple University Press, 1970.

Wallace, Alfred Russel. "On the Tendency of Varieties to Depart Indefinitely from the Original Type." *Journal of the Proceedings of the Linnaean Society: Zoology* 3 (1859), 53–62.

Young, Robert M. "Malthus and the Evolutionists: The Common Context and Biological and Social Theory." *Past and Present* 43 (May 1969). Reprinted in Young, *Nature's Place in Victorian Society*. Cambridge: Cambridge University Press, 1985, 23–55.

INDEX OF NAMES

INDEX OF THEMES